Sonia Malik
Editor

Biotechnology and Production of Anti-Cancer Compounds

Springer

Editor
Sonia Malik
Graduate Program in Health Sciences
Biological and Health Sciences Center
Federal University of Maranhao
Sao Luis, MA, Brazil

ISBN 978-3-319-53879-2 ISBN 978-3-319-53880-8 (eBook)
DOI 10.1007/978-3-319-53880-8

Library of Congress Control Number: 2017935979

Printed on acid-free paper

This Springer imprint is published by Springer Nature
The registered company is Springer International Publishing AG
The registered company address is: Gewerbestrasse 11, 6330 Cham, Switzerland

This book is affectionately dedicated to my dear mother.

Preface

Cancer is one of the most life-threatening diseases and a major cause of death worldwide. Plants act as an important source of anti-cancer compounds. A renaissance of public interest in plant-based products due to lesser side effects and better compatibility has led to an increased demand for anti-cancer drugs obtained from plants. To meet the ever increasing demand, plants producing anti-cancer compounds are harvested from their natural sources, which direct to their extinction. Biotechnology offers a tool to produce compounds of interest without harvesting the plants from nature. Biotechnological advancements provide the opening to make use of cells, tissue or organs of economically important plants by growing them under aseptic conditions and to genetically manipulate them to obtain the desired compounds.

This book provides up-to-date information on anti-cancer drugs obtained from plants, their market demand, value as well as the role of biotechnology in the improvement of plant-based anti-cancer compounds. Chapters discuss the recent developments and techniques to obtain anti-cancer drugs from plants, in vitro protocols for optimized production of these compounds, their mode of action, and biosynthetic pathways. Experiences and views of researchers working in this area have been shared. Future strategies and goals to find out the ways to obtain the highly demanded anti-cancer compounds in an eco-friendly, economic, and efficient way are highlighted. This book will be valuable to researchers/teachers and students working in the area of plant tissue culture, natural products, phytochemistry, pharmaceutical sciences, medicines, and drug discovery.

Sao Luís, Maranhao, Brazil Sonia Malik

Preface

Acknowledgments

The completion of this book could not have been possible without the consistent support of my beloved husband. He encouraged me to initiate this task and helped me to accomplish it. I would like to thank my wonderful son for understanding me when I was working on this book instead of playing with him. He has been a continuous source of inspiration for me. I would like to express my deepest gratitude to my great parents, grandparents, brothers and their families for their obstinate support and love.

Above all, to the Great Almighty for his countless love and blessings. God is always there to listen to our prayers and open doors if we are humble enough to knock and have faith.

Contents

Contributors

Jyothi Abraham Research Department of Botany, St. Thomas College, Palai, Arunapuram (P.O.), Kottayam, Kerala, India

Ala'a Al-Hrout Biology Department, College of Science, UAE University, Al-Ain, UAE

Flavia Maria Mendonça do Amaral Department of Pharmacy, Biological and Health Sciences Center, Federal University of Maranhão, Sao Luis, MA, Brazil

Amr Amin Biology Department, College of Science, UAE University, Al-Ain, UAE

Zoology Department, Faculty of Science, Cairo University, Giza, Egypt

Badriya Baig Biology Department, College of Science, UAE University, Al-Ain, UAE

J.F. Buyel Fraunhofer Institute for Molecular Biology and Applied Ecology IME, Aachen, Germany

Institute for Molecular Biotechnology, RWTH Aachen University, Aachen, Germany

Preeti Chaturvedi Department of Biological Sciences, College of Basic Sciences and Humanities, Uttarakhand, India

Ali Hilal-Alnaqbi Department of Mechanical Engineering, College of Engineering, UAE University, Al-Ain, UAE

Iliana Ionkova Department of Pharmacognosy, Medical University of Sofia, Sofia, Bulgaria

Katarzyna Kała Department of Pharmaceutical Botany, Jagiellonian University Medical College, Krakow, Poland

Jatinder Kaur Katnoria Department of Botanical and Environmental Sciences, Guru Nanak Dev University, Amritsar, India

Rajwant Kaur Department of Botanical and Environmental Sciences, Guru Nanak Dev University, Amritsar, India

Rajinder Kaur Department of Botanical and Environmental Sciences, Guru Nanak Dev University, Amritsar, India

Ajit Kumar Department of Horticulture, College of Agriculture, G.B. Pant University of Agriculture and Technology, Uttarakhand, India

Verity I.P. Loake Institute of Molecular Plant Sciences, School of Biological Sciences, University of Edinburgh, Edinburgh, UK

Sonia Malik Graduate Program in Health Sciences, Biological and Health Sciences Center, Federal University of Maranhão, Sao Luis, MA, Brazil

Ludmilla Santos Silva de Mesquita Department of Pharmacy, Biological and Health Sciences Center, Federal University of Maranhao, Sao Luis, MA, Brazil

Denise Fernandes Coutinho Moraes Department of Pharmacy, Biological and Health Sciences Center, Federal University of Maranhao, Sao Luis, MA, Brazil

Bożena Muszyńska Department of Pharmaceutical Botany, Jagiellonian University Medical College, Krakow, Poland

Avinash Kaur Nagpal Department of Botanical and Environmental Sciences, Guru Nanak Dev University, Amritsar, India

Marisol Ochoa-Villarreal Institute of Molecular Plant Sciences, School of Biological Sciences, University of Edinburgh, Edinburgh, UK

Vartika Pant Department of Biological Sciences, College of Basic Sciences and Humanities, Uttarakhand, India

Glaucia C. Pereira Biotechnology, Icelandic Institute for Intelligent Machines, Reykjavik, Iceland

Department of Computer Sciences, Polytechnic Institute, University Autonoma of Madrid, Madrid, Spain

Department of Bioengineering, Imperial College London, London, UK

Maria Nilce de Sousa Ribeiro Department of Pharmacy, Biological and Health Sciences Center, Federal University of Maranhao, Sao Luis, MA, Brazil

Pavlina Sasheva Department of Pharmacognosy, Medical University of Sofia, Sofia, Bulgaria

Ashita Sharma Department of Botanical and Environmental Sciences, Guru Nanak Dev University, Amritsar, India

Aly Farag El Sheikha Department of Biology, McMaster University, Hamilton, ON, Canada

Department of Food Science and Technology, Faculty of Agriculture, Minufiya University, Shibin El Kom, Minufiya Government, Egypt

Ruchi Singh Department of Biological Sciences, College of Basic Sciences and Humanities, Uttarakhand, India

Katarzyna Sułkowska-Ziaja Department of Pharmaceutical Botany, Jagiellonian University Medical College, Krakow, Poland

T. Dennis Thomas Department of Plant Science, Riverside Transit Campus, Central University of Kerala, Kasaragod, Kerala, India

Tripti Tewari Department of Biological Sciences, College of Basic Sciences and Humanities, Uttarakhand, India

Chapter 1
Medicinal Plants: Ethno-Uses to Biotechnology Era

Aly Farag El Sheikha

Abstract To date, medicinal plants form the backbone of primary healthcare for 70–95% of the population of the developing world. Therefore, medicinal plants help in alleviating human suffering and are widely used for traditional remedies, pharmaceutical materials, and trade. Cancer patients numbers are increasing worldwide, ranking this disease as the second disease cause of mortality for both sexes. Traditionally, medicinal plants have been used in the fight against cancer, then it is considered as the basis for medicines discovery, and nowadays more than 70% of anticancer drugs have a natural source. The biotechnological tools are necessary to select, multiply, improve, and analyze medicinal plants. This chapter highlights the history of using the medicinal plants indigenously worldwide, i.e., anticancer reservoirs and also answers to many questions, such as: Why the importance of using medicinal plants is increasing recently? What are the benefits of applying the biotechnology in medicinal plants? It then describes the new biotech technique of the traceability by using PCR-DGGE to determine the geographical origin of medicinal plants (a case study of Physalis fruits from four different countries) by analyzing the DNA fragments of microorganisms (yeasts) on plants. This method is based on the assumption that the microbial communities of environmental samples are unique to a geographic area.

Keywords Ethnomedicine history • Medicinal plants • Anticancer reservoirs • Benefits of biotechnology • Traceability • PCR-DGGE • Origin

---

A.F. El Sheikha (✉)
Department of Biology, McMaster University, 1280 Main St. West, Hamilton, ON, Canada, L8S 4K1

Department of Food Science and Technology, Faculty of Agriculture, Minufiya University, 32511 Shibin El Kom, Minufiya Government, Egypt
e-mail: elsheikha_aly@yahoo.com

S. Malik (ed.), *Biotechnology and Production of Anti-Cancer Compounds*,
DOI 10.1007/978-3-319-53880-8_1

Historical Background

Since the beginning of civilization, humanity has used plant materials suited for use in meeting the necessities of life. Historical contexts of using herbs traditionally depict that various medicinal plants were in use from several centuries BC by Egyptians, Chinese, Indians, Syrians, Babylonians, and Hebrews (Seid and Aydagnehum 2013; El Sheikha 2015b).

Archeologically, the Sumerians described well-established medicinal uses for such plants as laurel, caraway, and thyme (Falodun 2010); also studies have shown that the practice of herbal medicine dates as far back as 60,000 years ago in Iraq (Leroi-Gourhan 1975). The evolution of these plant-based medicine systems primarily based on plants within a local area. The great popular medicine systems are produced by several systems within Africa, the Chinese, and Tibetan of some parts of Asia, the Ayurvedic and Unani of the Indian subcontinent, the Native American of North America, and the Amazonian of South America (Mamedov 2012).

Historically, the use of herbs and spices in food preparation developed in part as a response to the threat of food-borne pathogens. Several studies illustrate that recipes are the most highly spiced in tropical regions where pathogens are the most abundant. Furthermore, the spices with the most potent antimicrobial activity tend to be selected (Billing and Sherman 1998). It is well known in all cultures; vegetables are spiced less than meat, presumably because they are more resistant to spoilage (Sherman and Hash 2001). Flowering plants were the source of most plant medicines. Stepp (2004) cited that many of the common weeds that populate human settlements, such as nettle, dandelion, and chickweed, have medicinal properties.

Ethnomedicines Uses of Herbs Between the Past and the Present

The usage of herbs for healing is the method of medicine as old as humankind itself. Man has been in a constant struggle with the disease since time immemorial, so he does not stop the search for new sources of the drug through the environment around him, and the evidence of this are many and varied including what is in the form of written documents, preserved monuments, and even original plant medicines. Increased awareness of the importance of the use of medicinal plants in the fight against diseases as a result of several years of accumulated experience due to which man learned to pursue drugs in many parts of plants, i.e., barks, seeds, fruit bodies, and other parts. Modern science has proved the functional role of plants in the treatment of diseases, and this has been evident through the incorporation of modern drug therapy for many of the drugs of plant origin which have already been used for thousands of years by the ancient civilizations. The continuous development is witnessed by the use of medicinal plants in the treatment of diseases, as well as increased awareness of the importance of the use of natural resources in the pharmaceutical industry. That has raised significantly the ability of pharmacists and physician to cope with the growing challenges in their battle against the diseases, which seek to provide a better life for humanity (Petrovska 2012).

Ancient Times

Ancient peoples had acquired some knowledge of medicinal plants and used it to try to find a way to cure diseases and ease the pain of it. Predominantly, these first trials at medicine were based on speculation and superstition. They believed that evil spirits are the cause of diseases. Therefore, the use of herbs could drive them out of the body that rendered the body an unsuitable host. Usually, the healers (men or women) in the tribe were aware of the plants used in the treatment. It must not lose sight of the role of those old beliefs in this enormous development in the field of medicine (Gonsalves 2010). Methods of folk treatments throughout the world commonly used medicinal plants as part of their traditions. As following, we briefly highlight how ancient civilizations identified the medicinal plants and how to use them to heal diseases.

In Egypt

From 3000 to 6000 years ago, the ancient Egyptians have invented an efficient pharmacological collection of numerous curing materials obtained from natural resources. Nunn (1996) states: "By far the most common form of treatment recommended in the medical papyri was the use of drugs, drawn from a very wide range of animal, mineral, and vegetable materials and administered in a variety of ways. The ancient Egyptians were renowned for their skill in this respect." The ancient Egyptians have written one of the earliest known records on Ebers Papyrus that dated to 1500 BC (Fig. 1.1), which contains information on over 850 plant medicines, including garlic, juniper, cannabis, castor bean, aloe, and mandrake (Sumner 2000). The Egyptian

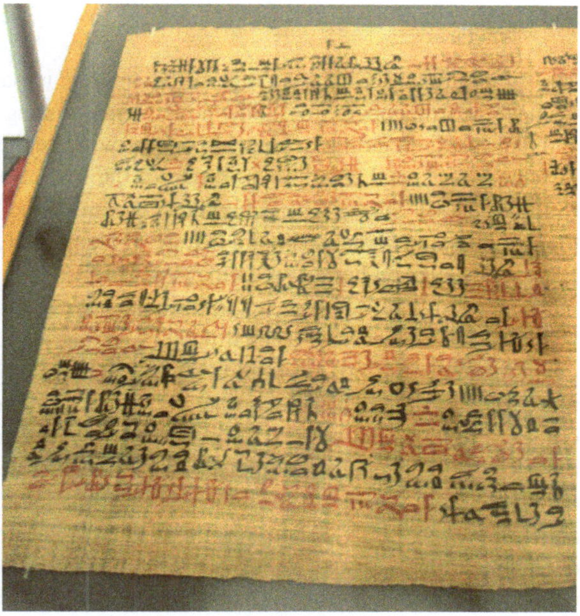

Fig. 1.1 Egyptian medical papyrus of medicinal plants dating to 1550 BC (*Source*: https://en.wikipedia.org/wiki/Ebers_Papyrus#/media/File:PEbers_c41-bc.jpg)

physicians prescribed sedatives, analgesics, gastrointestinal disorder remedies, and medicines for urinary tract diseases and the common cold (Nunn 1996; Oakes and Gahlin 2003). Plant extracts were prepared and taken internally, applied topically, and administered by fumigation and vapor inhalation. The Egyptians are also credited with the early medicinal use of wine, castor oil, marijuana, opium, mints, and beer made from barley and wheat (Shafik and Elseesy 2003). Oakes and Gahlin (2003) point out that "The Egyptians were the first people to use some drugs that modern studies have proved would have been medicinally useful."

In Greece

Plant-based therapeutic treatments continued to be augmented later by healthcare practitioners in ancient Greece 3000 through 1500 years ago. Only a few parts of these works have survived intact; the scientists have noted from what remains that there is a significant overlap with the Egyptian medicinal plants. Greek and Roman therapeutic practices were preserved through the writings of Hippocrates (e.g., *De herbis et curis*), especially Galen (e.g., *Therapeutics*), which were also as the headwaters for western medicine later (Robson and Baek 2009). Dioscorides, an authority on herbs who lived in the first century AD, is remarked for accumulating 24 detailed books on over 600 remedial plants and their proper uses under the title "De Materia Medica," the earliest known name of that terminology (Sigerist 1967; Von Staden 1989).

Following those developments, additional discoveries of useful medicinal plants resulted from experimentations in several early historic cultures 1000–2000 years ago in China and India.

In China

The mythological Chinese emperor Shennong is said to have written the first Chinese pharmacopeia, the "Shennong Ben Cao Jing." It lists 365 curative plants and their uses, including hemp, ephedra (provided ephedrine as a drug to contemporary medicine), and chaulmoogra (one of the first effective treatments for leprosy) (Sumner 2000). Succeeding generations grown on the Shennong Bencao Jing, as in the Yaoxing Lun (Treatise on the Nature of Healing Herbs), a seventh century Tang Dynasty treatise on herbal medicine (Wu 2005).

In ancient Chinese times "traditional Chinese medicine as medicine" and "Chinese herbal medicine as pharmacy" were already described as eminent specialties. More than 85% of Chinese medical materials originate from plants, but traditional Chinese medicine practitioners also prescribe animal/insects, minerals, and crude artificial compounds. Also, the term "Chinese herbal medicine" also encompasses some ethnic herbal medicines and folk medicines in China. In Chinese herbal medicine, there are 11,146 different kinds of plants, 1581 kinds of animals/animal and insects, 80 kinds of mineral medications, and more than 50 kinds of crude chemical preparations, as well as 5000 (total one million) clinically validated herbaceous formulations. Different from other herbal medicines and Western medicines,

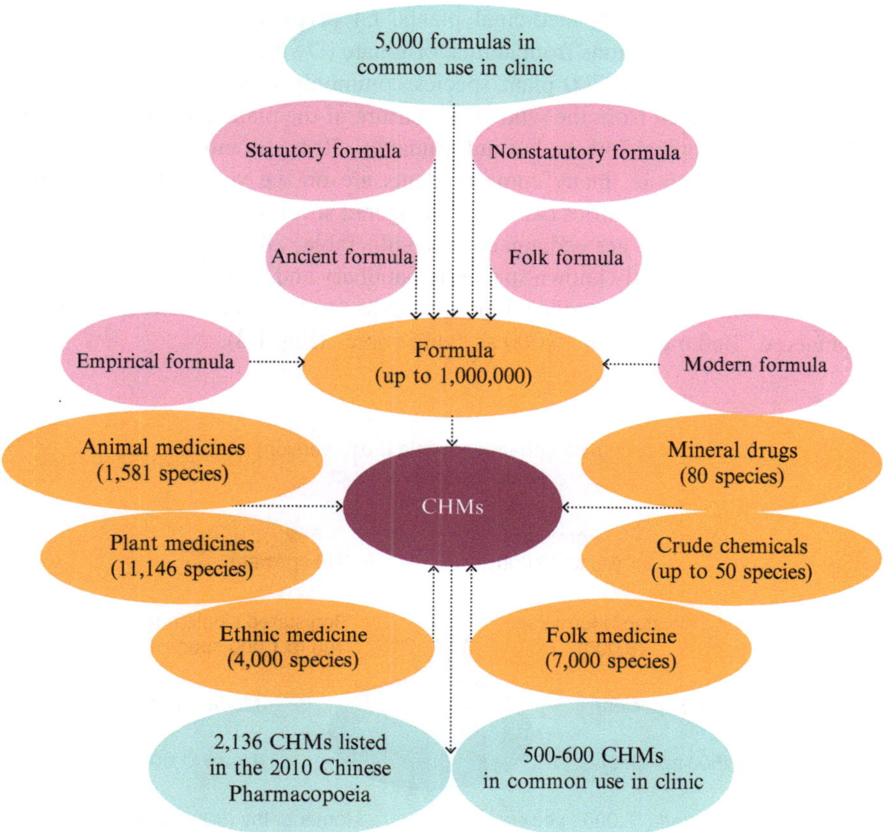

Fig. 1.2 Chinese herbal medicine: How to maximize the use of natural resources (Pan et al. 2014)

Chinese herbal medicines are often prescribed as formulas under the guidance of traditional Chinese medicines theories and practice. Each herbal medicine prescription (formula, *Fang-Ji* in Chinese) is a cocktail of many herbs tailored to the individual patient (Pan et al. 2014). It allows us to blend herbs to strengthen their benefits and reduce or eliminate any side effects when they are used each alone (Fig. 1.2).

In India

As early as 1900 BC, Ayurvedic medicine has used many medicinal plants (e.g., turmeric) (Aggarwal et al. 2007). How was the Ayurveda system basis established? It can be identified through the earliest Sanskrit writings, i.e., the Atharva Veda and Rig Veda are some of the available documents, which detail the medical knowledge (Sumner 2000). During the first millennium BC, many other curative plants, and minerals used in Ayurveda were later described by ancient Indian herbalists (e.g., Charaka and Sushruta). The Sushruta Samhita attributed to Sushruta in the sixth century BC describes 700 medicinal plants, 64 preparations based on mineral sources, and 57 preparations from animal originate (Dwivedi and Dwivedi 2007).

Ninety percent of the 700 plant species commonly used in the Indian herbal industry are collected from the wild. The treasure of the plant and animal diversity is found in the tropical forests, but unfortunately, 50% of them have already been destroyed. Furthermore, many curative plants are on the verge of extinction. In 1997, there were many entries (427) of endangered species into the Red Data Book of India, of which 124 are endangered, 81 vulnerable, 28 considered extinct, 100 rare, and 34 insufficiently known species (Chaudhary and Singh 2010). In 2012, the Red Data Book of India described 5766 species as "endangered," 3947 as "critically endangered," and more than 10,000 as "vulnerable" (Fig. 1.3).

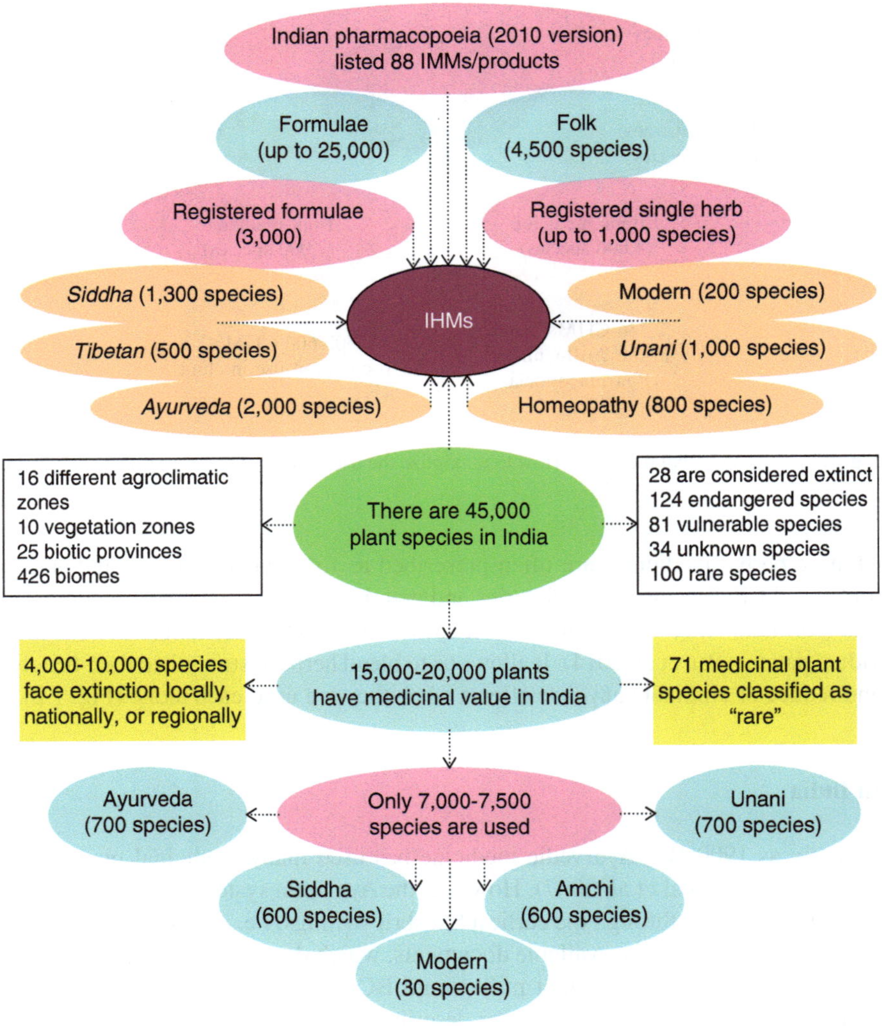

Fig. 1.3 Indian medicinal plants between the importance and the risk of extinction (Pan et al. 2014)

Middle Ages

During the Early Middle Ages, the primary source of medical knowledge in Europe and England were Benedictine monasteries. Arsdall (2002) reported that translating and copying ancient Greco-Roman and Arabic works were most of these religious scholars' efforts rather than creating essential new practices and information. The monasteries kept many Greek and Roman writings on medicine safe by hand copying of manuscripts. So these monasteries were considered as radiation centers of medical knowledge locally, and also their gardens provided many medicinal herbs, which use a simple remedy for common disorders. The traditional medicine at the same time in the home and village continued uninterrupted, supporting many wandering and settled herbalists. Hildegard of Bingen was one of the famous women in the herbal folk. A twelfth-century Benedictine nun wrote a medical text called "Causae et Curae" (Truitt 2009).

Arabic Achievements in Herbal Medicine

In the Middle Ages, the ancient Hippocratic-Greek medical know-how was adapted and improved by Arabian herbalists, pharmacologists, chemists, and physicians. Moreover, the majority of Arabs are Muslims, and Arabic culture and Islamic ideology are closely related. As such, Arabic medicine may also be called Greco-Arab or Islamic medicine. After the fall of the Roman Empire, the Arabic world became the center of scientific and medical knowledge for very long times (from 632 to 1258 CE).

During the middle ages, Arabic medicine contributed significantly to the evolution of modern medicine and pharmacy in Europe. For example, the European pharmacopoeia relied on Muslim writings and information therein until the late nineteenth century (see Fig. 1.4) (Saad et al. 2005; Azaizeh et al. 2010). The superiority of Muslims and Arabs in the field of medicine due to their implementation of the provisions of the Holy Qur'an, where serves as a good approach to both religion and life issues, and this also applies to the Sunnah of the Prophet Muhammad peace be upon him, where is the Koran full curriculum of religion and life in all its fields. The science proves every day some of the scientific facts mentioned by the Quran Karim from more than 1400 years and the rest comes.

More than 700 species of 2600 plant species in the Middle East region are noted for their use of herbs or botanical pesticides; however, only 200–250 plant species are still in use in Arab folk medicine for the treatment of different diseases (Al-Harbi et al. 1996). The western Mediterranean coastal region (from Alexandria to Sallum, Egypt) comprises 230 plant species belonging to 48 families; 89% of these species had medicinal importance, 62% of the species were common, approximately 24.9% were occasional, and 13% were scarce (Azaizeh et al. 2010). Arabic herbal medicine used in various forms include infusion, decoction, juice, syrup, roasted materials, oil, macerated plant parts, fresh salads or fruits, milky sap, poultice, and paste,

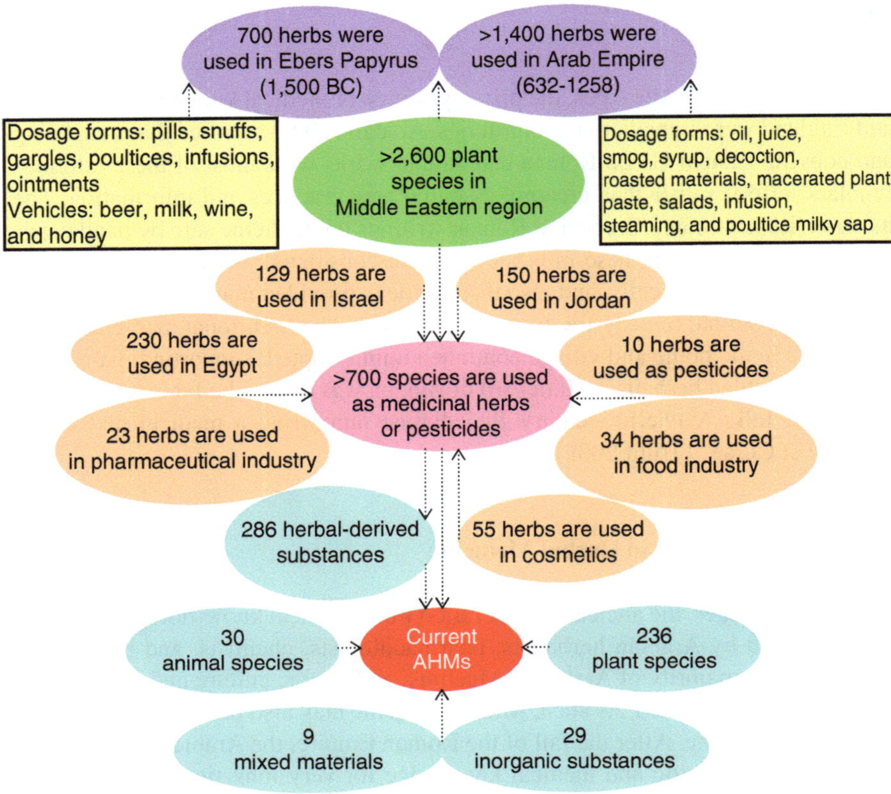

Fig. 1.4 Arabic herbal medicine: Extended history of accomplishments (Pan et al. 2014)

of which some formulations of herbal medicines are still used today. Although Arabic herbal medicine is the first choice for many people in dealing with ailments in the Middle East, most of the herbalists (i.e., those in Jordan), who acquire the expertise from their predecessors, are not adequately trained in herbal medicine (Azaizeh et al. 2010) (Fig. 1.5).

Al-Andalus was an important center for Arab herbalism between 800 and 1400, as was Baghdad. There are many works of famous Arab scholars in that period which specializes in herbal medicine. Examples of these books: "The Book of Simples" authored by Abulcasis (936–1013) in Cordoba, which is a primary source for later European herbals. "Corpus of Simples" authored by Ibn al-Baitar (1197–1248) in Malaga, a complete Arab herbal which introduced 200 new healing herbs, i.e., tamarind, nux vomica, and *Aconitum* (Castleman 2001). Avicenna wrote two books, the first is "The Canon of Medicine" in (1025), which lists 800 tested drugs, plants, and minerals (Jacquart 2008). The second one is dedicated to a discussion of the healing properties of herbs, including senna, sandalwood, nutmeg, cinnamon,

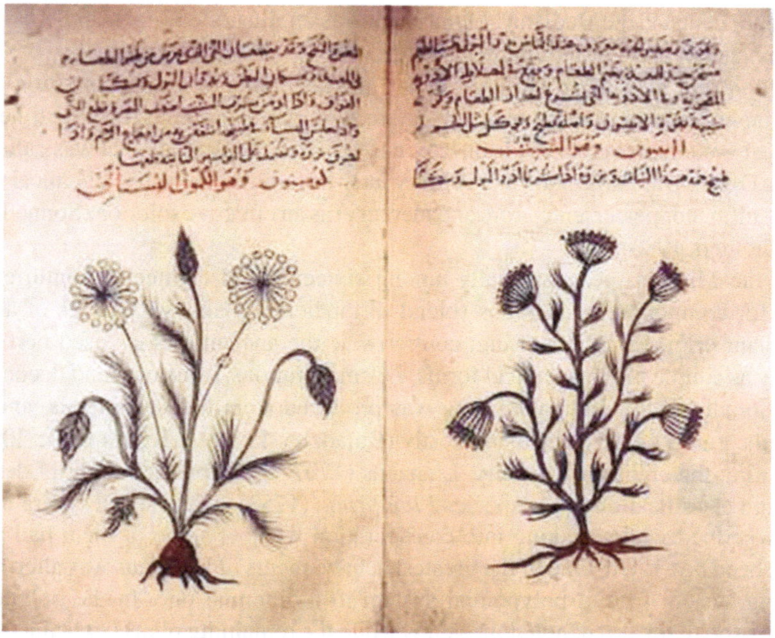

Fig. 1.5 Page from the guidebook of Arabic herbal medicine dating to 1334 CE in Arabic, describes medicinal features of cumin and dill (*Source*: http://en.wikipedia.org/wiki/Image:Arabic_herbal_medicine_guidebook.jpeg)

rosewater, myrrh, and rhubarb (Castleman 2001). Many other pharmacopoeia books were written in eleventh and twelfth centuries (and printed in Venice in 1491) by Abu-Rayhan Biruni and Ibn Zuhr (Avenzoar) (Krek 1979).

Modern Era

Early Modern Era

It is worth mentioning that all information about herbals was available in English for the first time in the fifteenth century. Grete Herball in 1526 published the first herbal in English. Furthermore, the two best-known herbals references in English were "The Herball or General History of Plants" by John Gerard in 1597 and "The English Physician Enlarged" by Nicholas Culpeper in 1653. Gerard's text was a translated version of a book by the Belgian herbalist and the source of his illustrations from a German botanical work. On the other side, Culpeper's book based on traditional medicine with astrology, magic, and folklore. The era of exploration and the Columbian Exchange introduced new medicinal plants to Europe. In the sixteenth century, the Badianus Manuscript was an illustrated Mexican herbal written in Nahuatl and Latin (Gimmel 2008).

In contrast, the second millennium saw the beginning of a slow corrosion of the prominent position held by plants as sources of curative effects. At that time, this medical system proved that it is utterly ineffective in the face of what was called the Black Death. But Paracelsus in a century later introduced the active chemical drugs (e.g., sulfur, copper, arsenic, iron, and mercury) as an effective solution (Sonnedecker and Kremers 1986).

In the Middle Ages, especially among sixteenth and eighteenth centuries, the need for compound medications (blend of medicinal plants with drugs of animal and plant origin) was increasing, contrariwise the ancient nations used herbs primarily as simple pharmaceutical forms, i.e., macerations, infusions, and decoctions. Additionally, if the compound drug was produced from herbs, minerals, and rare animals, it was highly valued and costly (Bojadzievski 1992; Toplak Galle 2005).

During the eighteenth century, Linnaeus (1707–1788) provided a brief description and classification of the species *Plantarium* (1753). These species were named and described without taking into consideration whether some of them had previously been described. One of the greatest achievements of Linnaeus was altering the naming system from a polynomial system to a binomial one. In the polynomial system, the first word denoted the genus while the remaining phrase explained other characteristics of the plant (e.g., the willow Clusius was named *Salix pumila angustifolia antera*). But the name of species in binomial system consisted of the genus name and the species name; the genus name started with a capital letter, and the species name began with a small letter (Jančić 2002).

Turning Points in the Use of Medicinal Plants

The early nineteenth century witnessed a significant turning point on the level of systematic scientific evaluation to explain the therapeutic role of medicinal plants. It has contributed to the successive discoveries of many of the active chemical compounds found in various species of medicinal plants such as tannins, glycosides, etheric oils, vitamins, and hormones (Dervendzi 1992).

Medicinal plants were exposed to the risk of exclusion from its use as a means of therapy at the end of the nineteenth century and early twentieth century. This is attributable to the preparation processes of medicinal plants which have a great effect on their pharmacological efficacy and then leading up to non-use of medicinal plants during this period. The drying process is one of the most influential operation on the properties of medicinal plants where the risk lies in the inhibiting effect on active compounds that have a vital role in the therapeutic functions of these plant. In order to avoid the negative impact of drying, the trend has been to isolate active compounds from medicinal plants as pure compounds and use directly in the pharmaceutical industry. Early twentieth century announced yet another turning point in the therapeutic use of the medicinal plants, where research efforts went toward the utilization of the fresh medicinal plant, especially the ones with labile medicinal substances. Besides, many efforts were invested in the study of manufacturing and cultivation conditions of medicinal plants (Lukic 1985; Kovacevic 2000).

It is evident that the active compounds extracted from herbs are considered natural products, most seamless laboratory. Therefore, it is given the fact that man is an integral part of nature, the degree of acceptance of the human body to those obtained drugs from natural sources is very high (Nelson and Cox 2005). Based on several studies (chemical, physiological, clinical), a lot of plants were restored to a pharmacy, which was forgotten before, such as *Secale cornutum*, *Filix mas*, *Punica granatum*, *Aconitum*, *Colchicum*, *Ricinus*, *Hyoscyamus*, *Stramonium*, *Opium*, and *Styrax*. There are many of examples of this kind; possibly they will push serious research into the ancient manuscripts on medicinal plants that would not be remarked out of curiosity about history but as potential sources of modern pharmacotherapy (Petrovska 2012).

Currently, almost all pharmacopoeias in the world deprive plant drugs of real medicinal value (British Pharmacopoeia Commission 2007; Council of Europe 2008; The United States Pharmacopoeial Convention 2008). There are countries (United Kingdom, Russia, Germany) that have separate herbal pharmacopoeias (Blumenthal et al. 1998; British Pharmacopoeia Commission 2007). Practically, a much higher number of unofficial drugs are always used. Their application is based on the experiences of popular medicine (folkloric medicine) or grounded on the results of recent research (conventional medicine). The use of medicinal plants is independent or in combination with synthetic drugs (complementary medicine) by the recommendation of the physician or pharmacist or through self-medication. Knowledge of the precise diagnosis of the illness is more necessary for the suitable and successfully applied therapy as medicinal plants, i.e., the pharmacological effect of their components is essential. In Germany as the major European producer and consumer of herbal preparations, rational phytotherapy is utilized grounded on applications of drugs whose efficiency depends on the dose and identified active components, and experimental and clinical tests have corroborated their effectiveness. Those preparations have been manufactured from standardized plant drug extracts, and they adhere to all requirements for pharmaceutical quality of medications (Petrovska 2012).

Importance of Medicinal Plants Has Increased Recently .... Why?

Why the Need Arises to Search for New Sources of Drugs?

Interestingly in recent years, there has been an evolution in the use of medicinal plants due to the side effects of synthetic drugs, lack of new pharmaceutical remedies for microbial resistance, many chronic diseases, as well as the unprecedented investment in medicinal research and development (Pan et al. 2010). Additionally, the high cost of medicines and the inability of many developing countries to purchase modern drugs have forced them to search for products in the form of medicinal plants that are proved to be cheap, efficient, safe, and culturally acceptable (Adefa and Abraha 2011).

According to the data retrieved from the World Health Organization (WHO), approximately 35,000–70,000 species has been used as drugs. This number corresponds to 14–28% of the 250,000 plants species estimated to occur around the world (Nair and Nathan 1998; Padulosi et al. 2002; Al-Sokari and El Sheikha 2015). In the global market today, much more 50 major medicaments originated from tropical plants. The variability in the biological and chemical components of plants represents a potentially limitless renewable fountain for the use in the development of novel pharmaceuticals (Mamedov 2012).

In this view, studies indicate that 25% of the modern drugs are derived from the extracts of medicinal plants. Moreover, documentation of remedial plants knowledge is incomplete as the result of a limited inventory of medicinal plants traditionally used by local people (Farnsworth 1994; El Sheikha et al. 2014).

Economic Returns for the Use of Medicinal Plants

Herbs market has expanded dramatically all over the world, an emanation of growing interest to the medical field. The income of market for sales of medicinal plants in the 1980s has climbed to about 3 billion USD/year in the North America (Glaser 1999). Also, in Canada, the use of medicinal plants has increased. Where the survey indicated that targeted 2500 people and carried out by Berger (2001) on the age groups starting at the age of 15 years to the older category, the results were 38% of the respondents are using herbal remedies, and this percentage is high compared to 1999, which was 28%. For 2007 in South America, Brazil is outstanding with 160 million USD (Tasheva and Kosturkova 2013).

Food and Drug Administration (FDA) have approved only about 1200 new drugs since 1950 (Munos 2009). As a result, in both the developing and the industrialized countries over the past 40 years, the use of herbs and its products for health objectives has increased in popularity worldwide (Humber 2002). Furthermore, with modern science/technology and ideas, the global pharmaceutical companies have begun to rediscover herbs as a possible source of new medicines and renewed their strategies to develop and discover the drugs originated from natural products (Li and Zhang 2008).

Generally in the European Union (EU), herbal products for which curative demands are made must be regulated and marketed as medications, however, those that do not make such claims be found in the cosmetic or food groups. Therefore, the need arises to unite scientific and regulatory standards that govern the marketing of herbal products, which now are progressively being to achieve this goal. At retail selling prices, the European herbal medicine market was worth over 2.8 billion USD in 1994. But at the local price level, in 1995, herbal medicinal products were estimated to be worth 5.6 billion USD (AESGP 1998). A similar increase was observed in Western Europe with 6 billion USD income for 2 years from 2003 to 2004. The sales increased in the Czech Republic by 22% from 1999 to 2001 and jumped twice in Bulgaria (Tasheva and Kosturkova 2013). The global herbal supplements and remedies market exhibited robust growth over the last decade, with little or no significant decline on account of the recent economic recession (GIA 2016). In 2012, global sales of Chinese herbal medicine reached 83 billion USD, up more than 20% from 2011 (WHO 2013). The global herbal supplements and remedies market is forecast to reach 107 billion USD by the year 2017, spurred by growing aging population and increasing consumer awareness about general health and well-being, according to a new report from Global Industry Analysts. Additionally, the fact that herbal supplements and remedies cause little or no side effects and provide greater efficacy is also proving to be a major factor aiding market growth (GIA 2016). By 2020, the global market of medicinal plant and its products could reach 115 billion USD, with Europe the largest and the Asia-Pacific the fastest growing markets. The demand is driven by women as the main consumers of dietary supplements, by growing emphasis on healthy living and concerns over the side effects of mainstream drugs (Shetty and Rinaldi 2015).

How to Use the Herbal Medicines?

Awareness in the Use of Medicinal Plants

It is interesting that the plants that are used in food (during the cooking process, for example) as well as those that are used for therapeutic purposes to be mostly safe. In spite of the recording of the few cases of allergies as a result of the use of certain herbs (very limited number) in foods and dietary supplements, and this, of course, never reduces the high degree of safety provided by the use of medicinal plants. But everyone should not hide the fact that the effect of toxic plants may reach the level of seriousness to become deadly (e.g., poison ivy) (AHPA-ERB 2008). Perhaps history shows us this scientific truth, as Socrates was executed for 2400 years by a lethal dose of poison hemlock. To ensure the safe use of herbal products, there should be a high level of awareness and responsibility among consumers, i.e., follow label directions or the recommendation of healthcare provider for any herbal product (AHPA-ERB 2008).

Regulation and Legislation of Herbal Medicines

WHO has issued guidelines for the estimation of herbal drugs. This document covered such topics as improving procedures for clinical trials using herbal remedies, guidelines for quality specifications of plant substances and preparations, guidelines for pharmacodynamic and general pharmacological studies of herbal medications and toxicity investigations of herbal medicines, and assessing the medicinal plants research (WHO 1996). Additionally, the WHO Traditional Medicines Strategy 2014–2023 was developed to review a framework. WHO and its partners aimed at enabling (Traditional Medicine/Complementary and Alternative Medicine) to play a greater role in reducing excess morbidity and mortality, especially among impoverished populations. Many of the efforts were made at the level of traditional

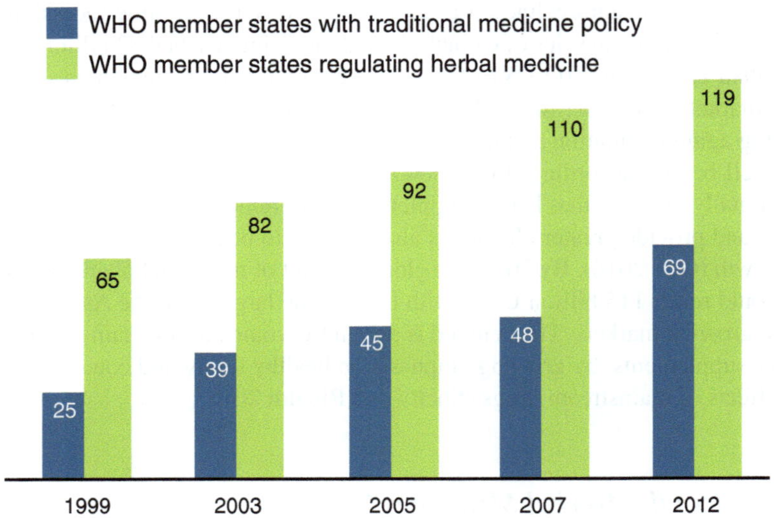

Fig. 1.6 Progress of the regulations on medicinal plants worldwide (WHO 2013)

medicines to cope with varying regulations. Each country has different rules for drug authority (WHO 2013). Over the past 15 years, the WHO has recorded a marked increase in countries with its national policies or regulations on traditional medicine about herbal medicines (see Fig. 1.6).

Based on the Council Directive 2004/24/EC (European Commission 2004), the medicinal products for human use fall under the general frame of the pharmaceutical products. It needs prior marketing approval before obtaining access to the market and investigates the requirements for the documentation of safety, quality, and efficacy, the file and expert reports (Calapai 2008). This framework has effectively been in operation by the European Agency for the Evaluation of Medicinal Plant Products (EMEA) that acts as a central agency for single European drugs marketing licenses operating a Herbal Medicinal Products Working Party (HMPWP). But, some EU Member States (UK, Germany, France, Italy, etc.) have taken different techniques in reviewing herbal remedies. To differentiate between what falls under each part of herbal medicinal products, and other products, Spain has developed a draft legislation that contains the definitions of "herbal medicinal products" and "phytotraditional products." The second one is not considered as "pharmaceutical specialties" and are therefore not classified as herbal medicinal products (IARC 2002).

In Canada, the herbal products not registered as drugs are sold as foods and are thus exempt from the medicines verification and review process, which evaluates product safety and effectiveness. In 1984, an Expert Advisory Committee on Herbs and Botanical Preparations was formed to be as advisers for the Health Protection Branch (HPB). The lists of hazardous herbal products were published by HPB, which the last version (1995) elicited a tremendous response from consumers and the herbal industry (Kozyrskyj 1997). The provincial government of British

Columbia in 2000 approved the regulations that established traditional Chinese medicine as an alternative form of essential healthcare, but the practitioners face several restrictions, including, for example, the Canadian Medicare doesn't cover the cost (Johnson 2001).

Act of Food, Drug, and Cosmetics in the USA distinguishes a product mainly by its intended use (21 Code of Federal Regulations (CFR) 201.128). Whatever the use of this product, it should be accompanied by labeling claims, as a food (including a dietary supplement), medicine (including a biological medicine), a medical device (e.g., gutta-percha) or cosmetic (FDA 2015).

Medicinal Plants Were Used Traditionally to Heal from Cancer: Why? and How?

Why People Use Herbal Medicine with Cancer?

Over the centuries, the herbal medicine has been used to heal many different health problems. The most of these uses promoted by significantly included the relaxation and cope with anxiety, depression, irritable bowel syndrome, menstrual problems, eczema, and hay fever. Concerning cancer, some studies have indicated that a significant proportion of cancer patients (60%) use medicinal plants as one of the complementary and alternative medicines (CAM) most commonly used, and it will be in parallel with conventional cancer drugs. Commonly used plants include garlic, echinacea, ginger, St John's wort, and ginseng (Langmead and Rampton 2001).

Cancer Research UK in early 2014 reported the behavior, beliefs, information, and requirements of people with cancer who take herbal medicines (Cancer Research UK 2014). The study indicated that patients mostly took herbal medicines to take back some control of their disease. They also thought that the remedies would not cause side effects.

Medicinal Plants: Cancer Remedies Between the Past and the Present

One of the major obstacles to overall public health is cancer. It is responsible for one in every four deaths in the USA alone. American Cancer Society in 2003 had estimated that there are one million new cases of invasive cancer detected with over half a million deaths from skin, squamous cell, and basal cancers. Cancer as a particular disease has a long history with herbal remedies (Hartwell 1982). Despite the suspicion that some may ask the effectiveness of these treatments as had previously been used in folklore and traditional medicine (Cragg and Newman 2005).

The natural products as secondary metabolites, which are derived from plants and microbes, have a vital contribution in the chemotherapy of cancer. The tropical rainforests are considered as a renewable source of novel anticancer molecules for several reasons including:

- Increasing the probability of exposure some of these tropical plants at risk of extinction shortly and thus will miss the natural sources of drugs (Cox 2000).
- Great biodiversity which is characterized by tropical forests, which also make these as variety sources of natural medicines (Burslem et al. 2001).

There are some proofs which mentioned that the herbal remedies might help to prevent or relieve the symptoms of cancer or its side effects. Many researchers and through many studies conducted on lung cancer patients have been pointing out that the Chinese medicinal plants during the chemotherapy phase might be useful, but there is still a need for further studies (Cancer Research UK 2016).

Herbal Medicines: Traditional Treatment of Cancer Worldwide

Herbal drugs are enjoying widespread popularity all over the world (Almeida et al. 2006). About 85% of the world population uses medicinal plants, and the demand is increasing in developed and developing countries to prevent and heal diseases (Abramov 1996). Moreover, the investigation of herbal drugs from plants to treat cancer has been reported (Lee 1999). However, only 10% of medicinal plant species is cultivated today while the larger majority being left under wild stands threat (Cunninghum 2001).

The following detailed table (Table 1.1) illustrates the medicinal plant species that are used indigenously for cancer therapy.

New Vision of Medicinal Plants by Biotechnological Lenses

How Will the Medicinal Plants Be Drugs?

Background of Knowledge

Regarding the derivative medicines from plants, it is important to keep in mind some of the conceptual differences. Plants can be used as curative resources in many ways. They can be used as herbal teas, crude extracts in pharmaceutical preparations, such as fluid, tinctures, powder, extracts, pills, and capsules when they are considered as phytopharmaceutical preparations or herbal drugs (Rates 2001).

OPS (Organización Panamericana de la Salud) was defined the medicinal plant as (1) any plant used to relieve, prevent, or treat a disease or to alter the pathological and physiological process, or (2) any plant species employed as a source of medicines (Arias 1999). Herbal medicine preparation is any manufactured medicine obtained exclusively from plants (aerial and non-aerial parts, resins, juices, oil),

Table 1.1 List of indigenous medicinal plants used for the treatment of cancer in folk medicine worldwide

Country	Plant species	Family name	Parts used	References
Jordan	*Narcissus tazetta* L.	*Amaryllidaceae*	Flowers	Talib and Mahasneh (2010)
	Arum dioscoridis	*Araceae*	Leaves	Ali-Shtayeh et al. (2000), Hudaib et al. (2008)
	Arum hygrophilum	*Araceae*	Leaves	Hudaib et al. (2008)
	Arum palaestinum	*Araceae*	Leaves	Said et al. (2002), Hudaib et al. (2008)
	Hedera helix L.	*Araliaceae*	Leaves, berries	Oran and Al-Eisawi (1998)
	Inula viscosa L.	*Asteraceae*	Flower heads	Talib and Mahasneh (2010)
	Calendula arvensis L.	*Asteraceae*	Dry flowering branches	Oran and Al-Eisawi (1998)
	Anthemis pseudocotula	*Asteraceae*	Flower heads	Oran and Al-Eisawi (1998)
	Luffa cylindrica L.	*Cucurbitaceae*	Seeds, aerial parts	Talib and Mahasneh (2010)
	Arbutus andrachne L.	*Ericaceae*	Leaves, fruits, roots	Said et al. (2002)
	Mercurialis annua L.	*Euphorbiaceae*	Leaves	Said et al. (2002)
	Quercus calliprinos	*Fagaceae*	Fruits, barks	Said et al. (2002)
	Globularia arabica L.	*Globulariaceae*	Leaves	Oran and Al-Eisawi (1998)
	Laurus nobilis L.	*Lauraceae*	Leaves	Said et al. (2002)
	Ononis sicula	*Leguminosae*	Aerial parts	Talib and Mahasneh (2010)
	Anagyris foetida L.	*Leguminosae*	Leaves	Oran and Al-Eisawi (1998)
	Urginea maritima L.	*Liliaceae*	Bulbs	Oran and Al-Eisawi (1998)
	Allium cepa	*Liliaceae*	Bulbs, leaves	Ali-Shtayeh et al. (2000)
	Viscum cruciatum	*Loranthaceae*	Pads, leaves	Ali-Shtayeh et al. (2000), Al-Qura'n (2009)
	Cocculus pendulus	*Menispermeaceae*	Leaves, branches	Oran and Al-Eisawi (1998)
	Triticum aestivum L.	*Poaceae*	Shoots	Said et al. (2002)
	Platanus orientalis L.	*Polypodiaceae*	Leaves	Oran and Al-Eisawi (1998)
	Clematis flammula L.	*Ranunculaceae*	Leaves	Oran and Al-Eisawi (1998)
	Zizyphus spina-christi L.	*Rhamnaceae*	Fruits, leaves	Dafni et al. (2005), Hudaib et al. (2008), Saied et al. (2008)

(continued)

Table 1.1 (continued)

Country	Plant species	Family name	Parts used	References
	Sarcopoterium spinosum L.	*Rosaceae*	Leaves, seeds, roots	Hudaib et al. (2008)
	Crataegus azarolus L.	*Rosaceae*	Flowers, fruits	Said et al. (2002)
	Urtica pilulifera L.	*Urticaceae*	Leaves	Said et al. (2002)
Lebanon	*Bongardia chrysogonum* L.	*Berberidaceae*	Whole plant	Baydoun et al. (2015)
	Carthamus tenuis	*Asteraceae*	Seeds	Baydoun et al. (2015)
	Cyclamen coum Mill.	*Primulaceae*	Whole plant, tubers	Baydoun et al. (2015)
	Clematis flammula L.	*Ranunculaceae*	Whole plant	Baydoun et al. (2015)
	Crataegus azarolus L.	*Rosaceae*	Flowers, fruits	Said et al. (2002, 2008)
Thailand	*Alpinia galanga*	*Zingiberaceae*	Rhizomes	Lee and Houghton (2005)
	Alpinia officinarum	*Zingiberaceae*	Rhizomes	Lee and Houghton (2005)
	Oroxylum indicum L.	*Bignoniaceae*	Fruits	Roy et al. (2007)
	Oroxylum indicum L.	*Bignoniaceae*	Stem barks	Costa-Lotufo et al. (2005)
	Rhinacanthus nasutus L.	*Acanthaceae*	Leaves, roots	Farnsworth and Bunyapraphatsara (1992)
	Dioscorea birmanica	*Dioscoreaceae*	Rhizomes	Boonyaratanakornkit and Chantaptavan (1993), Itharat et al. (2004)
	Dioscorea membranacea	*Dioscoreaceae*	Rhizomes	Itharat et al. (1999a)
	Hydnophytum formicarum	*Rubiaceae*	Rhizomes	Itharat et al. (1999b, 2002)
	Premna herbacea	*Verbenaceae*	Rhizomes	Boonyaratanakornkit and Chantaptavan (1993)
	Salacia chinensis L.	*Celastraceae*	Stems	Itharat et al. (1999b, 2002)
	Siphonodon celastrineus	*Celastraceae*	Stems, leaves	Chayamarit (1995)
	Smilax corbularia	*Smilacaceae*	Rhizomes	Boonyaratanakornkit and Chantaptavan (1993)
	Smilax glabra	*Smilacaceae*	Rhizomes	Boonyaratanakornkit and Chantaptavan (1993)
	Suregada multiflora	*Euphorbiaceae*	Stems	Chayamarit (1995), Itharat et al. (1999b, 2002)
Malaysia	*Alpinia galanga*	*Zingiberaceae*	Rhizomes	Ismail et al. (1999), Lee and Houghton (2005)
	Cayratia japonica	*Vitaceae*	Roots	Ismail et al. (1999), Lee and Houghton (2005)
	Jasminum sambac	*Oleaceae*	Roots	Ismail et al. (1999), Lee and Houghton (2005)

	Physalis minima L	*Solanaceae*	Stems, leaves	Ismail et al. (1999), Lee and Houghton (2005)
	Tabernaemontana divaricata L.	*Apocynaceae*	Roots	Ismail et al. (1999), Lee and Houghton (2005)
Taiwan	*Physalis angulata*	*Solanaceae*	Stems, leaves	El Sheikha et al. (2008a)
Bangladeshi	*Emblica officinalis*	*Euphorbiaceae*	Fruits	Rajeshkumar et al. (2003)
	Nigella sativa L.	*Ranunculaceae*	Seeds	Gali-Muhtasib et al. (2004)
	Oroxylum indicum L.	*Bignoniaceae*	Stem barks	Lambertini et al. (2004)
	Tribulus terrestris L.	*Zygophyllaceae*	Fruits	Bedir et al. (2002), Ali et al. (2001), Sun et al. (2003)
Pakistan	*Caraluma edulis*	*Apocynaceae*	Aerial parts, flowers	Nisar et al. (2014)
	Nerium oleander	*Apocynaceae*	Leaves, roots	Nisar et al. (2014)
	Sonchus asper	*Asteraceae*	Aerial parts	Nisar et al. (2014)
	Citrullus colocynthis	*Cucurbitaceae*	Dried pulp of fruit and root	Nisar et al. (2014)
	Cyperus papyrus	*Cyperaceae*	Whole plant	Nisar et al. (2014)
	Euphorbia helioscopia	*Euphobiaceae*	Leaves, stems, roots, seeds	Nisar et al. (2014)
	Vicia sativa	*Fabaceae*	Whole plant	Nisar et al. (2014)
	Acacia ampliceps	*Leguminosae*	Leaves	Nisar et al. (2014)
	Crotalaria burhia	*Leguminosae*	Whole plant	Nisar et al. (2014)
	Hibiscus rosa	*Malvaceae*	Leaves, flowers	Nisar et al. (2014)
	Ficus carica	*Moraceae*	Leaves, fruits, latex	Nisar et al. (2014)
	Jasminum grandiflorum	*Oleaceae*	Flowers, roots, stems, barks	Nisar et al. (2014)
	Arundo donax	*Poaceae*	Roots, leaves	Nisar et al. (2014)
	Fagonia cretica	*Zygophyllaceae*	Whole plant	Nisar et al. (2014)
	Hedera nepalensis	*Araliaceae*	Leaves	Ahmed et al. (2013)
	Viola serpense	*Violaceae*	Flowers, leaves	Ahmed et al. (2013)
South Africa	*Rhus leptodictya*	*Anacardiaceae*	Leaves	Fouche et al. (2008)
	Rhus chirindensis	*Anacardiaceae*	Stems	Fouche et al. (2008)
	Rhus lancea	*Anacardiaceae*	Fruits, stems	Fouche et al. (2008)
	Gomphocarpus physocarpus	*Apocynaceae*	Roots	Fouche et al. (2008)

(continued)

Table 1.1 (continued)

Country	Plant species	Family name	Parts used	References
	Acokanthera oppositifolia	*Apocynaceae*	Stems, roots	Fouche et al. (2008)
	Gomphocarpus fruticosus	*Apocynaceae*	Leaves, stems, fruits, roots	Fouche et al. (2008)
	Cussonia paniculata	*Araliaceae*	Leaves	Fouche et al. (2008)
	Asparagus aethiopicus L.	*Asparagaceae*	Roots	Fouche et al. (2008)
	Asparagus transvaalensis	*Asparagaceae*	Stems	Fouche et al. (2008)
	Brachylaena rotundata	*Asteraceae*	Leaves	Fouche et al. (2008)
	Xanthium strumarium L.	*Asteraceae*	Stems	Fouche et al. (2008)
	Arctotis arctotoides	*Asteraceae*	Whole plant	Fouche et al. (2008)
	Artemisia afra	*Asteraceae*	Leaves	Fouche et al. (2008)
	Athrixia elata	*Asteraceae*	Leaves, seeds	Fouche et al. (2008)
	Helichrysum nudifolium	*Asteraceae*	Roots	Fouche et al. (2008)
	Oncosiphon piluliferum	*Asteraceae*	Whole plant	Fouche et al. (2008)
	Schkuhria pinnata	*Asteraceae*	Stems	Fouche et al. (2008)
	Tithonia diversifolia	*Asteraceae*	Leaves	Fouche et al. (2008)
	Vernonia staehelinoides	*Asteraceae*	Leaves	Fouche et al. (2008)
	Kigelia africana	*Bignoniaceae*	Leaves, roots	Fouche et al. (2008)
	Cadaba aphylla	*Capparaceae*	Roots	Fouche et al. (2008)
	Gymnosporia tenuispina	*Celastraceae*	Flowers, leaves, stems	Fouche et al. (2008)
	Parinari capensis	*Chrysobalanaceae*	Whole plant	Fouche et al. (2008)
	Combretum zeyheri	*Combretaceae*	Leaves	Fouche et al. (2008)
	Ipomoea cairica L.	*Convolvulaceae*	Whole plant	Fouche et al. (2008)
	Cotyledon cuneata	*Crassulaceae*	Roots	Fouche et al. (2008)
	Kalanchoe paniculata	*Crassulaceae*	Roots	Fouche et al. (2008)
	Kalanchoe thyrsiflora	*Crassulaceae*	Leaves, roots	Fouche et al. (2008)
	Sansevieria pearsonii	*Dracaenaceae*	Roots	Fouche et al. (2008)
	Psoralea pinnata	*Fabaceae*	Stems	Fouche et al. (2008)

Country	Plant species	Family name	Parts used	References
	Erythrina lysistemon	*Fabaceae*	Whole plant	Fouche et al. (2008)
	Pelargonium acraeum	*Geraniaceae*	Whole plant	Fouche et al. (2008)
	Drimia robusta	*Hyacinthaceae*	Whole plant	Fouche et al. (2008)
	Moraea polystachya	*Iridaceae*	Whole plant	Fouche et al. (2008)
	Leucosidea sericea	*Rosaceae*	Leaves	Fouche et al. (2008)
	Physalis peruviana L.	*Solanaceae*	Whole plant	El Sheikha et al. (2008a), Fouche et al. (2008)
	Solanum acanthoideum	*Solanaceae*	Roots	Fouche et al. (2008)
	Solanum panduriforme	*Solanaceae*	Whole plant	Fouche et al. (2008)
	Solanum tomentosum L.	*Solanaceae*	Stems	Fouche et al. (2008)
	Withania somnifera L.	*Solanaceae*	Leaves	Fouche et al. (2008)
Ethiopia	*Plumbago zeylanica* L.	*Plumbaginaceae*	Leaves	Abera (2014)
	Rumex abyssinicus	*Polygonaceae*	ND[a]	Mekonnen et al. (2010)
Nigeria	*Adenia lobata*	*Passifloraceae*	Whole plant	Agoreyo et al. (2012)
Mexico	*Rollinia mucosa*	*Annonaceae*	ND	Shi et al. (1997)
	Annona muricata L.	*Annonaceae*	Leaves	Moghadamtousi et al. (2015)
	Bursera fagaroides	*Burseraceae*	Steam barks	Rojas-Sepúlveda et al. (2012)
	Hemiangium excelsum	*Celastraceae*	ND	Popoca et al. (1998)
	Phoradendron reichenbachianum	*Santalaceae*	ND	Ríos et al. (2001)
	Cuphea aequipetala	*Lythraceae*	ND	Vega-Ávila et al. (2004)
	Penstemon barbatus	*Plantaginaceae*	ND	Moreno-Escobar et al. (2011)
	Zea mays L.	*Poaceae*	ND	Kuga et al. (1993)
	Colubrina macrocarpa	*Rhamnaceae*	ND	Popoca et al. (1998)
Brazil	*Euphorbia tirucalli*	*Euphorbia*	ND	Betancur-Galvis et al. (2002)
Cameroon	*Rumex abyssinicus*	*Polygonaceae*	Bulbs	Tamokou et al. (2013)
	Rumex bequaertii	*Polygonaceae*	Bulbs, roots	Munavu et al. (1984), Tamokou et al. (2013)
Ireland	*Rumex obtusifolius*	*Polygonaceae*	ND	Harshaw et al. (2010)

[a]*ND* not determined

either in the crude form or as a pharmaceutical formulation (Rates 2001). A drug is a product prepared according to legal and technical procedures that is used for the diagnosis, cure, or prevention of disease and has been scientifically characterized regarding its efficacy, safety, and quality (WHO 2013). A drug is a pharmacologically active compound, which is a component of medicine, irrespective of its natural, biotechnological, or synthetic origin.

Plant Selection

The technique used for drug development from plant resources depends on the aim. There are several methods in which this can be done, including traditional use, chemical content, or a combination of several criteria. Different strategies will result in a herbal medicine or an extract bioactive compound. Therefore, the selection of a suitable plant for a pharmacological study is an important and decisive stage (Williamson et al. 1996).

Preparation and Isolation of Bioactive Substances

Once the plant is selected, the next stage is its collection and botanical identification; then it should be submitted to a stabilization process. It is important that plant recollection involves a professional botanist who can correctly identify the species and these plants and prepare part of the material for herbarium preservation to have a reference material. Preferably, the date and place of recollection should be listed and the information retained for further collection, if necessary (Rates 2001).

Figure 1.7 shows the diagram of methods used for obtaining the bioactive plant-derived substances.

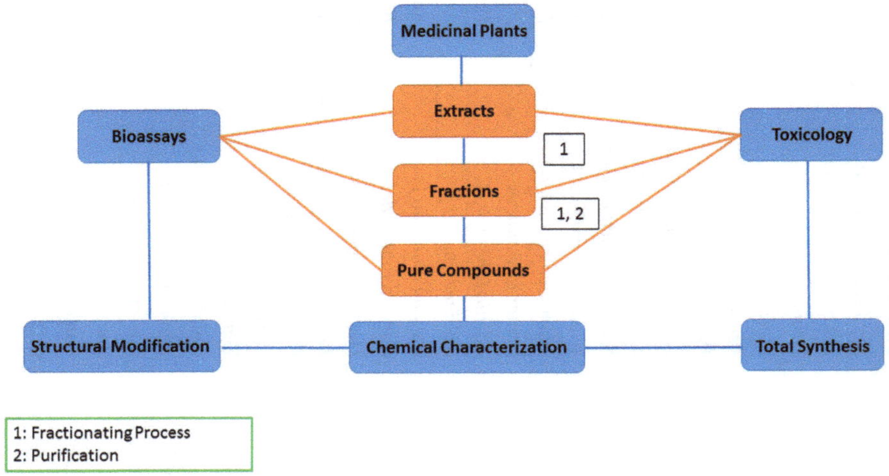

Fig. 1.7 How to extract the bioactive compounds from plants (Rates 2001). Reproduced with permission of Elsevier

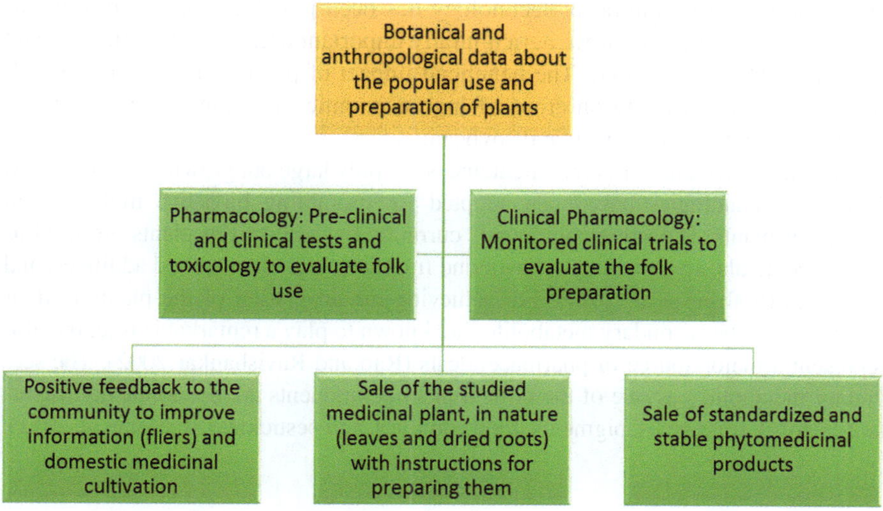

Fig. 1.8 Pharmacological validation methods of herbal medicines (Rates 2001). Reproduced with permission of Elsevier

The Need of Pharmacological Validation

As known, the drug is expensive. Furthermore, the study of medicinal plants also allows their use in pharmaceutical formulations, which are called phytomedicines or herbal medicines. This approach also requires toxicity and efficacy studies, but these are less time-consuming. On another hand, the steps of fractionation, purification, and bioassay are not required or are far less complex (Fig. 1.8).

These phyto-drugs when they become registered to become a medicine that needs to comply with the basic standards necessary for all drugs. Practically, there is high variability in the plant material; therefore, a minimum concentration or a concentration range is often used rather than an exact level. Most plant species have a wide therapeutic window (e.g., a toxic compound is considerably higher than the therapeutic dose). In this case, chemical structures are favored, as opposed to extracts. Standardized products provide more security and increase the level of trust people have in herbal drugs. Also, standardization allows comparison of the clinical effectiveness, pharmacological effects, and side effects of several products (e.g., against a placebo) (Gurib-Fakim 2006).

Benefits of Applying Biotechnology in Medicinal Plants

Why It Has Increased the Need to Apply Biotechnology to Medicinal Plants?

The definition of biotechnology is the application of scientific techniques to modify and improve microorganisms, plants, and animals to reinforce their value. The area of biotechnology concerning to agricultural applications is called agricultural biotechnology. Agricultural biotechnology has been practiced for a long time, as people have sought to improve agriculturally important organisms by selection and breeding (Wieczorek 2003). The beneficial impact of plant biotechnology worldwide has been especially on crops of high economic importance such as wheat, maize, rice, potato, sunflower, and soybean.

The global demand for herbal medicine is not only large but growing continuously. Different technologies have been adopted for promoting bioactive molecules in medicinal plants. Bioactive compounds currently extracted from plants are used as pharmaceuticals, agrochemicals, flavor and fragrance ingredients, food additives, and pesticides (Siahsar et al. 2011). For achieving the adaptation of the plants to their environment, the secondary metabolites are known to play a remarkable role, but also represent a major source of pharmaceuticals (Rao and Ravishankar 2002). The secondary metabolites as one of the critical plant components are economically important as drugs, fragrances, pigments, food additives, and pesticides (Siahsar et al. 2011).

What Are the Advantages of Applied Biotechnological Techniques?

Biotechnological tools are essential for the multiplication and genetic enhancement of the medicinal plants by adopting technologies such as in vitro regeneration and genetic transformation. It could also be harnessed for the production of secondary metabolites using plants as bioreactors (Khan et al. 2009).

Following are a few examples of advantages of applying some current biotechnological techniques to medicinal plants.

Cell culture technology: The evolving commercial importance of the secondary metabolites has in recent years resulted in a considerable interest, in secondary metabolism and particularly in the possibility to alter the production of bioactive metabolites using cell culture technology. One of the main advantages of cell culture technology is that it may provide a reliable and renewable source of plant remedies and could be used for the large scale, from which these metabolites can be extracted. Siahsar et al. (2011) reported the advantages of a cell culture technology over the conventional cultivation of plants. Those features could be summarized in the following points:

- Useful compounds can be produced under controlled conditions independent of any changings of soil or climatic conditions.
- The cultured cells by this technique are free from insects and microorganisms.
- The tropical or alpine plant cells could easily be multiplied to yield their specific metabolites.
- With the automated control of cell growth and rational regulation of metabolite processes would improve the productivity and reduce the labor costs.
- Organic substances are extractable from callus cultures.

Genetic transformation technology: Genetic transformation has been proved to be a powerful tool for the production of plants with desired traits in several crops. It promises to overcome some of the substantial agronomic and environmental problems that have not been solved using conventional plant breeding programs (Sawahel 1997).

Agrobacterium-*mediated gene transfer*: In plant transformation mediated by *Agrobacterium tumefactions*, the pathogenic bacterium originated from the soil has become the source of foreign genes introduced into plant cells and the subsequent regeneration of transgenic plants (Tzfira et al. 2004). The rapid progress in the area of crop biotechnology is mainly because of the development of efficient regeneration and suitable *Agrobacterium*-mediated transformation protocols for different crop species. Similar success could also be achieved in the medicinal plants, which in turn could be used for the enhancement of secondary metabolites content. Thus, *Agrobacterium* transformation has become a routine technique as the genetic transformation of many important medicinal plants (Siahsar et al. 2011).

Trace Medicinal Plants from Field to Fork by Biotech Approaches …. Is It Possible?

Determine the Geographical Origin of Herbal Materials Is Vital Requirement …. Why?

Globalization of trade is expanding natural product markets that sustain life and promote good health. But the financial climate is squeezing profit margins and exacerbating the propensity for contamination, fraudulent market substitution and the use of unlabeled fillers are challenging the growth of this trade. Consumers are becoming increasingly concerned about the authenticity of the products they purchase. It should be noted that international trade in herbal products is one of the keys and influential forces in the global economy, and the remarkable demand for these products continues to increase in both developing and developed countries. Currently, more than 1000 companies produce medicinal plant products with annual revenues more than 60 billion USD. Determination of the geographical origin of the raw materials of herbal plants "imported-exported" is a demand for the traceability system. Whether it will be used as food or drug (178/2002/EC, NTA Vol 2B Ed. July 2003), the WHO/2003 guideline should lead to the total transparency of herbal medicines production from plant to the final product (El Sheikha 2015a). Therefore, the use of an authentic herbal material is the first step in ensuring quality, safety, and efficacy of herbal medicines (Khan et al. 2009; Li et al. 2014; Lee et al. 2014).

Current Techniques Used for Determining the Origin of Medicinal Plants

Multi-element techniques: It is known that many herbal materials of the same species from different geographical districts behave differently in medical applications, and thus the substances require special treatment regarding commercial trade and safety management. *Marsdenia tenacissima* is one of the famous Chinese medicinal plants because of its high quality that attracts excellent prices, so there may be an economic incentive to adulterate these premium products with lower prices. Therefore, the need becomes essential to find a method to trace the geographic origin of *M. tenacissima*, so as to reflect positively on patients by reassuring, ensure fair competition and protect geographical indication. The multi-element marker may be useful for the geographical indication of the medicinal plants (Li et al. 2014).

Metabolomic approaches: Currently, the origins of medicinal plants are authenticated mostly by evaluating their phenotype. The pros of this technique are simple, rapid, and requires only an expert's knowledge and experience, but the cons are centered in the objectivity of scientific testing, and, therefore, it is not suitable for the quality control of medicinal plants. Moreover, phenotype evaluation is practically impossible when morphological features of the plants are destroyed by processing, such as grinding (Wang et al. 2012). How to overcome these limitations? The metabolomic approaches have been suggested to detect metabolic differences arising from environmental or genetic effects. Several analytical techniques have been used to obtain metabolomic profiles, including Fourier transform near-infrared (FT-NIR) spectroscopy, ^{1}H-nuclear magnetic resonance (^{1}H-NMR) spectroscopy, Gas chromatography-mass spectroscopy (GC-MS), Liquid chromatography-mass spectrometry (LC-MS) (Vergouw et al. 2008; López-Rituerto et al. 2012; Cevallos-Cevallos et al. 2012; Luo et al. 2013). FT-NIR allows for the rapid and nondestructive analysis of the metabolome, though quantitative data at the compound level is hard to obtain (Botros et al. 2008). NMR ensures reproducible and fast analysis; however, its low sensitivity prevents quantifying metabolites at low concentrations (Baek et al. 2012). Mass spectroscopy can be advantageous over ^{1}H-NMR and FT-NIR. Why? Because it provides higher sensitivity and broader detection range despite its relatively poor reproducibility (Moco et al. 2007). The researchers need to select an adequate approach based on their targets. For metabolomics techniques, choosing the appropriate analytical tool is essential for the authentication of medicinal plant origins because the choice will affect the chemical profile and detection range (Dunn and Ellis 2005). Lee et al. (2014) indicated that ^{1}H-NMR and LC-MS, which were the best techniques for *Gastrodia elata* and *Rehmannia glutinosa*, respectively, were preferable for origin discrimination over the others. Reasoned by integrating all the results, ^{1}H-NMR is the most prominent technique for discriminating the origins of two plants.

PCR-DGGE: Innovative Biological Barcode of Medicinal Plants

Principles of technique: PCR-DGGE is a molecular tool that could allow efficient identification of microbes in environmental samples, including foodstuffs. DGGE can separate the amplicons of roughly the same size but with different DNA sequences (Madigan et al. 2009). The differences in DNA sequences render them to have different denaturing and annealing properties (see Fig. 1.9). Specifically, the gel used for DGGE included denaturating agents (a mixture of urea and formamide) (Muyzer 1999), and the denatured DNA fragment stops moving when it reaches its melting point (threshold of enough denaturant). Due to this requirement, a partially melted double-stranded DNA can no longer migrate through the gel (Fischer and Lerman 1983). As GC content is one of the primary factors impacting the separation mechanism in DGGE, a GC clamp (~40 bp with high GC content) is added to the forward primer to anchor the PCR fractions together once they have denatured (Rettedal et al. 2010).

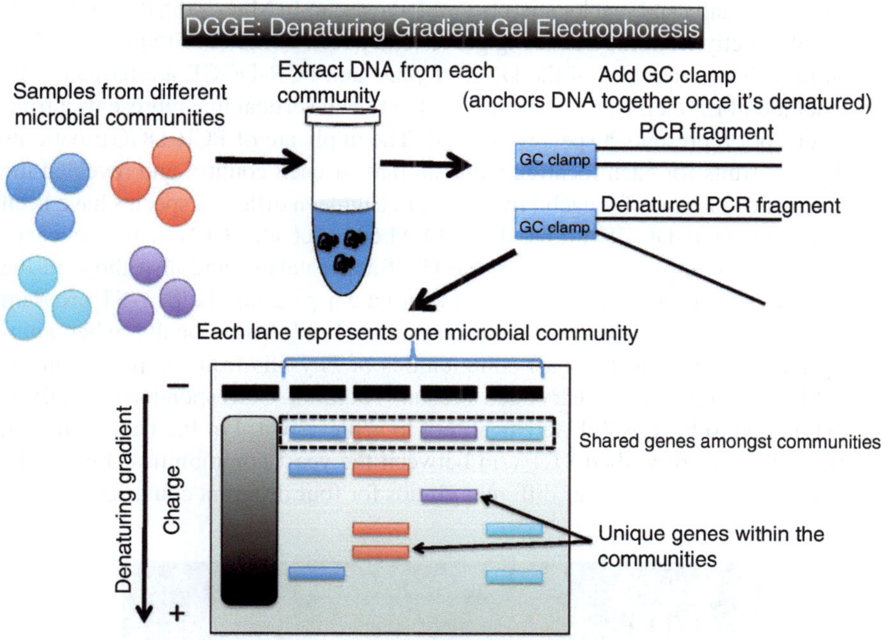

Fig. 1.9 Assessment of different microbial communities by DGGE (*Source*: https://en.wikipedia.org/wiki/File:Step-by-step_procedure_of_using_DGGE_analysis_in_microbiology.pdf)

Benefits of technique: Several studies have shown that PCR-DGGE is reliable, reproducible, rapid, inexpensive, and analyzing a large number of samples in just one step (Ercolini 2004; El Sheikha et al. 2011a; Vaz-Moreira et al. 2013; El Sheikha 2015c). The DGGE enables not only the study of microbial diversity but also can be coupled with other techniques such as cloning and subsequent sequencing to analyze the specific DNA sequences (Muyzer et al. 1993). In this method, the separation is not based on the size of the DNA fragment, but rather the melting property of the PCR product that property is considered one of the advantages of this approach. Therefore, DGGE can be more discriminating than other molecular biology-based approaches to study the microbial diversity. Also this technique could be used as a traceability and safety tool at the same time (El Sheikha and Montet 2016; El Sheikha and Xu 2017).

Applications to medicinal plants: DGGE is strongly preferred and is considered one of the best techniques for monitoring the microbial communities of plant samples (El Sheikha et al. 2009, 2011b; El Sheikha 2010a, 2011; El Sheikha and Montet 2011; Nganou et al. 2012; El Sheikha and Ray 2014).

Physalis possesses many medicinal properties as the whole plant, including antipyretic, depurative, diuretic, pectoral, and vermifuge. Furthermore, plant decoction is used in the treatment of abscesses, cough, fevers, or a sore throat (Duke and Ayensu 1985; El Sheikha 2004; El Sheikha et al. 2008b). Mature fruits of Physalis (*Physalis ixocarpa* Brot., *Physalis pruinosa* L., *Physalis peruviana* L.) were collected in a particular field in two different districts from three countries: Colombia,

Egypt, and Uganda and one location from Madagascar. In May 2009, the fruits were collected directly from the tree using gloves and put in sterile bags to preserve their initial flora. The extraction of the DNA of yeast and PCR-DGGE are based on the new method of El Sheikha (2010b). In Fig. 1.10, each vertical line represents a fruit, and each spot represents a species of yeast. The duplicate of PCR-DGGE patterns of Physalis fruits for each location were similar for each country and revealed the presence of 6–11 bands for each Physalis fruit. Eighteen different species have been identified by PCR-DGGE (Table 1.2) (El Sheikha et al. 2012). The sequences obtained were compared to those in the GenBank database and also those in the Ribosomal database (http://rdp.cme.msu.edu/index.jsp) using the BLAST program (http://ncbi.nlm.nib.gov/BLAST/) (Altschul et al. 1997). Additionally, when comparing the similarity of the yeast communities of Physalis fruit samples from the four different countries in harvested season, Factorial Correspondence Analysis (FCA) proved to be a useful statistical tool to achieve that. For the fruits samples, the two variances described 81.1% in between the yeast communities (Fig. 1.11). We can observe precisely four different groups for four different countries.

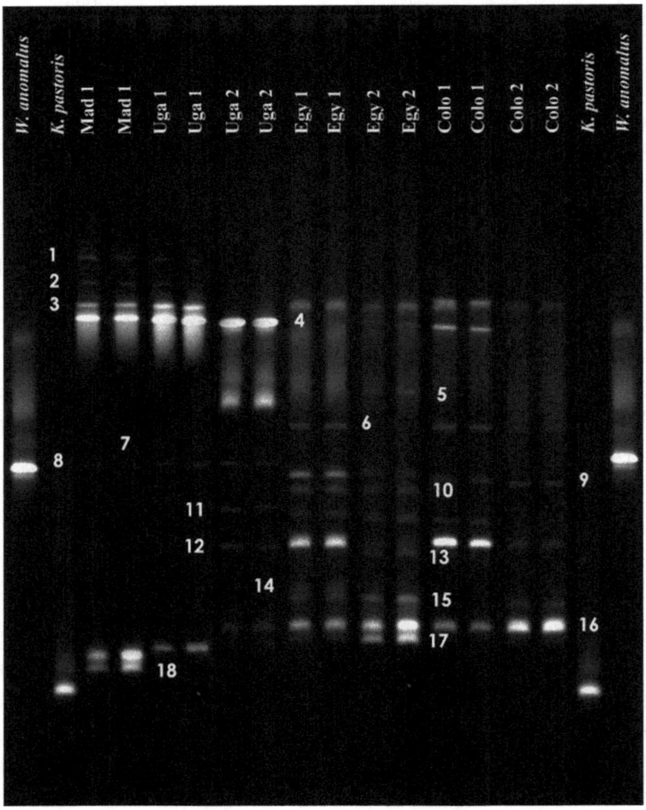

Fig. 1.10 PCR-DGGE pattern of yeast DNA of different varieties of Physalis from four countries. *Colo* Colombia, *Egy* Egypt, *Uga* Uganda, *Mad* Madagascar. (*1*, *2*) Two different locations (El Sheikha et al. 2012). Reproduced with permission of Elsevier

Table 1.2 Dominant yeast species on Physalis fruits identified by DGGE (*Source*: El Sheikha et al. 2012)

Band (s)	Closest relative	% Identity[a]	Source
1	*Glactomyces* sp.	85%	AB563187
2	*Pythium aphanidermatum*	97%	EU253558
3	*Trichosporon asahii*	99%	AB617985
4	*Hanseniaspora uvarum*	90%	EU441908
5	*Aureobasidium pullulans*	93%	HM461723
6	*Candida* sp.	97%	AY242304
7	*Candida zemplinina*	92%	HM191633
8	*Wickerhamomyces anomalus*	87%	FR717639
9	*Candida tropicalis*	94%	HM589856
10	*Rhodotorula graminis*	99%	FR717634
11	*Pichia guillermondii*	99%	HO211992
12	*Torulaspora delbrueckii*	98%	GU225752
13	*Pichia* sp.	95%	AB598080
14	*Issatchenkia orientalis*	98%	EF460648
15	*Saccharomyces cerevisae*	97%	CAA89576
16	*Pichia fermentans*	99%	HQ262367
17	*Pichia kluyveri*	96%	FN667994
18	*Pichia membranifaciens*	87%	FR691649

Reproduced with permission of Elsevier
[a]% Similarity with the reference strain

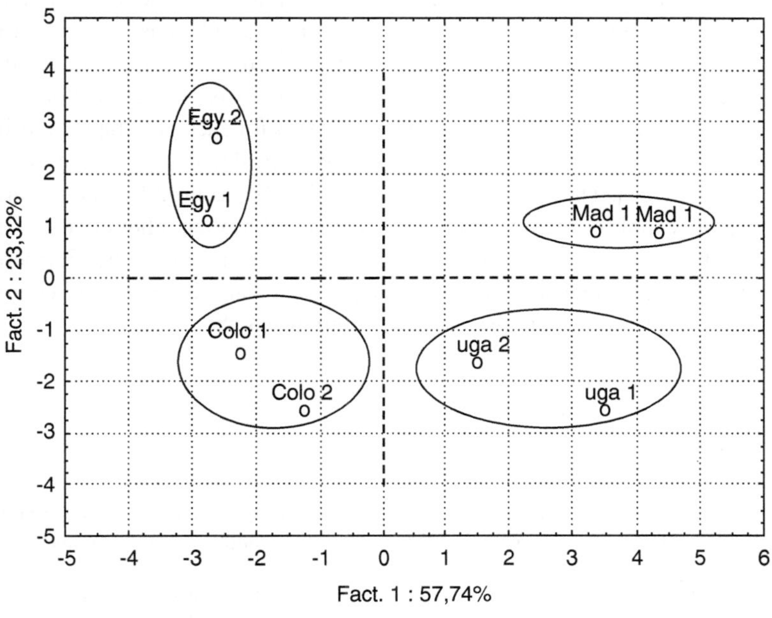

Fig. 1.11 Factorial analysis of yeast DNA of different varieties of Physalis from four countries. *Colo* Colombia, *Egy* Egypt, *Uga* Uganda, *Mad* Madagascar. (*1*, *2*) Two different locations (El Sheikha et al. 2012). Reproduced with permission of Elsevier

Conclusion and Future Prospects

During the last 10 years, an intense interest has emerged in "nutraceuticals" (or "functional foods") in which phytochemical constituents can have long-term health-promoting or medicinal qualities. The differentiation between the nutraceuticals and herbs can sometimes be vague. The primary characteristic of nutraceuticals is not only the nutritional role in the diet but also its benefits to health in the long term (i.e., chemoprevention). In contrast, many medicinal plants possess particular therapeutic advantages (over short or long intervals) without any nutritional role. There is universal interest in traditional and alternative medicine, and represent that interest in the growing demand for the use of medicinal plants in medical practices, and this is what the WHO reported recently. There is vast spread of herbal products in the global market (e.g., food additives, herbal teas, juices, and essential oils). Also the continuous development of the research and practice of their medicinal uses (i.e., patents) was a turning point of herbs from the centuries-old traditions until it became one of the certified therapeutic methods.

Perhaps the threat of extinction facing plant wealth in general and medicinal plants in particular is alarm to the world to make greater efforts toward:

- Conservation of natural resources to ensure to maintain a balance and biodiversity
- Maximizing the interest and benefit of medicinal plants as a natural resource for the treatment of the most severe diseases of this age, including cancer

Numerous tracking methodologies are now available for medicinal plants, but regardless of the approach, critical questions still need to be answered before their use. These include questions on sensitivity, accuracy, robustness, frequency of testing, and cost. Molecular biology methods will be increasingly used shortly. PCR-DGGE profiles have proven to be biological markers for each geographical origin of Physalis fruits (as a medicinal plant) and could be considered as a provider of unique biological barcodes. This global technique can thus be proposed as a rapid analytical traceability tool for medicinal plants (less than 24 h) than other microbial methods avoids the precise analysis of yeast by biochemistry or sequencing. Furthermore, the diversity studies of many others medicinal plants and its microbial ecology in which they occur provide another area for future research.

References

Abera B (2014) Medicinal plants used in traditional medicine by Oromo people, Ghimbi district, Southwest Ethiopia. J Ethnobiol Ethnomed 10:40. http://www.ethnobiomed.com/content/10/1/40. Accessed 16 July 2016

Abramov V (1996) Traditional medicine, vol 134. World Health Organization, Geneva, pp 1–3

Adefa M, Abraha B (2011) Ethnobotanical survey of traditional medicinal plants in Tehuledere district, South Wollo, Ethiopia. J Med Plant Res 5(26):6233–6242

AESGP (Association Européenne des Spécialités Pharmaceutiques Grand Public; The Association of the European Self-Medication Industry) (1998) Herbal medicinal products in the European Union. Study carried out on behalf of the European Commission, Brussels. http://pharmacos.eudra.org/F2/pharmacos/docs/doc99/Herbal%20Medecines%20EN.pdf. Accessed 16 July 2016

Aggarwal BB, Sundaram C, Malani N, Ichikawa H (2007) Curcumin: the Indian solid gold. In: Aggarwal BB, Surh Y-J, Shishodia S (eds) The molecular targets and therapeutic uses of curcumin in health and disease. Springer, New York, NY

Agoreyo BO, Okoro NC, Choudhary MI (2012) Preliminary phytochemical analysis of two varieties of *Adenia lobata* (Jacq) and the antioxidant activity of their various solvent fractions. Bayero J Pure Appl Sci 5(1):182–186. doi:10.4314/bajopas.v5i1.32

Ahmed E, Arshad M, Saboor A, Qureshi R, Mustafa G, Sadiq S, Chaudary KS (2013) Ethnobotanical appraisal and medicinal use of plants in Patriata, New Murree, evidence from Pakistan. J Ethnobiol Ethnomed 9:13. http://www.ethnobiomed.com/content/9/1/13. Accessed 16 July 2016

AHPA-ERB (AHPA Foundation for Education & Research on Botanicals) (2008) A field guide to herbal dietary supplements. AHPA-ERB Foundation, Silver Spring, MD. http://www.ahpa.org/Portals/0/Documents/fieldguide.pdf?ver=2014-05-14-132502-000. Accessed 16 July 2016

Al-Harbi MM, Qureshi S, Ahmed MM, Raza M, Baig MZA, Shah AH (1996) Effect of camel urine on the cytological and biochemical changes induced by cyclophosphamide in mice. J Ethnopharmacol 52(3):129–137

Ali NA, Julich WD, Kusnick C, Lindequist U (2001) Screening of Yemeni medicinal plants for antibacterial and cytotoxic activities. J Ethnopharmacol 74:173–179

Ali-Shtayeh MS, Yaniv Z, Mahajna J (2000) Ethnobotanical survey in the Palestinian area: a classification of the healing potential of medicinal plants. J Ethnopharmacol 73:221–232

Almeida CFCBR, Amorim ELC, Albuquerque UP, Maia MBS (2006) Medicinal plants popularly used in the Xingó region – a semi-arid location in Northeastern Brazil. J Ethnobiol Ethnomed 2:1–7

Al-Qura'n S (2009) Ethnopharmacological survey of wild medicinal plants in Showbak, Jordan. J Ethnopharmacol 123:45–50

Al-Sokari SS, El Sheikha AF (2015) In vitro antimicrobial activity of crude extracts of some medicinal plants from Al-Baha region in Saudi Arabia. J Food Nutr Sci 3(1–2):74–78

Altschul SF, Madden TL, Schäffer AA, Zhang J, Zhang Z, Miller W, Lipman DJ (1997) Gapped BLAST and PSI-BLAST: a new generation of protein databases search programs. Nucleic Acids Res 25(17):3389–3402

Arias TD (1999) Glosario de Medicamentos: desarollo, evaluacĩon y uso. Organizacĩon Panamricana de La Salud. Organizacĩon Mundial de La Salud, Washington, DC

Arsdall AV (2002) Medieval herbal remedies: the old English herbarium and Anglo-Saxon Medicine. Psychology Press, New York, NY

Azaizeh H, Saad B, Cooper E, Said O (2010) Traditional Arabic and Islamic medicine, a re-emerging health aid. Evid Based Complement Alternat Med 7(4):419–424

Baek SH, Bae ON, Park JH (2012) Recent methodology in ginseng analysis. J Ginseng Res 36:119–134

Baydoun S, Lamis C, Helena D, Nelly A (2015) Ethnopharmacological survey of medicinal plants used in traditional medicine by the communities of Mount Hermon, Lebanon. J Ethnopharmacol 173:139–156

Bedir E, Khan IA, Walker LA (2002) Biologically active steroidal glycosides from *Tribulus terrestris*. Pharmazie 57:491–493

Berger E (2001) The Canada Health Monitor surveys of health issues in Canada, Survey 22. Ottawa, ON, Health Canada

Betancur-Galvis LA, Morales GE, Forero JE, Roldan J (2002) Cytotoxic and antiviral activities of colombian medicinal plant extracts of the *Euphorbia* genus. Mem Inst Oswaldo Cruz 97:541–546

Billing J, Sherman PW (1998) Antimicrobial functions of spices: why some like it hot. Q Rev Biol 73(1):3–49
Blumenthal M, Busse WR, Goldberg A, Gruenwald J, Hall T, Riggins CW, Rister RS (eds) (1998) The complete German commission E monographs: therapeutic guide to herbal Medicines. American Botanical Council/Integrative Medicine Communications, Austin, TX/Boston, MA
Bojadzievski P (1992) The health services in Bitola through the centuries. Society of Science and Art, Bitola
Boonyaratanakornkit L, Chantaptavan V (1993) Identification and specification of Khao-Yen-Neua and Khao-Yen-Tai. Thai J Pharm Sci 1:79–90
Botros L, Sakkas D, Seli E (2008) Metabolomics and its application for non-invasive embryo assessment in IVF. Mol Hum Reprod 14:679–690
British Pharmacopoeia Commission (2007) British pharmacopoeia 2007. The Stationery Office, London. http://webarchive.nationalarchives.gov.uk/20080804194642/mhra.gov.uk/committees/medicinesadvisorybodies/britishpharmacopoeiacommission/index.htm. Accessed 16 July 2016
Burslem DFRP, Garwood NC, Thomas SC (2001) Tropical forest diversity—the plot thickens. Science 291:606–607
Calapai G (2008) European legislation on herbal medicines: a look into the future. Drug Saf 31(5):428–431
Cancer Research UK (2014) A study looking at the use of herbal medicine by people with cancer. http://www.cancerresearchuk.org/about-cancer/find-a-clinical-trial/a-study-use-herbal-medicines-cancer. Accessed 16 July 2016
Cancer Research UK (2016) About chemotherapy for lung cancer. http://www.cancerresearchuk.org/about-cancer/type/lung-cancer/treatment/chemotherapy/about-chemotherapy-for-lung-cancer. Accessed 16 July 2016
Castleman M (2001) The new healing herbs: the classic guide to nature's best medicines featuring the top 100 time-tested herbs. Rodale Inc., New York, NY
Cevallos-Cevallos JM, Futch DB, Shilts T, Folimonova SY, Reyes-De-Corcuera JI (2012) GC–MS metabolomic differentiation of selected citrus varieties with different sensitivity to citrus huanglongbing. Plant Physiol Biochem 53:69–76
Chaudhary A, Singh N (2010) Herbo mineral formulations (Rasaoushadhies) of Ayurvedia an amazing inheritance of Ayurvedic pharmaceutics. Anc Sci Life 30(1):18–26
Chayamarit K (1995) Thai medicinal plants, 5th edn. Department of Forestry, Bangkok
Costa-Lotufo LV, Khan MH, Ather A, Wilke DV, Jimenez PC, Pessoa C, Moraes ME (2005) Studies of the anticancer potential of plants used in Bangladeshi folk medicine. J Ethnopharmacol 99:21–30
Council of Europe, Strasburg. 2008. European pharmacopoeia. 6th edn., www.worldcat.org/title/european-pharmacopoeia/oclc/170932841. Accessed 16 July 2016
Cox PA (2000) Will tribal knowledge survive the millennium? Science 287:44–45
Cragg GM, Newman DJ (2005) Plants as sources of anti-cancer agents. J Ethnopharmacol 100:72–79
Cunninghum AB (2001) Applied ethnobotany: people, wild plant use and conservation. Earthscan publications Ltd., London
Dafni A, Levy S, Lev E (2005) The ethnobotany of Christ's Thorne jujube (*Ziziphus spina-christi*) in Israel. J Ethnobiol Ethnomed 1:8
Dervendzi V (1992) Contemporary treatment with medicinal plants. Tabernakul, Skopje
Duke JA, Ayensu ES (1985) Medicinal plants of China. Reference Publications, Inc., Algonac, MI
Dunn WB, Ellis DI (2005) Metabolomics: current analytical platforms and methodologies. Trac Trends Anal Chem 24:285–294
Dwivedi G, Dwivedi S (2007) History of medicine: Sushruta – the clinician – teacher par excellence. http://medind.nic.in/iae/t07/i4/iaet07i4p243.pdf. Accessed 16 July 2016
El Sheikha AF (2015a) New strategies of traceability for determining the geographical origin of medicinal plants: innovation of biological barcode by PCR-DGGE. ACMAP 5th Annual

Conference Program book FARGO-ND. J Med Act Plants 4(2): Supplement. http://scholarworks.umass.edu/jmap/vol4/iss2/3. Accessed 16 July 2016

El Sheikha AF (2004) Technological, chemical and microbiological studies on some packed foods. Dissertation, Minufiya University

El Sheikha AF (2010a) Determination of the geographical origin of fruits by using 26S rDNA fingerprinting of yeast communities by PCR-DGGE: an application to Shea tree fruits. J Life Sci 4(6):9–15

El Sheikha AF (2010b) Determination of geographical origin of Shea tree and Physalis fruits by using the genetic fingerprints of the microbial community by PCR/DGGE: analysis of biological properties of some fruit extracts. Dissertation, Montpellier University II

El Sheikha AF (2011) Détermination de l'origine géographique des fruits: exemples du karité et du Physalis par l'utilisation d'empreintes génétiques sur la communauté microbienne par PCR/DGGE. In: Leclerc C (ed). Éditions Universitaire Européennes, GmbH & Co. KG, Sarrebruck

El Sheikha AF (2015b) *Rumex nervosus*: an overview. Int J Innov Hortic 4(2):87–95

El Sheikha AF (2015c) New strategies for tracing foodstuffs: biological barcodes utilizing PCR-DGGE. Adv Food Technol Nutr Sci Open J 1(1):1–7

El Sheikha AF, Al-Sokari SS, Al-Ghamdi AY (2014) Ithrib (*Rumex nervosus* Vahl) as a famous ethno-medicinal plant of Saudi Arabia. In: 5th Annual Conference of the American Council for Medicinally Active Plants (ACMAP). 15–18 June 2014. Fargo, North Dakota, USA

El Sheikha AF, Bouvet J-M, Montet D (2011a) Biological bar-code for the determination of geographical origin of fruits by using 28S rDNA fingerprinting of fungal communities by PCR-DGGE: an application to Shea tree fruits. Qual Assur Saf Crops Foods 3(1):40–47

El Sheikha AF, Condur A, Métayer I, Le Nguyen DD, Loiseau G, Montet D (2009) Determination of fruit origin by using 26S rDNA fingerprinting of yeast communities by PCR-DGGE: preliminary application to Physalis fruits from Egypt. Yeast 26:567–573

El Sheikha AF, Durand N, Sarter S, Okullo JBL, Montet D (2012) Study of the microbial discrimination of fruits by PCR-DGGE: application to the determination of the geographical origin of Physalis fruits from Colombia, Egypt, Uganda and Madagascar. Food Control 24(1–2):57–63

El Sheikha AF, Métayer I, Montet D (2011b) A Biological bar-code for determining the geographical origin of fruit by using 28S rDNA fingerprinting of fungi communities by PCR-DGGE: an application to Physalis fruits from Egypt. Food Biotechnol 25(2):115–129

El Sheikha AF, Montet D (2011) Determination of fruit origin by using 28S rDNA fingerprinting of fungi communities by PCR-DGGE: an application to Physalis fruits from Egypt, Uganda and Colombia. Fruits 66(2):79–89

El Sheikha AF, Montet D (2016) How to determine the geographical origin of seafood? Crit Rev Food Sci Nutr 56(2):306–317

El Sheikha AF, Ray R (2014) Is PCR-DGGE innovative molecular tool for detection of microbial plant pathogens? In: Sharma N (ed) Biological controls for preventing food deterioration. John Wiley & Sons, Ltd., Chichester

El Sheikha AF, Sainz de Los Terreros A, Nabet C, Montet D, Larroque M, Pelissier Y (2008a). Le Coqueret du Pérou (*Physalis peruviana* L.). 1st Prize of Ethnopharmacology and Ethnopharama laboratory. 10th International Symposium of Aromatherapy and Medicinal Plants, Grasse, 11–13 April 2008

El Sheikha AF, Xu J (2017) Traceability as a key of seafood safety: reassessment and possible applications. Rev Fish Sci Aquac 25(2):158–170. doi:10.1080/23308249.2016.1254158

El Sheikha AF, Zaki M, Bakr A, El Habashy M, Montet D (2008b) Physico-chemical properties and biochemical composition of Physalis (*Physalis pubescens* L.) fruits. Global Science Books Ltd., UK. Food 2:124–130

Ercolini D (2004) PCR-DGGE fingerprinting: novel strategies for detection of microbes in food. J Microbiol Methods 56:297–314

European Commission (2004) Directive 2004/24/EC of the European Parliament and of the Council of 31 March 2004 amending, as regards traditional herbal medicinal products, Directive 2001/83/EC on the Community code relating to medicinal products for human use.

30/04/20104. Official Journal of the European Union: L136/85. http://ec.europa.eu/health/files/eudralex/vol-1/dir_2004_24/dir_2004_24_en.pdf. Accessed 16 July 2016
Falodun A (2010) Herbalmedicine in Africa-distribution, standardization and prospects. Res J Phytochem 4(3):154–161
Farnsworth NR (1994) Ethnopharmacology and drug development. In: Prance GT (ed) Ethnobotany and the search for new drugs. Wiley, Chichester
Farnsworth NR, Bunyapraphatsara N (1992) Thai medicinal plant: recommended for primary health care system. Prachachon company, Bangkok
FDA (Food and Drug Administration) (2015) Botanical drug development guidance for industry. http://www.fda.gov/downloads/drugs/guidancecomplianceregulatoryinformation/guidances/ucm458484.pdf. Accessed 16 July 2016
Fischer SG, Lerman LS (1983) DNA fragments differing by single base-pair substitutions are separated in denaturing electrophoreseis analysis of polymerase chain reaction-amplified genes coding for 16S rRNA. Proc Natl Acad Sci U S A 80(6):1579–1583
Fouche G, Cragg GM, Pillay P, Kolesnikova N, Maharaj VJ, Senabe J (2008) *In vitro* anticancer screening of South African plants. J Ethnopharmacol 119:455–461
Gali-Muhtasib H, Diab-Assaf M, Boltze C, Al-Hmaira J, Hartig R, Roessner A, Schneider-Stock R (2004) Thymoquinone extracted from black seed triggers apoptotic cell death in human colorectal cancer cells via a p53-dependent mechanism. Int J Oncol 25:857–866
GIA (Global Industry Analysts, Inc.) (2016) A worldwide business strategy & market intelligence source. http://www.strategyr.com/. Accessed 16 July 2016
Gimmel M (2008) Reading medicine in the Codex de la Cruz Badiano. J Hist Ideas 69(2):169–192
Glaser V (1999) Billion-dollar market blossoms as botanicals take root. Nat Biotechnol 17:17–18
Gonsalves J (2010) Economic botany and ethnobotany. International Scientific Academy, Delhi
Gurib-Fakim A (2006) Medicinal plants: traditions of yesterday and drugs of tomorrow. Mol Aspects Med 27(1):1–93
Harshaw D, Nahar L, Vadla B, Saif-E-Naser GM, Sarker SD (2010) Bioactivity of *Rumex obtusifolius* (Polygonaceae). Arch Biol Sci Belgrade 62(2):387–392
Hartwell JL (1982) Plants used against cancer. Quarterman, Lawrence, MA
Hudaib M, Mohammad M, Bustanji Y, Tayyeb R, Yousef M, Aburjeie M, Aburjai T (2008) Ethnopharmacological survey of medicinal plants in Jordan, Mujib Nature Reserve and surrounding area. J Ethnopharmacol 120:63–71
Humber JM (2002) The role of complementary and alternative medicine: accommodating pluralism. JAMA 288(13):1655–1656
IARC (International Agency for Research on Cancer) (2002) Some traditional herbal medicines, some mycotoxins, naphthalene and styrene. IARC Monogr Eval Carcinog Risks Hum 82:1–556
Ismail Z, Ismail N, Lassa J (1999) Malaysian herbal monograph, vol 1. Malaysian Monograph Committee, Kuala Lumpur
Itharat A, Houghton PJ, Eno-Amooquaye E, Burke PJ, Sampson J, Raman A (2004) In vitro cytotoxic activity of Thai medicinal plants used traditionally to treat cancer. J Ethnopharmacol 90:33–38
Itharat A, Singchangchai P, Subchareon P, Supavita T, Ubonkaw P, Ratanasuwan P, Kummee S, Srisuchatum P (2002) Folk wisdom about treatment of traditional doctors in Southern Thailand: focus groups discussion. Prince of Songkla University, Songkla
Itharat A, Supavita T, Jusanit P, Singchangchai P, Subchareon P, Deevisad G, Ubonkaw P, Ratanasuwan P, Kajpunyapong V, Muangsrinun R (1999b) Survey of medicinal plants and local folk wisdom in Southern Thailand: case study at Talebun National Parks. Prince of Songkla University, Songkla
Itharat A, Supavita T, Singchangchai P, Kajjarun U, Ratanasuwan P, Pongpeerachote S, Chantaptawan V (1999a) Identification and specification of Khao-Yen-Neua and Khao-Yen-Tai from traditional doctors in Thailand. Prince of Songkla University, Songkla

Jacquart D (2008) Islamic pharmacology in the middle ages: theories and substances. Eur Rev 16(2):219–227

Jančić R (2002) Botanika farmaceutika. Public company Sl. List SRJ, Beograd

Johnson T (2001) News. Chinese medicine now part of primary care scene in BC. Can Med Assoc J 164:1195

Khan MY, Aliabbas S, Kumar V, Rajkumar S (2009) Recent advances in medicinal plant biotechnology. Indian J Biotechnol 8:9–22

Kovacevic N (2000) Fundamentals of pharmacognosy. Personal edition, Beograd

Kozyrskyj A (1997) Herbal products in Canada. How safe are they? Can. Family Phys Med Famille Can 43:697–702

Krek M (1979) The enigma of the first Arabic book printed from movable type. J Near East Stud 38(3):203–212

Kuga H, Ejima A, Mitui I, Sato K, Ishihara N, Fukuda K, Saito F, Uenakai K (1993) Isolation and characterizacion of cytotoxic compounds from corn. Biosci Biotechnol Biochem 57(6):1020–1021. doi:10.1271/bbb.57.1020

Lambertini E, Piva R, Khan MTH, Lampronti I, Bianchi N, Borgatti M, Gambari R (2004) Effects of extracts from Bangladeshi medicinal plants on in vitro proliferation of human breast cancer cell lines and expression of estrogen receptor alpha gene. Int J Oncol 24:419–423

Langmead L, Rampton DS (2001) Review article: herbal treatment in gastrointestinal and liver disease-benefits and dangers. Aliment Pharmacol Ther 15:1239–1252

Lee CC, Houghton P (2005) Cytotoxicity of plants from Malaysia and Thailand used traditionally to treat cancer. J Ethnopharmacol 100:237–243

Lee D-K, Lim DK, Um JA, Lim CJ, Hong JY, Yoon YA, Ryu Y, Kim HJ, Cho HJ, Park JH, Seo YB, Kim K, Lim J, Kwon SW, Lee J (2014) Evaluation of four different analytical tools to determine the regional origin of *Gastrodia elata* and *Rehmannia glutinosa* on the basis of metabolomics study. Molecules 19:6294–6308

Lee KH (1999) Antitumor agents 195 Anticancer drug design based on plant derived natural products. J Biomed Sci 6:236–250

Leroi-Gourhan A (1975) The flowers found with Shanidar IV, a Neanderthal burial in Iraq. Science 190(4214):562–564

Li C, Yang S-C, Guo Q-S, Zheng K-Y, Shi Y-F, Xiao X-F, Long G-Q (2014) Determining the geographical origin of the medicinal plant Marsdenia tenacissima with multi-element analysis and data mining techniques. Chemom Intel Lab Syst 136:115–120

Li X-J, Zhang H-Y (2008) Western-medicine-validated anti-tumor agents and traditional Chinese medicine. Trends Mol Med 14(1):1–2

López-Rituerto E, Savorani F, Avenoza A, Busto JH, Peregrina JM, Engelsen SB (2012) Investigations of La Rioja terroir for wine production using ^{1}H NMR metabolomics. J Agric Food Chem 60:3452–3461

Lukic P (1985) Pharmacognosy. SSO Faculty of Pharmacy, Beograd

Luo F, Lu R, Zhou H, Hu F, Bao G, Huang B, Li Z (2013) Metabolic effect of an exogenous gene on transgenic *Beauveria bassiana* using Liquid Chromatography–Mass Spectrometry-based metabolomics. J Agric Food Chem 61:7008–7017

Madigan MT, Martinko JM, Dunlap PV, Clark DP (2009) Brock biology of microorganisms, 12th edn. Pearson Education Inc., San Francisco, CA

Mamedov N (2012) Medicinal plants studies: history, challenges and prospective. Med Aromat Plants 1(8):e133. doi:10.4172/2167-0412.1000e133

Mekonnen T, Urga K, Engidawork E (2010) Evaluation of the diuretic and analgesic activities of the rhizomes of *Rumex abyssinicus* Jacq in mice. J Ethnopharmacol 127:433–439

Moco S, Vervoort J, Bino RJ, de Vos RCH, Bino R (2007) Metabolomics technologies and metabolite identification. Trac Trends Anal Chem 26:855–866

Moghadamtousi SZ, Rouhollahi E, Karimian H, Fadaeinasab M, Mohammad F, Ameen Abdulla M, Abdul Kadir H (2015) The chemopotential effect of *Annona muricata* leaves against azoxymethane-induced colonic aberrant crypt foci in rats and the apoptotic effect of acetogenin annomuricin E in HT-29 cells: a bioassay-guided approach. PLoS One 10(4):e0122288. doi:10.1371/journal.pone.0122288

Moreno-Escobar JA, Bazald S, Villarreal ML, Bonilla-Barbosa JR, Mendoza S, Rodríguez-López V (2011) Cytotoxic and antioxidant activities of selected Lamiales species from Mexico. Pharm Biol 49(12):1243–1248. doi:10.3109/13880209.2011.589454

Munavu RM, Mudamba LO, Ogur JA (1984) Isolation and characterization of the major anthraquinone pigments from *Rumex abysinica*. Planta Med 50(1):111. doi:10.1055/s-2007-969640

Munos B (2009) Lessons from 60 years of pharmaceutical innovation. Nat Rev Drug Discov 8:959–968

Muyzer G (1999) DGGE/TGGE a method for identifying genes from natural ecosystems. Curr Opin Microbiol 2(3):317–322

Muyzer G, De Waal EC, Uitterlinden AG (1993) Profiling of complex microbial populations by denaturing gradient gel electrophoresis analysis of polymerase chain reaction-amplified genes coding for 16S rRNA. Appl Environ Microbiol 59:695–700

Nair M, Nathan G (1998) Medicinal plants: cure for the 21th century; Biodiversity, conservation and utilization of Medicinal plants: Proceedings of the seminar UPM, Serdang, Malaysia

Nelson D, Cox M (2005) Lehninger principles of biochemistry, 4th edn. W.H. Freeman and Company, New York, NY

Nganou DN, Durand N, Tatsadjieu NL, Meile JC, El Sheikha AF, Montet D, Mbufung CM (2012) Determination of coffee origin by using 28S rDNA fingerprinting of fungal communities by PCR-DGGE: application to the Cameroonian coffee. Int J Biosci 2:18–30

Nisar MF, Jaleel F, Waseem M, Ismail S, Toor Y, Haider SM, Zhong JL (2014) Ethno-medicinal uses of plants from district Bahawalpur, Pakistan. Curr Res J Biol Sci 6(5):183–190

Nunn JF (1996) Ancient Egyptian medicine. Trans Med Soc Lond 113:57–68

Oakes L, Gahlin L (2003) Ancient Egypt: an illustrated reference to the myths, religions, pyramids and temples of the land of the pharaohs. Barnes & Noble, New York, NY

Oran SA, Al-Eisawi DM (1998) Check list of medicinal plants in Jordan. Dirasat 25:84–112

Padulosi S, Leaman D, Quek P (2002) Challenges and opportunities in enhancing the conservation and use of medicinal and aromatic plants. J Herbs Spices Med Plants 9:243–267

Pan S-Y, Litscher G, Gao S-H, Zhou S-F, Yu Z-L, Chen H-Q, Zhang S-F, Tang M-K, Sun J-N, Ko K-M (2014) Historical perspective of traditional indigenous medical practices: the current renaissance and conservation of herbal resources. Evid Based Complement Alternat Med 2014:525340. doi:10.1155/2014/525340

Pan S-Y, Pan S, Yu Z-L, Ma D-L, Chen S-B, Fong W-F, Han Y-F, Ko K-M (2010) New perspectives on innovative drug discovery: an overview. J Pharm Pharm Sci 13(3):450–471

Petrovska BB (2012) Historical review of medicinal plants' usage. Pharmacogn Rev 6(11):1–5

Popoca J, Aguilar A, Alonso D, Villarreal ML (1998) Cytotoxic activity of selected plants used as antitumorals in Mexican traditional medicine. J Ethnopharmacol 59:173–177. doi:10.1016/S0378-8741(97)00110-4

Rajeshkumar NV, Pillai MR, Kuttan R (2003) Induction of apoptosis in mouse and human carcinoma cell lines by *Emblica officinalis* polyphenols and its effect on chemical carcinogenesis. J Exp Clin Res 22:201–212

Rao SR, Ravishankar GA (2002) Plant cell cultures: chemical factories of secondary metabolites. Biotechnol Adv 20:101–153

Rates SMK (2001) Plants as source of drugs. Toxicon 39:603–613

Rettedal EA, Clay S, Brözel VS (2010) GC-clamp primer batches yield 16S rRNA gene amplicon pools with variable GC clamps, affecting denaturing gradient gel electrophoresis profiles. FEMS Microbiol Lett 312(1):55–62

Ríos MY, González-Morales A, Villarreal ML (2001) Sterols, triterpenes and bioflavonoids of *Viburnum jucundum* and cytotoxic activity of ursolic acid. Planta Med 67:683–684. doi:10.1055/s-2001-17357

Robson B, Baek OK (2009) The engines of hippocrates: from the dawn of medicine to medical and pharmaceutical informatics. John Wiley & Sons, Hoboken, NJ

Rojas-Sepúlveda A, Mendieta-Serrano M, Antúnez Mojica M, Salas-Vidal E, Marquina S, Villarreal ML, Puebla AM, Delgado J, Alvarez L (2012) Cytotoxic podophyllotoxin type-lignans from the steam bark of *Bursera fagaroides* var. *fagaroides*. Molecules 17:9506–9519. doi:10.3390/molecules17089506

Roy MK, Nakahara K, Na Thalang V, Trakoontivakorn G, Takenaka M, Isobe S, Tsushida T (2007) Baicalein, a flavonoid extracted from a methanolic extract of *Oroxylum indicum* inhibits proliferation of a cancer cell line in vitro via induction of apoptosis. Pharmazie 62:149–153

Saad B, Azaizeh H, Said O (2005) Tradition and perspectives of Arab herbal medicine: a review. Evid Based Complement Alternat Med 2(4):475–479

Said O, Fulder S, Khalil K, Eli K, Bashar S (2008) Maintaining a Physiological Blood Glucose Level with 'Glucolevel', a combination of four anti-diabetes plants used in the traditional Arab herbal medicine. eCAM 5(4):421–428

Said O, Khalil K, Fulder S, Azaizeh H (2002) Ethnopharmacological survey of medicinal herbs in Israel, the Golan Heights and the West Bank region. J Ethnopharmacol 83:251–265

Saied AS, Gebauer J, Hammer K, Buerkert A (2008) *Ziziphus spina-christi* (L.) Willd: a multipurpose fruit tree. Genet Resour Crop Evol 55:929–937

Sawahel W (1997) Plant genetic transformation technology. Daya Publishing House, Delhi

Seid MA, Aydagnehum SG (2013) Medicinal plants biodiversity and local healthcare management system in Chencha district; Gamo Gofa, Ethiopia. J Pharmacogn Phytochem 2(1):284–293

Shafik A, Elseesy W (2003) Medicine in ancient Egypt. In: Seli H, Shapiro H (eds) Medicine across cultures. Kluwer, Boston, MA

Sherman PW, Hash GA (2001) Why vegetable recipes are not very spicy. Evol Hum Behav 22(3):147–163

Shetty P, Rinaldi A (2015) Traditional medicine for modern times: facts and figures. http://www.scidev.net/global/medicine/feature/traditional-medicine-modern-times-facts-figures.html. Accessed 16 July 2016

Shi G, MacDougal JM, McLaughlin JL (1997) Bioactive annonaceous acetogenins from *Rollinia mucosa*. Phytochemistry 45:719–723. doi:10.1016/S0031-9422(97)00028-9

Siahsar B, Rahimi M, Tavassoli A, Raissi A (2011) Application of biotechnology in production of medicinal plants. Am Eur J Agric Environ Sci 11(3):439–444

Sigerist HE (1967) A history of medicine, I. Primitive and archaic medicine. Oxford University Press, New York, NY

Sonnedecker G, Kremers E (1986) Kremers and Urdang's history of pharmacy. American Institute of the History of Pharmacy, Madison, WI

Stepp JR (2004) The role of weeds as sources of pharmaceuticals. J Ethnopharmacol 92(2–3):163–166

Sumner J (2000) The natural history of medicinal plants. Timber Press, Portland, OR

Sun B, Qu W, Bai Z (2003) The inhibitory effect of saponins from *Tribulus terrestris* on Bcap-37 breast cancer cell line in vitro. Zhong Yao Cai 26:104–106

Talib WH, Mahasneh AM (2010) Antiproliferative activity of plant extracts used against cancer in traditional medicine. Sci Pharm 78:33–45

Tamokou J d D, Chouna JR, Fischer-Fodor E, Chereches G, Barbos O, Damian G, Benedec D, Duma M, APN E, Wabo HK, Kuiate JR, Mot A, Silaghi-Dumitrescu R (2013) Anticancer and antimicrobial activities of some antioxidant-rich Cameroonian medicinal plants. PLoS One 8(2):e55880. doi:10.1371/journal.pone.0055880

Tasheva K, Kosturkova G (2013) Role of biotechnology for protection of endangered medicinal plants. In: Petre M (ed) Environmental biotechnology - new approaches and prospective applications. InTech, Rijeka

The United States Pharmacopoeial Convention. Washington: 2008. USP 31 the United States Pharmacopoeia. www.usp.org/sites/default/files/usp_pdf/EN/products/usp2008p2supplement3.pdf. Accessed 16 July 2016

Toplak Galle K (2005) Domestic medicinal plants. Mozaic book, Zagreb

Truitt ER (2009) The virtues of balm in late medieval literature. Early Sci Med 14(6):711–736

Tzfira T, Li J, Lacroix B, Citovsky V (2004) *Agrobacterium* tDNA integration: molecules and models. Trends Genet 20:375–383
Vaz-Moreira I, Egas C, Nunes OC, Manaia CM (2013) Bacterial diversity from the source to the tap: a comparative study based on 16S rRNA gene-DGGE and culture-dependent methods. FEMS Microbiol Ecol 83(2):361–374
Vega-Ávila E, Tapia Aguilar R, Jiménez Estrada M, Villareal Ortega M, Román-Ramos R (2004) Cytotoxic activity of *Cuphea aequipetala*. Proc West Pharmacol Soc 47:129–133. doi:10.1016/j.jep.2010.11.055
Vergouw CG, Botros LL, Roos P, Lens JW, Schats R, Hompes PGA, Burns DH, Lambalk CB (2008) Metabolomic profiling by near-infrared spectroscopy as a tool to assess embryo viability: a novel, non-invasive method for embryo selection. Hum Reprod 23:1499–1504
Von Staden H (1989) Herophilus: the art of medicine in early Alexandria: edition, translation, and essays. Cambridge University Press, New York, NY
Wang J-B, Zeng L-N, Zang Q-C, Gong Q-F, Li B-C, Zhang X-R, Chu X-H, Zhang P, Zhao Y-L, Xiao X-H (2012) Colorimetric grading scale can promote the standardization of experiential and sensory evaluation in quality control of traditional Chinese medicines. PLoS One 7:e48887. doi:10.1371/journal.pone.0048887
WHO (2013) WHO traditional medicine strategy: 2014-2023. http://www.who.int/medicines/publications/traditional/trm_strategy14_23/en/. Accessed 16 July 2016
WHO (World Health Organization) (1996) Annex II. Guidelines for the assessment of herbal medicines. (WHO Technical Report Series No. 863), Geneva
Wieczorek A (2003) Use of biotechnology in agriculture—benefits and risks. Governance for industrial transformation. Proceedings of the 2003 Berlin conference on the human resources. Honolulu, HI: University of Hawaii Press. http://hdl.handle.net/10125/3349. Accessed 16 July 2016
Williamson E, Okpako DT, Evans FJ (1996) Selection, preparation and pharmacological evaluation of plant material. Wiley, Chichester
Wu J-N (2005) An illustrated Chinese materia medica. Oxford University Press, New York, NY

Chapter 2
How Plants Can Contribute to the Supply of Anticancer Compounds

J.F. Buyel

Abstract Plants were the first sources of medicines used by humankind, with evidence of herbal remedies dating back at least 60,000 years. Many plants have been used medicinally because they produce secondary metabolites with pharmacological properties, including compounds such as paclitaxel (Taxol) that inhibit cell division and can therefore be used as a treatment for cancer. With the advent of recombinant DNA and molecular biotechnology in the 1970s, plants have also been modified genetically to produce more of their native pharmaceutically active substances, or even nonnative compounds. The scope of medicinal plants has also expanded beyond secondary metabolites to include pharmaceutical recombinant proteins, such as human antibodies. This chapter provides an overview of the anticancer compounds naturally produced in plants and how gene technology has been used to facilitate their production. It also considers how plant-based expression systems can help to supply modern healthcare systems with protein-based anticancer compounds such as monoclonal antibodies, lectins, and anticancer vaccines.

Keywords Lectins • Monoclonal antibodies • Plant molecular pharming • Plant secondary metabolites • Therapeutic anticancer vaccines

Abbreviations

ADCC	Antibody-dependent cellular cytotoxicity
ADCs	Antibody-drug conjugates
APIs	Active pharmaceutical ingredients
ATPS	Aqueous two-phase systems

J.F. Buyel (✉)
Fraunhofer Institute for Molecular Biology and Applied Ecology IME,
Forckenbeckstraße 6, 52074 Aachen, Germany

Institute for Molecular Biotechnology, RWTH Aachen University,
Worringerweg 1, 52074 Aachen, Germany
e-mail: johannes.buyel@ime.fraunhofer.de

S. Malik (ed.), *Biotechnology and Production of Anti-Cancer Compounds*,
DOI 10.1007/978-3-319-53880-8_2

CDC	Complement-dependent cytotoxicity
CHO	Chinese hamster ovary
DoE	Design-of-experiments
EBA	Expanded-bed adsorption
EBV	*Epstein-Barr virus*
FDA	Food and Drug Administration
GMP	Good manufacturing practice
HBsAg	Hepatitis B-soluble antigen
HBV	*Hepatitis B virus*
HCPs	Host cell proteins
HPV	*Human papillomavirus*
mAbs	Monoclonal antibodies
ML1	Mistletoe lectin 1
NK	Natural killer
PAT	Process analytical technology
PEG	Polyethylene glycol
QbD	Quality-by-design
R&D	Research and development
RIP	Ribosome-inactivating protein
RuBisCO	Ribulose-1,5-bisphosphate carboxylase/oxygenase
T_h1	T-helper
T-DNA	Transfer DNA
VEGF	Vascular endothelial growth factor
VFUs	Vertical farming units
VLPs	Virus-like particles

Introduction

Cancer is one of the major challenges in modern medicine and healthcare systems (Yabroff et al. 2011) due to the severe physiological and psychological burden suffered by patients and their families (Faller et al. 2013; Linden and Girgis 2012). Cancer also has a negative impact on the economy in general, costing an estimated $895.2 billion in healthcare-related payments and reduced productivity in 2008 (American Cancer Society 2010).

There were 6.2 million cancer-related deaths in 2003, equivalent to approximately 13% of all deaths worldwide (McGuire 2016). The cancer-related mortality rate is higher in developing countries than industrialized countries because of the socioeconomic conditions that restrict access to anticancer therapies (Sankaranarayanan 2014). More than 14 million new cancer cases are reported each year, and this is expected to increase by 26% over the next 35 years due to demographic changes and improved diagnostics (Pritzkuleit et al. 2010; Rottenberg et al. 2010).

Cancer is not a narrowly defined condition with a single cause, but the collective term for more than 100 different yet related diseases (American Cancer Society 2015). There are gender-specific differences in the incidence of certain cancers that may reflect developmental differences between the sexes; for example women are more frequently diagnosed with breast cancer than men, or lifestyle factors; for example men are more frequently diagnosed with lung cancer than women (American Cancer Society 2015; McGuire 2016). The latter is also the single most frequent cause of cancer-related deaths (1.59 million per year) (McGuire 2016). The common link between all cancer types is that a subset of cells acquires the ability to proliferate in a rapid and uncontrolled manner. In most tissues, cancer cells initially form a localized malignant tumor, but cells eventually break away from the primary tumor and spread through the blood and/or lymphatic systems to form secondary tumors, a process known as metastasis (Alberts et al. 2002). Some tumors remain benign and noninvasive, and these are not classified as cancers (Silverstein et al. 2006). A set of six characteristics has been proposed to define cancer, i.e., (1) self-sufficiency in growth signals, (2) insensitivity to antigrowth signals, (3) evasion of apoptosis, (4) limitless replicative potential, (5) sustained angiogenesis, and (6) tissue invasion and metastasis (Hanahan and Weinberg 2000). An expansion of this set has been proposed more recently, including (7) deregulated metabolism, (8) evasion of the immune system, (9) genome instability, and (10) inflammation (Hanahan and Weinberg 2011). However, these definitions may focus too much on the cellular rather than the tissue level of the diseases (Sonnenschein and Soto 2013).

A predisposition to cancer can be inherited, but it often occurs spontaneously due to exposure to environmental risk factors such as smoking, high-energy radiation, or carcinogenic chemical substances. Cancer can also be caused by infections with certain bacteria, e.g., *Helicobacter pylori* (Hong et al. 2012). More often, cancer can be caused by viruses (Cummins and Tangney 2013; Weinberg 2006), including *Human papillomavirus* (HPV) (Chen et al. 2015), the major cause of cervical cancer, and presumably *Epstein-Barr virus* (EBV), which is linked to Burkitt's lymphoma (Brady et al. 2007). Ultimately, all triggers result in mutations and/or epigenetic changes in DNA structure that inactivate tumor-suppressor genes such as *TP53* (Bieging et al. 2014) or activate proto-oncogenes such as *HER-2* (Chial 2008). The trigger may in some cases be directly mutagenic, e.g., (1) the induction of point mutations by alkylating agents, nucleoside analogs, or intercalating chemicals; (2) the incorrect repair of DNA double-strand breaks induced predominantly by radiation; or (3) the integration of foreign DNA, disrupting the original genetic context and causing aberrant gene expression as observed for some viruses (Akagi et al. 2014). The effects can also be indirect, e.g., the induction of chronic inflammation or infections that promote the proliferation of a subset of cells, e.g., B-lymphocytes, increasing the likelihood of uncontrolled growth as assumed for EBV in Burkitt's lymphoma.

Due to the heterogeneous nature of cancer there is no universal treatment. Four different general approaches are available, which can be followed alone or in combination (Sudhakar 2009). Surgery involves the physical removal of malignant tumor tissue. This can in theory effect a complete cure in a single procedure, but a small number of cancer cells may remain at the excision site eventually leading to the formation of a new tumor. Therefore, healthy tissue adjacent to the tumor is often removed to create a safety margin, which is an undesirable side effect and does not necessarily increase the survival rate (Hernandez et al. 2009; Kubota 2011). Furthermore, in advanced cancers, small secondary tumors may be difficult to locate and/or remove. Therefore, surgery is often combined with chemotherapy or radiotherapy to increase the likelihood of a cure (Salama and Chmura 2014).

Radiotherapy uses X-ray or gamma radiation to damage the DNA of rapidly dividing cancer cells beyond repair, causing the cells to undergo apoptosis. Even though healthy cells typically have more efficient DNA repair mechanisms than cancer cells and can thus better withstand the radiation (Bernstein and Bernstein 2015; Gajecka et al. 2005), they can be affected by radiotherapy as well which can give rise to severe side effects such as the depletion of hematopoietic precursor cells and (in the longer term) even the induction of other forms of cancer (Berrington de Gonzalez et al. 2011; Mauch et al. 1995). The efficacy of high-energy radiation is dependent on the oxygen concentration in the tumor tissue, because the free radicals formed when oxygen interacts with the radiation cause further damage to the DNA of the cancer cells. However, hypoxia is often observed in large solid tumors, which means that radiotherapy is less effective against larger tumors (Harrison et al. 2002).

Chemotherapy is the treatment of cancer with drugs. This approach is advantageous because it can kill residual cancer cells and small, undetectable secondary tumors (Polireddy and Chen 2016). Chemotherapy can also be combined with radiotherapy to increase the therapeutic efficacy (Shahid 2016). One drawback is that the efficacy of chemotherapy depends on the way drugs are distributed in tissues, and poor results are often observed with larger solid tumors due to the limited vascularization, which prevents effective tumor penetration (Minchinton and Tannock 2006). The active pharmaceutical ingredients (APIs) used for chemotherapy are often small molecules, such as paclitaxel (Chabner and Roberts 2005). Such molecules can circulate relatively freely and reach the tumor site(s) even if their precise location is unknown. The first generation of chemotherapeutics were developed to disrupt the metabolism and/or mitotic activity of rapidly dividing cells, whereas the second generation instead targeted signaling components, such as protein kinases or growth factor receptors (Chabner and Roberts 2005). For example, paclitaxel is a first-generation drug that disrupts mitosis by preventing tubulin depolymerization, whereas gefitinib is a second-generation drug that inhibits signaling via the epidermal growth factor receptor (Chabner and Roberts 2005; Wani and Horwitz 2014). Whereas some cancer drugs have a simple structure suitable for total chemical synthesis (Neidle and Thurston 2005), most are complex molecules that must be produced using biotechnology (Baldi et al. 2008; Howat et al. 2014). Paclitaxel provides a useful example of the latter scenario. This compound was originally isolated from the bark of the Pacific yew tree (*Taxus brevifolia*) (Wani and Horwitz 2014), but is now produced in transgenic plant cell suspension cultures at the 75,000-L scale (Zhong 2002). Some cancer drugs demonstrate limited selectivity but most also affect rapidly dividing healthy cells, such as hair follicle cells and B-lymphocytes, resulting in the common side effects of chemotherapy: hair loss and a compromised immune system (Sfikakis et al. 2005; Trueb 2010).

Immunotherapy harnesses the immune system against cancer, and is the most selective treatment approach and therefore the treatment associated with the least severe side effects (Caspi 2008; Schuster et al. 2006). Immunotherapy can take several forms, including the use of vaccines to prevent cancer, as seen with the vaccine against HPV to prevent cervical cancer (De Vincenzo et al. 2013; Poljak 2012), introduction of cytokines to manipulate the immune response, or antibody therapy to target cancer cells in the same way that antibodies normally target pathogens (Schuster et al. 2006). In the latter case, monoclonal antibodies (mAbs) are directed against cancer-specific cell surface structures including receptors and other surface proteins that are overexpressed in tumors, or glycan structures that are more common in cancer cells—these tumor-selective targets are collectively described as tumor markers (Christiansen et al. 2014; Chung and Christianson 2014; Duffy et al. 2014; Weiner et al. 2012). After binding to cancer cells, the mAbs can elicit antibody-dependent cellular cytotoxicity (ADCC) or complement-dependent cytotoxicity (CDC) through their constant domains, causing natural killer (NK) cells to force the cancer cells into apoptosis (Mellstedt 2003). Alternatively, the antibodies may block the binding of growth factors (Gong et al. 2004) or carry a toxic conjugate such as monomethyl auristatin E (Polakis 2016), which is taken up into the tumors. Such antibody-drug conjugates (ADCs) can combine the ADCC of regular antibodies with the additional toxic effect of a conjugated cytotoxic effector (Peters and Brown 2015; Scott et al. 2012). Despite these benefits and several dozen approved products (Scott et al. 2012), antibody therapy requires large doses of expensive and highly pure mAb, e.g., 3 mg per kg body mass (Ben-Kasus et al. 2007) or ~750 mg per square meter of body surface area (Cheson and Leonard 2008), sometimes meaning the equivalent of 6–12 g per patient (Chames et al. 2009) (Table 2.1). This is why immunotherapy can cost several thousand euros per year per patient. The situation can be even worse if the antibody has a short serum half-life or induces an immune response in the patient, because this reduces the effective concentration of the protein and increases the required doses (Glassman and Balthasar 2014; Senter 2009). Furthermore, the comparably large size of an antibody (~150 kDa) can inhibit tumor penetration and reduce the effectiveness of the treatment against large solid tumors (Beckman et al. 2011).

These drawbacks can be circumvented by anticancer vaccines. The benefit of such vaccines is that, in theory, they require only a single low-dose treatment for each patient (e.g., 15 μg), which significantly reduces the costs of production and administration (Table 2.1). Anticancer vaccines exploit the ability of the human adaptive immune system to raise neutralizing antibodies against foreign epitopes (Murphy et al. 2008), in this case epitopes unique to or more abundant on cancer cells or the pathogens that cause cancer. For example, GSK and Merck & Co. have developed vaccines that form viruslike particles (VLPs) based on the L1 surface protein of the most virulent HPV strains, to reduce the risk of infection and the resulting HPV-related cervical cancer (De Vincenzo et al. 2013; Poljak 2012). These vaccines act in a preventive manner, whereas the HPV E6 and E7 proteins are being developed as therapeutic vaccines, i.e., vaccines that prevent the onset of cervical cancer even when an HPV infection has been established (Buyel et al. 2012; Massa et al. 2007; Venuti et al. 2009). Immunotherapy can also be combined with chemotherapy (Bang et al. 2010).

Table 2.1 Drug types used for cancer therapy with the corresponding amounts of API required per treatment

Drug type	API	Disease	Treatment duration [weeks]/ number of doses [–] (for vaccines)	ø API per patient [mg]	Reference
Small molecule	Paclitaxel	Breast cancer	25	5994	Committee for Medicinal Products for Human Use (CHMP (2015a))
Vaccine	Inactivated pertussis toxoid	Acellular pertussis	3	11	Thierry-Carstensen et al. (2013)
	HBsAg[a]	Hepatitis B	4	440	Aziz et al. (2006)
	HBsAg	Hepatitis B	3	6	Baldy et al. (2003)
	HPV L1 protein	HPV infection	3	180	McCormack (2014)
	HPV L1 protein	HPV infection	3	30	Committee for Medicinal Products for Human Use (CHMP (2016))
	Hemagglutinin	Influenza	1	9	Kenney et al. (2004)
	Hemagglutinin	Influenza	1	10	Chi et al. (2010)
	Hemagglutinin	Influenza	1	12	Beran et al. (2009)
	Pf[b] epitope fusion to HBV[c] core particles	Malaria	3	35	Nardin et al. (2004)
mAb	Trastuzumab	Breast cancer	36–52	6569	Ben-Kasus et al. (2007)
	Bevacizumab	Colon carcinoma	240	67,140	Committee for Medicinal Products for Human Use (CHMP (2015b))
	Cetuximab	Colon carcinoma	240	134,940	Committee for Medicinal Products for Human Use (CHMP (2004))
	Aflibercept[d]	Colorectal cancer	6	2238	Committee for Medicinal Products for Human Use (CHMP (2012))
	Panitumumab	Colorectal cancer	6	1902	Committee for Medicinal Products for Human Use (CHMP (2007))
	Rituximab	Lymphoma	4	1306	Cheson and Leonard (2008)
	Rituximab	Lymphoma	8–16	103,800	Chames et al. (2009)
	Pembrolizumab	Metastatic melanoma	20	995	Committee for Medicinal Products for Human Use (CHMP (2015c))

[a] *HBsAg* hepatitis B-soluble antigen
[b] *Pf Plasmodium falciparum*
[c] *HBV Hepatitis B virus*
[d] This is a fusion protein used as a decoy receptor for vascular endothelial growth factor (VEGF)

Lectins are another class of molecules that can be used for immunotherapy or chemotherapy (Jiang et al. 2015). These plant-derived proteins bind to various carbohydrate structures on the cell surface and can induce immunomodulatory effects or apoptosis (Souza et al. 2013). For example, viscumin, also known as mistletoe lectin 1 (ML1) because it is the most abundant lectin in mistletoe (*Viscum album*), is a type II ribosome-inactivating protein (RIP) (Olsnes et al. 1982) that can be used to treat solid tumors (Zwierzina et al. 2011).

Ultimately, the choice of treatment depends on the type and grade of the cancer, its histology and location in the body, the stage of the disease, as well as the geographical region and healthcare options available to the patient (Manegold 2014; Merrett 2014). The difficult task is to identify a treatment that will effectively clear malignant tumors from the body with minimal side effects, e.g., the number of cycles of chemotherapy required (Heydarnejad et al. 2011). In the future, these issues could be reduced by individualized cancer treatments for each patient. The concept of individualized medicine has attracted increasing interest over the last decade because implementation is more feasible given the recent improvements in cancer diagnostics, molecular biology, high-throughput screening, donor cell cultivation, process scale-down, and single-use production systems (Klutz et al. 2015; Schilsky 2010). Personalized medicine requires a precise diagnosis for each patient, followed by the identification of patient-specific tumor markers based on blood or biopsy samples and then the production, selection, or even de novo development of highly selective small-molecule anticancer compounds, mAbs, or ADCs and their subsequent application with constant monitoring for therapeutic progress (Millner and Strotman 2016). Personalized therapy would require not only a paradigm change in the way clinical trials are designed today (Schork 2015), but also a platform that can manufacture complex APIs in comparably large amounts (gram range) in a short time, which will be difficult to achieve with the current, mostly cell culture-based expression systems.

Even without considering the specific challenges of personalized medicine, there remain major challenges for the economically feasible production of anticancer drugs with sufficient purity and at the necessary volume (Siddiqui and Rajkumar 2012). Chemical synthesis is often impractical for small-molecule drugs due to the complexity of anticancer compounds and not economical for entire proteins. Prokaryotic expression systems are generally unsuitable for mAbs and other therapeutic proteins due to the lack of some posttranslational modifications (e.g., *N*-linked glycosylation, which is often required for therapeutic proteins to function properly) and the inefficiency of others (e.g., disulfide bond formation, which is necessary for proteins to fold correctly). The inefficient formation of disulfide bonds causes insoluble proteins to accumulate as inclusion bodies that need to be solubilized and refolded in vitro, which reduces the yield and adds to the production costs (Eiberle and Jungbauer 2010). On the other hand, expression platforms based on mammalian cells, such as Chinese hamster ovary (CHO) cells, may be incompatible with the production of anticancer small-molecule drugs and proteins that target cell division.

The following sections of the chapter focus on plants as either natural sources of anticancer compounds or as hosts for the production of recombinant anticancer biopharmaceutical proteins. Some additional benefits of plants for the production of anticancer therapeutics, such as their scalability and sustainability, are discussed in concert with their potential merits as a platform for the manufacture of personalized medicines.

Plant Secondary Metabolites as Anticancer Drugs

Natural Sources of Plant Metabolites with Anticancer Activity

Plants have been used as medicines for at least 60,000 years (Fabricant and Farnsworth 2001) reflecting their ability to produce cocktails of secondary metabolites with a broad range of pharmacological properties, including anticancer activity (Kuttan et al. 1997; Zarkovic et al. 1998). Phenolic compounds, such as flavonoids, are the most promising plant-derived secondary metabolites for the treatment of cancer (Asensi et al. 2011; Wahl et al. 2011). Historically, plant-derived medicines were administered orally, either by the direct consumption of plant tissues or by the preparation of crude extracts, which is advantageous because these methods are both simple and inexpensive. Furthermore, the administration process is painless for the patient and safe because no syringe-based injection or other invasive procedures are required. The number of publications reporting the anticancer activity of plant extracts is expanding rapidly: a literature search in PubMed (http://www.ncbi.nlm.nih.gov/pubmed) on August 16th, 2016, resulted in 4122 hits including 462 review articles when using the search term "plant extract anti cancer" and a massive increase in publication activity can be observed in the field since 2000 (Fig. 2.1a). It is beyond the scope of this chapter to review all the plant-derived substances that have potential anticancer activity, their mode of action, and efficacy, and it is unlikely that all these extracts will be able to replicate their anticancer effects in larger scale tests (Fritz et al. 2013; Ioannidis 2005; Nuzzo 2014). Furthermore, even if the anticancer effects are genuine (Kwon et al. 2016), the crude mixtures will need to be tested for safety and efficacy because they may contain, as well as the API, additional metabolites with undesirable effects, and the relative concentrations of the API and other ingredients can be difficult to control. For example, a raw extract of poppy (*Papaver somniferum*) seeds not only contains sanguinarine (an alkaloid with anticancer activity) (Selvi et al. 2009) but also hallucinogenic and addictive opiates.

In the light of the risks presented by crude extracts, APIs with anticancer activity are usually isolated from plants allowing them to be tested for a specific mode of action at defined concentrations without side effects caused by other metabolites. The most prominent example is paclitaxel, a taxane found in the bark of the Pacific yew tree (Camidge 2001; Wani and Horwitz 2014). Paclitaxel is used to treat ovarian, breast, and pancreatic cancers among others (Committee for Medicinal Products for Human Use (CHMP) 2015a; Wani and Horwitz 2014), and was approved by the US Food and Drug Administration (FDA) in 1992. It was shown that in combination with gemcitabine, paclitaxel increased the median survival time of patients suffering

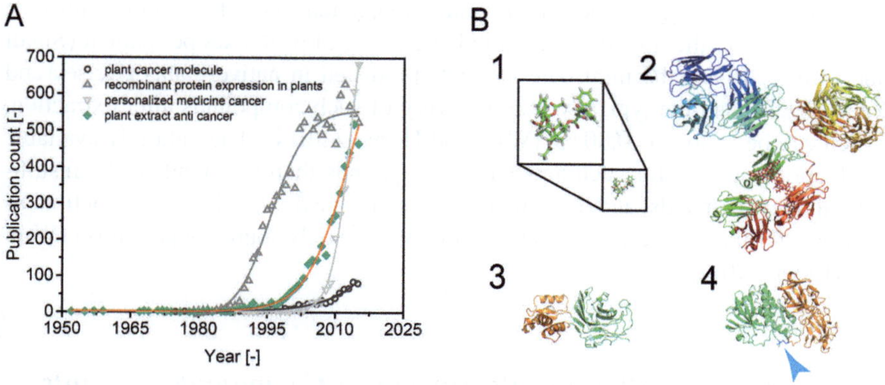

Fig. 2.1 Publication activity and anticancer compounds. (**a**) Publication history for four search queries related to anticancer compounds derived from or produced in plants. The most relevant query is shown as *green diamonds*. A Boltzmann function was fitted to each data series to illustrate the general trends in publication frequency. (**b**) Schematic illustration of the structures of molecules typically used for the treatment of cancer, all shown at the same size scale. (*1*) The small-molecule paclitaxel (sdf file: TA1_ideal_taxol; rcsb database; http://www.rcsb.org) with a zoomed-in box. (*2*) IgG2a monoclonal antibody from mice (pdb file: 1IGT) (Harris et al. 1997). The F_c part is shown in *red* and *green*, and the F_v regions are shown in *blue* and *orange*. (*3*) Therapeutic HPV vaccine candidate LicKM-E7-GGG ("3Djigsaw" (https://bmm.crick.ac.uk/~3djigsaw/) homology model using automated template selection). The HPV-E7 fusion part is shown in *orange*, and the lichenase part is shown in *green* (Buyel et al. 2012). (*4*) Mistletoe lectin viscumin (pdb file: 1M2T) (Krauspenhaar et al. 2002). The toxic A-chain is shown in *green*, the lectin-like B-chain is shown in *orange*, and the linking disulfide bond is highlighted by a *blue arrow*

from pancreatic cancer by >25% compared to treatment with gemcitabine alone (Von Hoff et al. 2013). Paclitaxel is an inhibitor of tubulin depolymerization, thereby stabilizing the microtubule cytoskeleton and blocking mitosis in a concentration-dependent manner (Wani and Horwitz 2014).

Other small-molecule anticancer compounds naturally found in plants include genistein (an isoflavone angiogenesis inhibitor), lycopene (a carotenoid found in many fruits), and resveratrol (a stilbene found in grape berries and therefore also in wines) which are currently undergoing clinical trials to test their efficacy against breast and oral cancer (Bosviel et al. 2012; King-Batoon et al. 2008; Schnekenburger et al. 2014; Zlotogorski et al. 2013). Additional examples of plant-derived metabolites that have shown anticancer activity in vitro include several pregnane glycosides (e.g., desmiflavaside D) and alkaloids (e.g., mahanine) which selectively inhibit the growth of breast and prostate cancer cell lines, respectively (Jagadeesh et al. 2007; Raees et al. 2016). Antitumor activity has also been reported for non-psychoactive cannabinoids (McAllister et al. 2015). The number of known plant secondary metabolites with potential anticancer activity is steadily increasing (Schnekenburger et al. 2014) and diverse ecosystems such as tropical rainforests and coral reefs are thought to harbor a plethora of yet unknown molecules with similar or even improved efficacy as well as completely new modes of action (Mukherjee et al. 2001; Pereira et al. 2012).

Despite the potency of such compounds, large amounts of plant biomass must be harvested due to the high doses required, e.g., ~3 g of paclitaxel per patient (Small and Catling 1999), the miniscule amounts produced in native plants (Kelsey and Vance 1992), and the typically poor recovery of such compounds after extraction, which can be as low as 0.004% (Wani and Horwitz 2014). The naturally available production capacity for such anticancer compounds therefore tends to be insufficient to meet demands, and the source species may be driven close to extinction, as observed for the Pacific yew, which was initially the only source of paclitaxel (Wani and Horwitz 2014).

Production of Small-Molecule Anticancer Compounds in Plants

The unsustainable depletion of natural plant populations can be avoided by using genetic engineering to extend an existing metabolic pathway or to introduce an entirely new metabolic pathway in a more suitable host plant (Wilson and Roberts 2012). This allows the selection of hosts with better growth rates, shorter generation times, greater biomass productivity, or more suitability for biomass processing. For example, the extraction of paclitaxel from Pacific yew trees (Wani and Horwitz 2014) requires a large cultivation area to grow a sufficient number of trees over a period of up to 200 years. The trees must be logged and stripped of bark to extract paclitaxel, killing the trees in the process. The size of the trees and the rigidity of the bark mean that expensive large-scale equipment is required. Even with optimal cultivation and processing infrastructure, the area-wise productivity is low. Assuming that a 100-year-old yew tree is covered with ~3 kg of bark containing paclitaxel at a concentration of 0.14 g kg^{-1} (Small and Catling 1999) and that such a tree requires 30 m^2 for cultivation, the productivity is only 0.14 mg a^{-1} m^{-2}. In contrast, tobacco (*Nicotiana tabacum*) produces biomass at the rate of 10 kg a^{-1} m^{-2} (Stoger et al. 2002). Even if genetic engineering in tobacco achieves 10% of the yields in yew trees (0.014 g kg^{-1} biomass), the productivity is still 140 mg a^{-1} m^{-2} and thus 1000-fold higher than the natural source. The production of secondary metabolites can also be increased by external factors, e.g., the lighting conditions (Buyel et al. 2015a).

Small-molecule anticancer compounds can also be produced in cultured plant cells (Tabata 2004). Despite the higher investment costs for fermentation infrastructure and the need for a sterile environment, cell cultures can be advantageous because it is possible to derive suspension cells directly from the native species or a close relative, benefiting from the intrinsic metabolic capability. Genetic engineering can then be used to increase yields if necessary. Paclitaxel provides a key example of this approach. Phyton Biotech (Delta BC, Canada) has implemented a 75,000-L-scale process compliant with good manufacturing practice (GMP) using *Taxus* sp. cells to produce the API for Bristol-Myers Squibb (New York, USA) (Zhong 2002). The product has been marketed since 1992 with annual sales of $1.592 billion in 2000 (Exposito et al. 2009). Since then, several approaches have been pursued to increase the paclitaxel titers during cultivation, e.g., using elicitors such as methyl jasmonate

(Cusido et al. 2014). Other approved compounds with anticancer activity include berberine and shikonin, which are manufactured in *Coptis japonica* and *Lithospermum erythrorhizon* cells, respectively (Fujita 2007). The typical scale of plant cell cultures is ~1000 L and above (Georgiev and Weber 2014).

Protein-Based Anticancer Compounds from Plants

The use of small-molecule anticancer compounds is limited by their poor selectivity, leading to side effects caused by nonspecific effects on the enzymes, metabolism, and cell cycle components of healthy cells. Proteins are more complex molecules that offer alternative mechanisms of action and can also be used to target cancer cells, thus reducing side effects. Small-molecule drugs and anticancer proteins are compared in Fig. 2.1b.

Anticancer Activity of Lectins

Lectins are a heterogeneous group of glycoproteins produced by many different plant species and have probably evolved as part of the molecular defense repertoire against pests and herbivores (Lannoo and Van Damme 2014; Van Damme 2014; Vandenborre et al. 2011). Lectins may comprise several polypeptide chains, forming, e.g., homotetramers like ArtinM from jackfruit (*Artocarpus heterophyllus*) (Souza et al. 2013) or heterodimers like the mistletoe lectin viscumin (Kourmanova et al. 2004). The common feature of lectins is their affinity towards carbohydrate structures, with different lectins favoring different oligosaccharides. This selectivity allows some lectins to bind more or less specifically to carbohydrates displayed on tumor cells, resulting in immunomodulatory or anticancer activity (Souza et al. 2013).

Viscumin is the most prominent example of a plant lectin with potential anticancer applications (Zwierzina et al. 2011). Viscumin is synthesized as a single polypeptide precursor which is activated by proteolytically removing a central amino acid linker sequence. The active form of the protein comprises an A-chain (former N-terminus) with N-glycosidase activity and a B-chain (former C-terminus) which binds to carbohydrates on the cell surface (Walsh et al. 2013). The two chains are covalently linked by a disulfide bond (Olsnes et al. 1982). The A-chain features β-sheet secondary structures but is dominated by α-helices (Fig. 2.1b) (Krauspenhaar et al. 2002). The mode of action of the A-chain involves the cleavage of the *N*-glycosidic bond at position A4324 in the 28*S* rRNA of the large subunit of eukaryotic ribosomes, hence the classification of viscumin as a type II RIP (Endo et al. 1988). The toxicity of purified viscumin (intravenous LD_{50} in mice) is 2.4 μm kg^{-1} (Olsnes et al. 1982). In comparison, the structurally related ricin toxin from the seeds of *Ricinus communis* (castor bean) has a toxicity of 30 μm kg^{-1} (Audi et al. 2005).

In contrast to the A-chain, the carbohydrate-binding B-chain of viscumin is mostly composed of β-sheets (Fig. 2.1b) and has three intra-chain disulfide bonds. Viscumin was initially shown to bind to β-galactosides, especially terminal galactose in the oligosaccharides of glycoproteins (Gabius et al. 1992; Lee et al. 1994; Olsnes et al. 1982) but it also binds selectively to terminal sialic acids, e.g., IV^6Neu5Ac-nLc4Cer residues in a mammalian context (Muthing et al. 2004), especially on glycoproteins and gangliosides (Zwierzina et al. 2011). The ganglioside CD75s carries an IV^6Neu5Ac-nLc4Cer residue and is associated with several types of solid tumor, including pancreatic cancer (Distler et al. 2008). The antitumor activity of viscumin therefore appears to reflect the ability of the B-chain to bind selectively to tumor cells allowing the toxic A-chain to be internalized and block protein synthesis. Previous studies have shown that viscumin also stimulates T-helper (T_h1) cells, thereby inducing the cellular component of the adaptive immune system to attack cancer cells (Zwierzina et al. 2011).

N-glycans are added to the viscumin B-chain as it passes through the secretory pathway to the apoplast (Niwa et al. 2003). These posttranslational modifications are not required for its antitumor activity, because aglycosylated recombinant viscumin purified from *Escherichia coli* inclusion bodies was also efficacious in phase I clinical trials (Zwierzina et al. 2011). However, authentic *N*-linked glycosylation may increase the potency of viscumin. Furthermore, the resolubilization and refolding of protein from inclusion bodies are laborious and inefficient (Eiberle and Jungbauer 2010). The expression of recombinant lectins in plants may be therefore ideal to achieve high yields, straightforward purification, and more potent APIs.

Recombinant Protein Expression in Plants

Benefits and Challenges

In addition to the production of native proteins with anticancer activity by some plant species, many species of plants (or the cells and tissues derived from them) have been used as expression systems for recombinant therapeutic proteins. The use of plants for the production of recombinant proteins is known as "molecular farming," and to emphasize the medical relevance when these are therapeutic proteins the alternative spelling "molecular pharming" is also used (Fischer et al. 1999, 2013; Ma et al. 2005; Menkhaus et al. 2004; Wilken and Nikolov 2012). In the context of biopharmaceutical manufacturing, plants are beneficial because they can synthesize complex proteins with authentic posttranslational modifications (e.g., glycosylation, disulfide bond formation), combined with low-cost upstream production, inherent process safety based on the inability of human pathogens to replicate in plants (Commandeur et al. 2003), and potential for flexible and very-large-scale production (Buyel et al. 2016c). The last two aspects are particularly important when comparing plants to mammalian cells because disastrous contamination with human pathogens is unlikely (Bethencourt 2009; Zimran et al. 2011) and the manufacturing capacity can be rapidly adapted to market demands. Two general types of expression strategies are available for plants: (1) transient expression using viral or bacterial vectors or combinations thereof and (2) expression in transgenic plants or plant cells (Fischer and Schillberg 2006; Paul and Ma 2011; Twyman et al. 2003).

In contrast to the production of proteins in microbes and animal cells, where a limited number of host systems have been standardized by the biomanufacturing industry (The CMC Biotech Working Group 2009), a plethora of plant species and expression strategies have been considered by academic and industrial development teams. In terms of intact plants, tobacco and its close relative *N. benthamiana* are now emerging as standard platforms for the production of recombinant proteins by stable transformation and transient expression, respectively, and cereals are regarded as promising hosts for stable expression because the recombinant proteins can be stored in seeds (Spiegel et al. 2016). Standardized platforms are also emerging based on tobacco, rice, and carrot cell suspension cultures (Santos et al. 2016; Schillberg et al. 2013). Scalable unit operations and more structured purification schemes are now being implemented for plant-derived products (Buyel and Fischer 2014c, e; Buyel et al. 2015c). The initial extraction method for plant-derived APIs depends on the tissue and subcellular localization used for expression. Products secreted into hydroponic medium (intact plants) or the fermentation broth (cell culture) can be recovered directly, similar to other cell culture-based processes (Drake et al. 2009). Recombinant proteins expressed in leaf or seed tissues are typically extracted by blade-based homogenizers/presses or mills, respectively (Bals and Dale 2011; Buyel and Fischer 2014d, e; Farinas et al. 2005; Hassan et al. 2014; Hassan et al. 2008; Kim et al. 2013). Aqueous buffers within the pH range of 7.0–8.0 are typically used for extraction (Buyel et al. 2015c).

Product Recovery and Purification

During downstream processing, the process schemes rely on the same methods used in cell culture-based systems, e.g., filtration and solid-liquid chromatography, but must be adapted to suit the characteristics of plants, e.g., a high particle burden (Buyel et al. 2014b) and a high concentration of host cell proteins (HCPs) such as ribulose-1,5-bisphosphate carboxylase/oxygenase (RuBisCO) if the proteins are not secreted into the medium (Buyel and Fischer 2014c), and product concentrations below $0.5\ g\ L^{-1}$ in the raw extracts (Buyel 2015) in contrast to $5–10\ g\ L^{-1}$ achieved in cultured mammalian cells (Li et al. 2010; Shukla and Thommes 2010), which can be challenging during initial product recovery (Winkelnkemper and Schembecker 2010). These modifications have been major cost drivers for plant-derived APIs in the past, accounting for up to 80% of the process costs (Buyel et al. 2015c; Wilken and Nikolov 2012). In particular, the clarification of the raw plant extracts has been characterized by low filter capacities ($<100\ L\ m^{-2}$) increasing the consumable costs of production. However, flocculants and other filter aids have recently been shown to improve particle retention on low-cost bag filters and can increase the downstream filter capacity to $>1000\ L\ m^{-2}$ (Buyel 2016; Buyel and Fischer 2014b; Buyel et al. 2014b, 2015b).

A number of genetic modifications have been used to facilitate the purification of proteins expressed in plants from the clarified extract. These modifications include oleosins as direct (Kapchie et al. 2011; Markley et al. 2006; Napier et al. 1996) or indirect fusions (McLean et al. 2012), zein (Geli et al. 1994; Torrent et al. 2009), elastin-like polypeptides (Conley et al. 2009; Tian and Sun 2011; Urry 1988), soybean agglutinin (Tremblay et al. 2011), or the frequently used hexa-histidine (His_6) tag (Buyel et al. 2012). All of these modifications confer specific properties that facilitate product purification, e.g., by two-phase extraction or affinity chromatography, but may not be suitable for anticancer compounds that are used for long-term treatment because the same tags may be immunogenic, thus reducing therapeutic efficacy (Fischer et al. 2012; Li 2011), or their commercial use may be restricted (Conley et al. 2011).

Process-based solutions for the challenging purification of plant-derived APIs have therefore been developed that are compatible with the requirements of biopharmaceutical proteins. Heat precipitation at ~70 °C has proven useful to remove >90% of all HCPs from extracts containing heat-stable products at a very early process stage, reducing the likelihood of product oxidation or proteolytic degradation (Buyel et al. 2014a, 2016b; Menzel et al. 2016). The low-pH (<5.0) precipitation of HCPs is an alternative step for heat-sensitive products (Azzoni et al. 2002; Buyel and Fischer 2014c). Additionally, abundant HCPs such as RuBisCO can be precipitated with polyethylene glycol (PEG) (Arfi et al. 2016). The subsequent purification and polishing steps to achieve the required final purity use chromatography modes such as affinity chromatography (e.g., protein A chromatography for the purification of mAbs) followed by orthogonal combinations of anion- and cation-exchange chromatography, hydrophobic interaction, and mixed-mode chromatography or size-exclusion chromatography (Nfor et al. 2011). For example, a two-step process consisting of protein A and ceramic hydroxyapatite chromatography was used to isolate an mAb from clarified tobacco extracts for phase I clinical studies (Ma et al. 2015). Because the product concentration can fall below 0.05 g L^{-1} (Menkhaus and Roseland 2008), the use of membrane absorbers instead of packed-bed columns can be advantageous, also reducing the process times due to the higher flow rates (Orr et al. 2013). Alternatively, ultrafiltration/diafiltration setups can be used to reduce the process volume (Lightfoot and Moscariello 2004; Lightfoot et al. 2008). Integrated approaches such as expanded-bed adsorption (EBA) chromatography and aqueous two-phase systems (ATPS) have been tested for the purification of recombinant proteins from plants but were either more expensive than conventional processes (Menkhaus and Glatz 2005) or attracted regulatory scrutiny due to the presence of fusion tags (Reuter et al. 2014).

Process Design and Monitoring

Non-platform processes designed to accommodate the properties of individual products require time and investment, and therefore contribute to the steadily increasing R&D costs in the biopharmaceutical industry (PhRMA 2014). Rational protein design (Boes et al. 2011) and characterization of HCPs (Buyel et al. 2013b) using open-source (UniProt-Consortium 2012) and/or commercial software (Chemical Computing Group, Montreal, Canada) can be used to established more standardized processes for plant-derived APIs. Furthermore, the concept of quality-by-design (QbD) and its associated tools, namely design-of-experiments (DoE) and process analytical technology (PAT) (De Beer et al. 2008; Landgrebe et al. 2010; Rathore 2009; Rathore et al. 2008, 2010; Read et al. 2010a; b), are being integrated at the stage of process development but must be adapted to the large number of individual plants per batch in contrast to the smaller number of fermenters in conventional processes (D'Este et al. 2012; Juca et al. 2011; Sainz et al. 2013). For example, new PAT monitoring has been implemented for plant cell suspension cultures during upstream production (Buyel et al. 2016a; Holland et al. 2013) and the complex mechanisms of protein expression (Buyel and Fischer 2014a) and flocculation (Buyel 2016) have been described using DoE.

Single-use technologies, such as filters, sensors, fermenters, and chromatography columns, are becoming increasingly popular in the biopharmaceutical industry (Allison and Richards 2014; Eibl and Eibl 2011; Laukel et al. 2011; O'Brien et al. 2012; Shukla and Gottschalk 2013; Whitford 2010) and are likely to become an integral part of future manufacturing plants (Klutz et al. 2015) because they facilitate process validation, accelerate change-over tasks, and avoid the need to clean equipment that comes into contact with the product (Fischer et al. 2012). Single-use technologies are also typically implemented in plant-based production processes (Ma et al. 2015) and the plants themselves can be regarded as single-use, biodegradable bioreactors.

Today, protein-based APIs derived from plants are still exotic products in the pharmaceutical industry. The only product of this class currently approved as a pharmaceutical for general human use is taliglucerase alfa (marketed as Elelyso) which was developed by Protalix Biotherapeutics and is produced in carrot cell suspension cultures (Mor 2015; Pastores et al. 2014). However, more than ten further product candidates are already in the clinical pipeline with still more in preclinical development (Gleba et al. 2014). These industry-derived products combined with further APIs developed in publically funded projects (Paul et al. 2013; Sparrow et al. 2007) have encouraged the regulatory authorities in the USA and EU to release draft guidelines for the production of plant-derived biopharmaceuticals (CPMP 2002; FDA 2002; Spok 2007), filling the legal vacuum that has thus far discouraged the big players in the pharmaceutical industry from making substantial investments into this sector (Das et al. 2008; Fischer et al. 2012; Ma et al. 2005), thus resulting in the focus on plant cell cultures instead of intact plants (Pastores et al. 2014). Plant-based production systems are now entering the biopharmaceutical arena as a competitive platform for the manufacture of biopharmaceutical proteins, including anticancer antibodies and lectins.

Transgenic Plants and the Potential Large-Scale Production of Anticancer Compounds

Scalability and Product Quality

In contrast to microorganisms and animal/plant cell cultures, transgenic plants can be cultivated as intact, multicellular organisms. One advantage is that sterile cultivation conditions and/or antibiotics are not required for upstream production, and another is that the pharmaceutical crops can be grown on the same agricultural scale as food and feed crops, thus allowing the production of multiple tons of recombinant proteins to fulfil high-demand markets (Buyel et al. 2016c; Fischer and Schillberg 2006). For example, 179.7 million hectares of genetically modified crops were grown in 2015, which is a 100-fold increase compared to 1996 (James 2015).Yet another advantage of transgenic plants is that the modification is much more stable than the genes carried by engineered microbes or mammalian cell lines (Baneyx and Mujacic 2004; Gerngross 2004; Hellwig et al. 2004; Wurm 2004). This ensures a defined and stable genetic status and facilitates the generation of well-defined master and working seed banks (Fischer et al. 2012). Transgenic lines are typically established by transformation using *Agrobacterium tumefaciens* or physical methods such as particle bombardment to introduce the gene(s) of interest into the plant genome, before regeneration into intact plants (Lorence and Verpoorte 2004; Newell 2000; Rivera et al. 2012). The initial transformants are typically subjected to 3–8 breeding cycles resulting in homozygous transgenic plant lines. The transformation and breeding process can take 8–24 months depending on the plant species and genetic construct, and can require extra time to prepare the seed banks required for large-scale production (Twyman et al. 2003). Although the advent of genome editing tools such as CRISPR/Cas9 can reduce this time (Bortesi and Fischer 2015; Puchta and Fauser 2014), transgenic plants are probably inadequate for urgent demands, as would be required for personalized cancer therapy. In contrast, transgenic plants function better as a bulk production system for proteins that are required for standard therapies in a large number of patients, e.g., anticancer vaccines or mAbs such as rituximab.

One matter of concern when considering plants for the production of vaccines and antibodies is that the glycans added to proteins by plants are structurally different to those produced in humans, which means that injected APIs can be immunogenic. Although antibodies against plant glycans have been detected in human serum, no adverse responses against plant-derived recombinant therapeutic proteins have been reported (Bardor et al. 2003; Chargelegue et al. 2000; Koprivova et al. 2004). In some cases, the nonhuman glycan profiles produced by plants can even be advantageous, e.g., modulating the effector functions of some plant-derived mAbs due to the carbohydrates attached to the Fc region, and simplifying the production of the enzyme glucocerebrosidase which is used to treat Gaucher's disease. In the latter case, the enzyme produced in carrot cells (Elelyso, see above) is devoid of terminal sialic acid residues, allowing efficient uptake by mannose receptors on circulating macrophages. The same protein produced in mammalian cells contains terminal sialic acid residues that must be cleaved off in vitro, thus increasing production costs (Grabowski et al. 2014). Hence, plants can be regarded not only as an alternative production system but also as a potential source of "biobetters." Where authentic human-type glycosylation is necessary for therapeutic efficacy, plants can be modified by mutating endogenous genes to remove plant-specific residues such as core α1,3-fucose and β1,2-xylose, and by introducing human genes needed for the synthesis of multi-antennary glycans, core α1,6-fucose, β1,4-galactose, and terminal sialic acids (Bakker et al. 2006; Strasser 2013, 2016; Strasser et al. 2008). These glyco-engineered lines can then be used as a platform for the production of recombinant proteins with authentic glycans.

Monoclonal Antibodies

Monoclonal antibodies can be used as highly specific and efficacious APIs for cancer therapy (Chiarella 2011; Gaughan 2015; Scott et al. 2012). In 2013, a total of almost 10 tonnes of mAbs was manufactured with an estimated market value of $US 75 billion (Ecker et al. 2015). Based on the current forecast of a 10% annual growth rate (Research 2013), this market will cross the $US 200 billion mark before 2025. The demand for some major single products such as rituximab and etanercept is already about 1 tonne per year (Kelley 2007). This amount is likely to increase massively if the supply of these products can be expanded to developing countries, which currently have limited access to such high-quality cancer therapeutics due to the manufacturing costs. For example, assuming that only every second person among the 14.1 million new cancer patients every year (Pritzkuleit et al. 2010; Rottenberg et al. 2010) requires mAb therapy, with a typical dose of ~10 g of protein pre patient (Chames et al. 2009), the annual demand for mAbs related to cancer therapy will increase to 70.5 tonnes. This calculation does not take into account the increasing incidence of cancer expected in the future due to demographic changes and improved diagnostics, and also ignores the likelihood that per-patient mAb consumption may increase due to improved survival rates.

Multi-tonne production scales for mAbs are difficult to achieve using conventional expression systems alone, even though mAb titers of 5–10 g L^{-1} are now possible on a regular basis (Li et al. 2010; Shukla and Thommes 2010). For example, despite such high titers, the investment costs for bioreactors would be enormous because even very-large-scale fermenters with volumes of 10,000–25,000 L (Kelley 2007; Li et al. 2010) would only produce ~5.0 tonnes of mAb per year, assuming a duration of 12 days per batch (Kelley 2007), and would thus cover less than 10% of the anticipated demand. Furthermore, the many benefits of single-use technologies are lost at such massive production scales because multiple disposable production trains would need to operate simultaneously, and at this point a fixed stainless-steel facility costs less to operate even when the up-front costs are included (Eibl et al. 2010; Shukla and Gottschalk 2013).

In contrast, plants provide their own in-built single-use bioreactors and the cost does not increase significantly regardless of the production scale, because plants can be numbered up as required and the only cost factor is the additional land or greenhouse space. The per se high biomass yield of some crops, e.g., 10,000 tonnes a^{-1} km^{-2} for tobacco (Stoger et al. 2002), can be increased to 91,000 tonnes a^{-1} km^{-2} if so-called vertical farming units (VFUs) are used for production instead of open-field cultivation (Buyel et al. 2016c; Holtz et al. 2015). Combining these biomass yields with antibody expression levels of 2 g kg^{-1} (Zischewski et al. 2015) and a typical mAb recovery of 70% during purification from transgenic plants (Ma et al. 2015) implies that ~0.55 km^{-2} of VFU area will be sufficient to produce 70 tonnes of purified mAb.

Additional benefits of VFUs include the improved containment and more tightly controlled growth conditions, which can be advantageous for GMP production processes (Fischer et al. 2012), as well as the potential for fully automated handling, even in a GMP environment (Wirz et al. 2012). To achieve the same output by fermentation, a bioreactor volume of >350,000 L would be required. On top of the investment costs, such a setup would carry an immense monetary risk given the frequency of batch failures, e.g., due to contamination (Lolas 2013): assuming CHO cell media costs of $US 55–90 L^{-1}, each failed batch would generate losses corresponding to $US 19–32 million only for the media. A more detailed discussion of the benefits of plants for the very-large-scale production of mAbs can be found elsewhere (Buyel et al. 2016c), but independent from the upstream production step, the downstream processing and scale-up considerations would be similar to those for cell culture-based systems (Ma et al. 2015; The CMC Biotech Working Group 2009), e.g., continuous manufacturing and a preference for single-use technologies (Angarita et al. 2015; Baur et al. 2015; Klutz et al. 2015). Plant-based systems may also benefit from the experiences gathered in the food processing industry (Goody 1997). In summary, plants as a production platform could tap into the full potential of mAbs as anticancer agents for a broad range of patients in industrialized as well as developing countries.

Lectins

Lectins such as viscumin could potentially be developed as anticancer drug candidates and a GMP-compliant process for the production of viscumin has already been established using a standard *E. coli* expression system, allowing viscumin to be tested in clinical trials (Zwierzina et al. 2011). However, the A- and B-chains were produced by two different bacterial strains in separate fermentations and both polypeptides formed inclusion bodies, so laborious resolubilization and refolding were necessary leading to poor recoveries of ~5% based on the amount of unfolded polypeptide educts, which is typical for such refolding processes (Eiberle and Jungbauer 2010). Additionally, the product was not glycosylated, which may affect its stability and efficacy as discussed above (Li et al. 2012). Other plant lectins have been expressed in the yeast *Pichia pastoris* at levels of 6–20 mg L^{-1} (Lannoo et al. 2007; Oliveira et al. 2008) but glycoproteins produced by yeast often bear predominantly high-mannose rather than complex-type glycans (Strasser 2016). Mammalian cells are unlikely to express viscumin at high titers because the lectin would be toxic to the host cells (Endo et al. 1988). Such issues can be overcome by using plants as expression hosts because lectins are native to plants and heterologous lectins are generally not toxic in plant systems. In the case of viscumin, the Fraunhofer Institute for Molecular Biology and Applied Ecology IME (IME) has started to develop a plant-based expression approach which has thus far achieved yields of ~5 mg kg^{-1} (our unpublished data). This is comparable to the 1.5 mg mL^{-1} achieved in *E. coli* after refolding, but the refolding step is not required. Purifying viscumin from the more complex plant matrix is unlikely to pose a major challenge because lactosyl-Sepharose can be used as an affinity resin, and the purification of proteins from plant extracts by affinity chromatography does not seem to be inhibited by the larger quantity of released HCPs compared to products secreted into the fermentation broth by CHO cells (Ma et al. 2015). Furthermore, plants have been successfully used to produce griffithsin, a lectin with the potential to block the transmission of *human immunodeficiency virus* (HIV) (Fuqua et al. 2015; O'Keefe et al. 2009).

Anticancer Vaccines

Vaccines are probably the most cost-effective and least invasive way to treat or even prevent cancer. In theory, a single dose of microgram amounts of an antigen can be sufficient to achieve lifelong protection against a specific form of cancer. In practice, several doses are often required and the protection can decline after a few years, so booster vaccinations may be necessary (Ouattara and Laurens 2015). Vaccines attract a higher regulatory burden in terms of safety than cancer therapeutics because they are given to healthy individuals and the risk/benefit ratio does not tolerate severe side effects (Kwok 2011). Therefore, subunit vaccines are preferred over live or attenuated pathogens because the former contain discrete antigenic components, typically surface proteins from the pathogen, and are therefore unable to revert to virulence (Baxter 2007). However, small protein entities tend to induce a weaker immune response than larger and more complex vaccines (Bachmann and Jennings 2010). Adjuvants such as alum can be used to enhance the immune response and thus increase protection, but VLPs are even more immunogenic. In simple terms, VLPs are empty viruses; that is, they are composed of typical viral structural proteins but do not carry any nucleic acid and therefore cannot replicate and cause disease (Wang and Roden 2013). The structural proteins can be produced in microbes, but as discussed above the lack of posttranslational modifications in bacteria and the high-mannose glycans produced by yeast are significant drawbacks, so higher eukaryotes are preferable. Mammalian cells produce correctly modified subunits but there is a risk of contamination with endogenous or adventitious viruses, which can necessitate the disposal of whole batches and even the closure of a facility (Bethencourt 2009). To ensure that not even trace quantities of viruses end up in the final drug product, virus filtration is an integral part of all mammalian cell culture-based production processes. This presents a unique challenge during the production of VLPs because the product is typically in the same size range as any live virus contaminants. Plants do not support the proliferation of human-trophic viruses and virus filtration is not necessary (Commandeur and Twyman 2005). Several VLPs have therefore been produced successfully in plants, including those based on HPV (Scotti and Rybicki 2013). Plants have also been used to produce subunit vaccines, e.g., a fusion protein including a modified version of the HPV protein E7 (Buyel et al. 2012). This vaccine candidate was not only protective against cervical cancer in the same manner as the commercial vaccines Cervarix and Gardasil, but could also be used as a therapeutic vaccine to prevent the development of cervical cancer in patients already carrying HPV (Massa et al. 2007; Venuti et al. 2009). Plants have also been used to produce a large number of vaccines not related to cancer, e.g., vaccines against malaria (Spiegel et al. 2015).

Transient Expression Systems for Individualized Cancer Therapies

The antibodies and lectins discussed above are administered to a large number of patients (or at-risk individuals in the case of vaccines) and transgenic plants are the most suitable plant-based expression platform. However, mAbs can also be used as personalized medicines designed to act against the unique features of a tumor in a specific patient (Duffy 2015). The total demand for such products will therefore be much lower than the general products described earlier, i.e., in the lower gram range rather than the kilogram to tonne scale necessary for the most in-demand mAbs (Kelley 2007). The regulatory approval process for personalized medicines will also need to differ from conventional procedures (Schork 2015). More importantly, time is often of the essence in cancer therapy, and neither conventional cell culture-based approaches nor transgenic plants with lead development times of several months are suitable expression systems for such applications. In contrast to these long development cycles, transient expression in plants can be achieved within a few weeks after obtaining the DNA sequence of a personalized antibody, which is a much more appropriate response time (Rosenberg et al. 2013; Shoji et al. 2012).

Transient expression can be achieved by infecting plants with replicating viruses carrying the genes for a recombinant protein, resulting in a systemic infection. This has been reported with several different viruses, and personalized antibodies expressed using a *tobacco mosaic virus* vector in tobacco have been tested in clinical trials of patients with non-Hodgkin's lymphoma (McCormick et al. 1999). Transient expression can also be achieved by the infiltration of wild-type plants (typically *N. benthamiana*) with genetically modified *A. tumefaciens* carrying the relevant transgene as part of the transfer DNA (T-DNA) on a resident plasmid which is most often part of a binary vector system (Glick et al. 2010). A type IV secretion system of the bacteria is responsible for T-DNA transfer into the plant cell nucleus, where the T-DNA is maintained transiently as an episomal structure facilitating the expression of any foreign genes (Valentine 2003). The bacterial infiltration and virus-based systems have also been combined so that the bacterial cells deliver a partial virus genome that can spread locally but not systemically (Gleba et al. 2014). Even though the levels of recombinant protein peak after several days and decline afterwards (Buyel et al. 2013a), transient expression typically achieves higher product yields than transgenic plants in the short term. This is because many plant cells temporarily contain multiple copies of the episomal expression vector allowing large amounts of protein to accumulate. One potential drawback is that earlier transient expression studies indicated relevant batch-to-batch variability in the product yield (O'Neill et al. 2008), which might cause regulatory concerns regarding the reproducibility of the process (Fischer et al. 2012). However, this issue can be overcome by improved process parameter controls, especially the incubation temperature and time (Buyel and Fischer 2012, 2013; Buyel et al. 2013a; Larsen and Curtis 2008). As discussed above, improved process control can also be achieved by VFUs and in fact most of the large-scale production facilities that have been constructed recently are built around transient expression technologies (Holtz et al. 2015; Wirz et al. 2012).

The concept has also been successfully applied to other scenarios, such as the production of emergency drugs, e.g., to counter the threat of a rapidly spreading influenza pandemic (Shoji et al. 2012; Stoger et al. 2014), which has attracted massive governmental support especially in the USA (Paul et al. 2013). Hence, molecular pharming based on transient expression in plants combines the speed and scalability that are required to supply sufficient amounts of anticancer compounds for personalized medicines in a timely manner.

Conclusions

The number of cancer diagnoses can be expected to increase due to demographic changes in industrialized countries along with improved diagnostics, but also because developing countries are seeking access to modern healthcare systems which are reliant on biotechnology. This demand cannot be met by established production systems such as microbes and mammalian cell cultures, not only because of the costs of production but also due to the inherent complexity and toxicity of anticancer drugs and proteins. Plants have the potential to fill this gap because they naturally produce a number of substances with anticancer activity, including small molecules such as paclitaxel or proteins such as viscumin. This potential can be increased further by applying recent advances in plant biotechnology and plant molecular pharming to oncology products, e.g., exploiting the rapid and flexible-scale production that can be achieved using transient expression systems to manufacture individualized medicines, or the massively scalable production that can be achieved using transgenic plants to manufacture bulk APIs.

Acknowledgements The author acknowledges Dr. Richard M Twyman for editorial assistance. This work was funded in part by the European Research Council Advanced Grant "Future-Pharma," proposal number 269110, and the Fraunhofer-Gesellschaft Internal Programs under Grant No. Attract 125-600164. The author has no conflict of interest to declare.

References

Akagi K, Li J, Broutian TR, Padilla-Nash H, Xiao W, Jiang B, Rocco JW, Teknos TN, Kumar B, Wangsa D, He D, Ried T, Symer DE, Gillison ML (2014) Genome-wide analysis of HPV integration in human cancers reveals recurrent, focal genomic instability. Genome Res 24:185–199

Alberts B, Johnson A, Lewis J, Raff M, Roberts K, Walter P (2002) The molecular basis of cancer-cell behavior. In: Alberts B et al (eds) Molecular biology of the cell, 4th edn. Garland Science, New York, NY

Allison N, Richards J (2014) Current status and future trends for disposable technology in the biopharmaceutical industry. J Chem Technol Biotechnol 89:1283–1287

American Cancer Society (2010) The global economic cost of cancer. American Cancer Society, Atlanta, GA

American Cancer Society (2015) Cancer facts and figures 2015. American Cancer Society, Atlanta, GA
Angarita M, Muller-Spath T, Baur D, Lievrouw R, Lissens G, Morbidelli M (2015) Twin-column CaptureSMB: a novel cyclic process for protein A affinity chromatography. J Chromatogr A 1389:85–95
Arfi ZA, Hellwig S, Drossard J, Fischer R, Buyel JF (2016) Polyclonal antibodies for specific detection of tobacco host cell proteins can be efficiently generated following RuBisCO depletion and the removal of endotoxins. Biotechnol J 11:507–518
Asensi M, Ortega A, Mena S, Feddi F, Estrela JM (2011) Natural polyphenols in cancer therapy. Crit Rev Clin Lab Sci 48:197–216
Audi J, Belson M, Patel M, Schier J, Osterloh J (2005) Ricin poisoning: a comprehensive review. JAMA 294:2342–2351
Aziz A, Aziz S, Li DS, Murphy L, Leone N, Kennedy M, Dhillon S, Van Thiel DH (2006) Efficacy of repeated high-dose hepatitis B vaccine (80 mu g) in patients with chronic liver disease. J Viral Hepat 13:217–221
Azzoni AR, Kusnadi AR, Miranda EA, Nikolov ZL (2002) Recombinant aprotinin produced in transgenic corn seed: extraction and purification studies. Biotechnol Bioeng 80:268–276
Bachmann MF, Jennings GT (2010) Vaccine delivery: a matter of size, geometry, kinetics and molecular patterns. Nat Rev Immunol 10:787–796
Bakker H, Rouwendal GJA, Karnoup AS, Florack DEA, Stoopen GM, Helsper JPFG, Van Ree R, Van Die I, Bosch D (2006) An antibody produced in tobacco expressing a hybrid beta-1,4-galactosyltransferase is essentially devoid of plant carbohydrate epitopes. Proc Natl Acad Sci U S A 103:7577–7582
Baldi A, Bisaria VS, Srivastava AK (2008) Biotechnological approaches for the production of some promising plant-based chemotherapeutics. In: Kayser O, Quax WJ (eds) Medicinal plant biotechnology. Wiley-VCH Verlag GmbH, Hoboken, NJ
Baldy JL, Elisbao Mdo C, Anzai ET, Pontello R, Reiche EM, Zaha-Inouye MM, Matsuo T, Tonani PC, Ferelle A, Henriques JN, Neves J (2003) Intradermal vaccination of adults with three low doses (2 micrograms) of recombinant hepatitis B vaccine. I. Seroconversion rate and adverse effects. Mem Inst Oswaldo Cruz 98:1101–1107
Bals B, Dale BE (2011) Economic comparison of multiple techniques for recovering leaf protein in biomass processing. Biotechnol Bioeng 108:530–537
Baneyx F, Mujacic M (2004) Recombinant protein folding and misfolding in Escherichia coli. Nat Biotechnol 22:1399–1408
Bang YJ, Van Cutsem E, Feyereislova A, Investigators TT (2010) Trastuzumab in combination with chemotherapy versus chemotherapy alone for treatment of HER2-positive advanced gastric or gastro-oesophageal junction cancer (TOGA): a phase 3, open-label, randomised controlled trial (vol 376, pg 687, 2010). Lancet 376:1302–1302
Bardor M, Faveeuw C, Fitchette AC, Gilbert D, Galas L, Trottein F, Faye L, Lerouge P (2003) Immunoreactivity in mammals of two typical plant glyco-epitopes, core alpha(1,3)-fucose and core xylose. Glycobiology 13:427–434
Baur D, Angarita M, Muller-Spath T, Morbidelli M (2015) Pareto-optimal design of the twin-column CaptureSMB process improves capacity utilization and productivity in protein A affinity capture. Biotechnol J 11:135–145
Baxter D (2007) Active and passive immunity, vaccine types, excipients and licensing. Occup Med (Lond) 57:552–556
Beckman RA, von Roemeling R, Scott AM (2011) Monoclonal antibody dose determination and biodistribution into solid tumors. Ther Deliv 2:333–344
Ben-Kasus T, Schechter B, Sela M, Yarden Y (2007) Cancer therapeutic antibodies come of age: targeting minimal residual disease. Mol Oncol 1:42–54
Beran J, Ambrozaitis A, Laiskonis A, Mickuviene N, Bacart P, Calozet Y, Demanet E, Heijmans S, Van Belle P, Weber F, Salamand C (2009) Intradermal influenza vaccination of healthy adults using a new microinjection system: a 3-year randomised controlled safety and immunogenicity trial. BMC Med 7:1–15

Bernstein C, Bernstein H (2015) Epigenetic reduction of DNA repair in progression to gastrointestinal cancer. World J Gastrointest Oncol 7:30–46
Berrington de Gonzalez A, Curtis RE, Kry SF, Gilbert E, Lamart S, Berg CD, Stovall M, Ron E (2011) Proportion of second cancers attributable to radiotherapy treatment in adults: a cohort study in the US SEER cancer registries. Lancet Oncol 12:353–360
Bethencourt V (2009) Virus stalls Genzyme plant. Nat Biotechnol 27:681–681
Bieging KT, Mello SS, Attardi LD (2014) Unravelling mechanisms of p53-mediated tumour suppression. Nat Rev Cancer 14:359–370
Boes A, Spiegel H, Delbruck H, Fischer R, Schillberg S, Sack M (2011) Affinity purification of a framework 1 engineered mouse/human chimeric IgA2 antibody from tobacco. Biotechnol Bioeng 108:2804–2814
Bortesi L, Fischer R (2015) The CRISPR/Cas9 system for plant genome editing and beyond. Biotechnol Adv 33:41–52
Bosviel R, Dumollard E, Dechelotte P, Bignon YJ, Bernard-Gallon D (2012) Can soy phytoestrogens decrease DNA methylation in BRCA1 and BRCA2 oncosuppressor genes in breast cancer? OMICS 16:235–244
Brady G, MacArthur GJ, Farrell PJ (2007) Epstein-Barr virus and Burkitt lymphoma. J Clin Pathol 60:1397–1402
Buyel JF (2015) Process development strategies in plant molecular farming. Curr Pharm Biotechnol 16:966–982
Buyel JF (2016) Procedure to evaluate the efficiency of flocculants for the removal of dispersed particles from plant extracts. J Vis Exp (11):e53940.
Buyel JF, Bautista JA, Fischer R, Yusibov VM (2012) Extraction, purification and characterization of the plant-produced HPV16 subunit vaccine candidate E7 GGG. J Chromatogr B 880:19–26
Buyel JF, Buyel JJ, Haase C, Fischer R (2015a) The impact of Pseudomonas syringae type III effectors on transient protein expression in tobacco. Plant Biol 17:484–492
Buyel JF, Fischer R (2012) Predictive models for transient protein expression in tobacco (Nicotiana tabacum L.) can optimize process time, yield, and downstream costs. Biotechnol Bioeng 109:2575–2588
Buyel JF, Fischer R (2013) Processing heterogeneous biomass: overcoming the hurdles in model building. Bioengineered 4:21–24
Buyel JF, Fischer R (2014a) Characterization of complex systems using the design of experiments approach: transient protein expression in tobacco as a case study. J Vis Exp 1:e51216
Buyel JF, Fischer R (2014b) Flocculation increases the efficacy of depth filtration during the downstream processing of recombinant pharmaceutical proteins produced in tobacco. Plant Biotechnol J 12:240–252
Buyel JF, Fischer R (2014c) Generic chromatography-based purification strategies accelerate the development of downstream processes for biopharmaceutical proteins produced in plants. Biotechnol J 9:566–577
Buyel JF, Fischer R (2014d) A juice extractor can simplify the downstream processing of plant-derived biopharmaceutical proteins compared to blade-based homogenizers. Process Biochem 50:859–866
Buyel JF, Fischer R (2014e) Scale-down models to optimize a filter train for the downstream purification of recombinant pharmaceutical proteins produced in tobacco leaves. Biotechnol J 9:415–425
Buyel JF, Gruchow HM, Boes A, Fischer R (2014a) Rational design of a host cell protein heat precipitation step simplifies the subsequent purification of recombinant proteins from tobacco. Biochem Eng J 88:162–170
Buyel JF, Gruchow HM, Fischer R (2015b) Depth filters containing diatomite achieve more efficient particle retention than filters solely containing cellulose fibers. Front Plant Sci 6:1–11
Buyel JF, Gruchow HM, Tödter N, Wehner M (2016a) Determination of the thermal properties of leaves by non-invasive contact free laser probing. J Biotechnol 217:100–108
Buyel JF, Hubbuch J and Fischer R (2016b) Comparison of tobacco host cell protein removal methods by blanching intact plants or by heat treatment of extracts. J Vis Exp (9):e54343

Buyel JF, Kaever T, Buyel JJ, Fischer R (2013a) Predictive models for the accumulation of a fluorescent marker protein in tobacco leaves according to the promoter/5′UTR combination. Biotechnol Bioeng 110:471–482

Buyel JF, Opdensteinen P, Fischer R (2014b) Cellulose-based filter aids increase the capacity of depth filters during the downstream processing of plant-derived biopharmaceutical proteins. Biotechnol J 10:584–591

Buyel JF, Twyman RM, Fischer R (2015c) Extraction and downstream processing of plant-derived recombinant proteins. Biotechnol Adv 33:902–913

Buyel JF, Twyman RM, Fischer R (2016c) Very-large-scale production of monoclonal antibodies in plants. In: Gottschalk U (ed) Process scale purification of antibodies, 2. Wiley-Interscience, New York, NY

Buyel JF, Woo JA, Cramer SM, Fischer R (2013b) The use of quantitative structure-activity relationship models to develop optimized processes for the removal of tobacco host cell proteins during biopharmaceutical production. J Chromatogr A 1322:18–28

Camidge R (2001) The story of taxol: nature and politics in the pursuit of an anti-cancer drug. Br Med J 323:115–115

Caspi RR (2008) Immunotherapy of autoimmunity and cancer: the penalty for success. Nat Rev Immunol 8:970–976

Chabner BA, Roberts TG Jr (2005) Timeline: chemotherapy and the war on cancer. Nat Rev Cancer 5:65–72

Chames P, Van Regenmortel M, Weiss E, Baty D (2009) Therapeutic antibodies: successes, limitations and hopes for the future. Br J Pharmacol 157:220–233

Chargelegue D, Vine ND, van Dolleweerd CJ, Drake PMW, Ma JKC (2000) A murine monoclonal antibody produced in transgenic plants with plant-specific glycans is not immunogenic in mice. Transgenic Res 9:187–194

Chen H, Chen XZ, Waterboer T, Castro FA, Brenner H (2015) Viral infections and colorectal cancer: a systematic review of epidemiological studies. Int J Cancer 137:12–24

Cheson BD, Leonard JP (2008) Monoclonal antibody therapy for B-cell non-Hodgkin's lymphoma. N Engl J Med 359:613–626

Chi RC, Rock MT, Neuzil KM (2010) Immunogenicity and safety of intradermal influenza vaccination in healthy older adults. Clin Infect Dis 50:1331–1338

Chial H (2008) Proto-oncogenes to oncogenes to cancer. Nat Educ 1:33

Chiarella P (2011) Production, novel assay development and clinical applications of monoclonal antibodies. Recent Pat Anticancer Drug Discov 6:258–267

Christiansen MN, Chik J, Lee L, Anugraham M, Abrahams JL, Packer NH (2014) Cell surface protein glycosylation in cancer. Proteomics 14:525–546

Chung C, Christianson M (2014) Predictive and prognostic biomarkers with therapeutic targets in breast, colorectal, and non-small cell lung cancers: a systemic review of current development, evidence, and recommendation. J Oncol Pharm Pract 20:11–28

Commandeur U, Twyman RM (2005) Biosafety aspects of molecular farming in plants. In: Fischer R, Schillberg S (eds) Molecular farming. Wiley-VCH Verlag GmbH & Co. KGaA, Hoboken, NJ

Commandeur U, Twyman RM, Fischer R (2003) The biosafety of molecular farming in plants. AgBiotechNet 5:1–9

Committee for Medicinal Products for Human Use (CHMP) (2004) Erbitux - scientific discussion. European Union, London

Committee for Medicinal Products for Human Use (CHMP) (2007) Vectibix - scientific discussion. European Union, London

Committee for Medicinal Products for Human Use (CHMP) (2012) Zaltrap. European Union, London

Committee for Medicinal Products for Human Use (CHMP) (2015a) Assessment report Abraxane. European Union, London

Committee for Medicinal Products for Human Use (CHMP) (2015b) Avastin - Annex I: Summary of product characteristics. European Union, London

Committee for Medicinal Products for Human Use (CHMP) (2015c) Keytruda. European Union, London

Committee for Medicinal Products for Human Use (CHMP) (2016) Cervarix - Annex I: Summary of product characteristics. European Union, London

Conley AJ, Joensuu JJ, Jevnikar AM, Menassa R, Brandle JE (2009) Optimization of elastin-like polypeptide fusions for expression and purification of recombinant proteins in plants. Biotechnol Bioeng 103:562–573

Conley AJ, Joensuu JJ, Richman A, Menassa R (2011) Protein body-inducing fusions for high-level production and purification of recombinant proteins in plants. Plant Biotechnol J 9:419–433

CPMP (2002) Points to consider on quality aspects of medicinal products containing active substances produced by stable transgene expression in higher plants. EMEA, London

Cummins J, Tangney M (2013) Bacteria and tumours: causative agents or opportunistic inhabitants? Infect Agents Cancer 8:1–8

Cusido RM, Onrubia M, Sabater-Jara AB, Moyano E, Bonfill M, Goossens A, Angeles Pedreno M, Palazon J (2014) A rational approach to improving the biotechnological production of taxanes in plant cell cultures of Taxus spp. Biotechnol Adv 32:1157–1167

D'Este C, de Souza P, Sharman C, Allen S (2012) Relocatable, automated cost-benefit analysis for marine sensor network design. Sensors (Basel) 12:2874–2898

Das V, Phillips PWB, Khachatourians GG (2008) Commercialization of plant-derived vaccines in Canada: a distant dream? Health Law Rev 16:25–29

De Beer TRM, Bodson C, Dejaegher B, Walczak B, Vercruysse P, Burggraeve A, Lemos A, Delattre L, Heyden YV, Remon JP, Vervaete C, Baeyens WRG (2008) Raman spectroscopy as a process analytical technology (PAT) tool for the in-line monitoring and understanding of a powder blending process. J Pharm Biomed Anal 48:772–779

De Vincenzo R, Ricci C, Conte C, Scambia G (2013) HPV vaccine cross-protection: highlights on additional clinical benefit. Gynecol Oncol 130:642–651

Distler U, Souady J, Hulsewig M, Drmic-Hofman I, Haier J, Denz A, Grutzmann R, Pilarsky C, Senninger N, Dreisewerd K, Berkenkamp S, Schmidt MA, Peter-Katalinic J, Muthing J (2008) Tumor-associated CD75s- and iso-CD75s-gangliosides are potential targets for adjuvant therapy in pancreatic cancer. Mol Cancer Ther 7:2464–2475

Drake PMW, Barbi T, Sexton A, McGowan E, Stadlmann J, Navarre C, Paul MJ, Ma JKC (2009) Development of rhizosecretion as a production system for recombinant proteins from hydroponic cultivated tobacco. FASEB J 23:3581–3589

Duffy MJ (2015) Personalized treatment for patients with colorectal cancer: role of biomarkers. Biomark Med 9:337–347

Duffy MJ, Lamerz R, Haglund C, Nicolini A, Kalousova M, Holubec L, Sturgeon C (2014) Tumor markers in colorectal cancer, gastric cancer and gastrointestinal stromal cancers: European group on tumor markers 2014 guidelines update. Int J Cancer 134:2513–2522

Ecker DM, Jones SD, Levine HL (2015) The therapeutic monoclonal antibody market. MAbs 7:9–14

Eiberle MK, Jungbauer A (2010) Technical refolding of proteins: do we have freedom to operate? Biotechnol J 5:547–559

Eibl R, Eibl D (2011) Single-use technology in biopharmaceutical manufacture. John Wiley & Sons, Hoboken, NJ

Eibl R, Kaiser S, Lombriser R, Eibl D (2010) Disposable bioreactors: the current state-of-the-art and recommended applications in biotechnology. Appl Microbiol Biotechnol 86:41–49

Endo Y, Tsurugi K, Franz H (1988) The site of action of the A-chain of mistletoe lectin I on eukaryotic ribosomes. The RNA N-glycosidase activity of the protein. FEBS Lett 231:378–380

Exposito O, Bonfill M, Moyano E, Onrubia M, Mirjalili MH, Cusido RM, Palazon J (2009) Biotechnological production of taxol and related taxoids: current state and prospects. Anticancer Agents Med Chem 9:109–121

Fabricant DS, Farnsworth NR (2001) The value of plants used in traditional medicine for drug discovery. Environ Health Perspect 109:69–75

Faller H, Schuler M, Richard M, Heckl U, Weis J, Kuffner R (2013) Effects of psycho-oncologic interventions on emotional distress and quality of life in adult patients with cancer: systematic review and meta-analysis. J Clin Oncol 31:782–793

Farinas CS, Leite A, Miranda EA (2005) Aqueous extraction of recombinant human proinsulin from transgenic maize endosperm. Biotechnol Prog 21:1466–1471

FDA (2002) Guidance for Industry: drugs, biologics, and medical devices derived from bioengineered plants for use in humans and animals. Food and Drug Administration, Rockville, MD

Fischer R, Schillberg S (2006) Molecular farming: plant-made pharmaceuticals and technical proteins. Wiley-Blackwell, Hoboken, NJ

Fischer R, Schillberg S, Buyel JF, Twyman RM (2013) Commercial aspects of pharmaceutical protein production in plants. Curr Pharm Des 19:5471–5477

Fischer R, Schillberg S, Hellwig S, Twyman RM, Drossard J (2012) GMP issues for recombinant plant-derived pharmaceutical proteins. Biotechnol Adv 30:434–439

Fischer R, Vaquero-Martin C, Sack M, Drossard J, Emans N, Commandeur U (1999) Towards molecular farming in the future: transient protein expression in plants. Biotechnol Appl Biochem 30:113–116

Fritz H, Seely D, Kennedy DA, Fernandes R, Cooley K, Fergusson D (2013) Green tea and lung cancer: a systematic review. Integr Cancer Ther 12:7–24

Fujita Y (2007) Industrial production of shikonin and berberine. In: Bock G, Marsh J (eds) Ciba Foundation Symposium 137 - applications of plant cell and tissue culture. John Wiley & Sons, Ltd., Hoboken, NJ

Fuqua JL, Hamorsky K, Khalsa G, Matoba N, Palmer KE (2015) Bulk production of the antiviral lectin griffithsin. Plant Biotechnol J 13:1160–1168

Gabius S, Joshi S, Kayser K, Gabius H (1992) The galactoside-specific lectin from mistletoe as biological response modifier. Int J Oncol 1:705–708

Gajecka M, Rydzanicz M, Jaskula-Sztul R, Wierzbicka M, Szyfter W, Szyfter K (2005) Reduced DNA repair capacity in laryngeal cancer subjects. A comparison of phenotypic and genotypic results. Adv Otorhinolaryngol 62:25–37

Gaughan CL (2015) The present state of the art in expression, production and characterization of monoclonal antibodies. Mol Divers 20:255–270

Geli MI, Torrent M, Ludevid D (1994) Two structural domains mediate two sequential events in [gamma]-Zein targeting: protein endoplasmic reticulum retention and protein body formation. Plant Cell 6:1911–1922

Georgiev MI, Weber J (2014) Bioreactors for plant cells: hardware configuration and internal environment optimization as tools for wider commercialization. Biotechnol Lett 36:1359–1367

Gerngross TU (2004) Advances in the production of human therapeutic proteins in yeasts and filamentous fungi. Nat Biotechnol 22:1409–1414

Glassman PM, Balthasar JP (2014) Mechanistic considerations for the use of monoclonal antibodies for cancer therapy. Cancer Biol Med 11:20–33

Gleba YY, Tuse D, Giritch A (2014) Plant viral vectors for delivery by agrobacterium. Curr Top Microbiol Immunol 375:155–192

Glick BR, Pasternak JJ, Patten CL (2010) Molecular biotechnology: principles and applications of recombinant DNA. John Wiley & Sons, Hoboken, NJ

Gong SJ, Jin CJ, Rha SY, Chung HC (2004) Growth inhibitory effects of trastuzumab and chemotherapeutic drugs in gastric cancer cell lines. Cancer Lett 214:215–224

Goody J (1997) Industrial food. Towards the development of a world cuisine. In: Counihan C, van Esterik P (eds) Food and culture. Routledge, London

Grabowski GA, Golembo M, Shaaltiel Y (2014) Taliglucerase alfa: an enzyme replacement therapy using plant cell expression technology. Mol Genet Metab 112:1–8

Hanahan D, Weinberg RA (2000) The hallmarks of cancer. Cell 100:57–70

Hanahan D, Weinberg RA (2011) Hallmarks of cancer: the next generation. Cell 144:646–674

Harris LJ, Larson SB, Hasel KW, McPherson A (1997) Refined structure of an intact IgG2a monoclonal antibody. Biochemistry 36:1581–1597

Harrison LB, Chadha M, Hill RJ, Hu K, Shasha D (2002) Impact of tumor hypoxia and anemia on radiation therapy outcomes. Oncologist 7:492–508

Hassan S, Keshavarz-Moore E, Ma J, Thomas C (2014) Breakage of transgenic tobacco roots for monoclonal antibody release in an ultra-scale down shearing device. Biotechnol Bioeng 111:196–201

Hassan S, van Dolleweerd CJ, Ioakeimidis F, Keshavarz-Moore E, Ma JK (2008) Considerations for extraction of monoclonal antibodies targeted to different subcellular compartments in transgenic tobacco plants. Plant Biotechnol J 6:733–748

Hellwig S, Drossard J, Twyman RM, Fischer R (2004) Plant cell cultures for the production of recombinant proteins. Nat Biotechnol 22:1415–1422

Hernandez J, Mullinax J, Clark W, Toomey P, Villadolid D, Morton C, Ross S, Rosemurgy A (2009) Survival after pancreaticoduodenectomy is not improved by extending resections to achieve negative margins. Ann Surg 250:76–80

Heydarnejad MS, Hassanpour DA, Solati DK (2011) Factors affecting quality of life in cancer patients undergoing chemotherapy. Afr Health Sci 11:266–270

Holland T, Blessing D, Hellwig S, Sack M (2013) The in-line measurement of plant cell biomass using radio frequency impedance spectroscopy as a component of process analytical technology. Biotechnol J 8:1231–1240

Holtz BR, Berquist BR, Bennett LD, Kommineni VJ, Munigunti RK, White EL, Wilkerson DC, Wong KY, Ly LH, Marcel S (2015) Commercial-scale biotherapeutics manufacturing facility for plant-made pharmaceuticals. Plant Biotechnol J 13:1180–1190

Hong SN, Lee SM, Kim JH, Lee TY, Kim JH, Choe WH, Lee SY, Cheon YK, Sung IK, Park HS, Shim CS (2012) Helicobacter pylori infection increases the risk of colorectal adenomas: cross-sectional study and meta-analysis. Dig Dis Sci 57:2184–2194

Howat S, Park B, Oh IS, Jin YW, Lee EK, Loake GJ (2014) Paclitaxel: biosynthesis, production and future prospects. N Biotechnol 31:242–245

Ioannidis JPA (2005) Why most published research findings are false. PLoS Med 2:696–701

Jagadeesh S, Sinha S, Pal BC, Bhattacharya S, Banerjee PP (2007) Mahanine reverses an epigenetically silenced tumor suppressor gene RASSF1A in human prostate cancer cells. Biochem Biophys Res Commun 362:212–217

James C (2015) Global status of commercialized biotech/GM Crops: 2015. International Service for the Acquisition of Agri-biotech Applications, Ithaca, NY

Jiang QL, Zhang S, Tian M, Zhang SY, Xie T, Chen DY, Chen YJ, He J, Liu J, Ouyang L, Jiang X (2015) Plant lectins, from ancient sugar-binding proteins to emerging anti-cancer drugs in apoptosis and autophagy. Cell Prolif 48:17–28

Juca SC, Carvalho PC, Brito FT (2011) A low cost concept for data acquisition systems applied to decentralized renewable energy plants. Sensors (Basel) 11:743–756

Kapchie VN, Hauck CC, Wang H, Murphy PA (2011) Process improvement for semipurified oleosomes on a pilot-plant scale. J Food Sci 76:C853–C860

Kelley B (2007) Very large scale monoclonal antibody purification: the case for conventional unit operations. Biotechnol Prog 23:995–1008

Kelsey RG, Vance NC (1992) Taxol and cephalomannine concentrations in the foliage and bark of shade-grown and sun-exposed taxus brevifolia trees. J Nat Prod 55:912–917

Kenney RT, Frech SA, Muenz LR, Villar CP, Glenn GM (2004) Dose sparing with intradermal injection of influenza vaccine. N Engl J Med 351:2295–2301

Kim TH, Yoo CG, Lamsal BP (2013) Front-end recovery of protein from lignocellulosic biomass and its effects on chemical pretreatment and enzymatic saccharification. Bioprocess Biosyst Eng 36:687–694

King-Batoon A, Leszczynska JM, Klein CB (2008) Modulation of gene methylation by genistein or lycopene in breast cancer cells. Environ Mol Mutagen 49:36–45

Klutz S, Magnus J, Lobedann M, Schwan P, Maiser B, Niklas J, Temming M, Schembecker G (2015) Developing the biofacility of the future based on continuous processing and single-use technology. J Biotechnol 213:120–130

Koprivova A, Stemmer C, Altmann F, Hoffmann A, Kopriva S, Gorr G, Reski R, Decker EL (2004) Targeted knockouts of Physcomitrella lacking plantspecific immunogenic N-glycans. Plant Biotechnol J 2:517–523
Kourmanova AG, Soudarkina OJ, Olsnes S, Kozlov JV (2004) Cloning and characterization of the genes encoding toxic lectins in mistletoe (Viscum album L). Eur J Biochem 271:2350–2360
Krauspenhaar R, Rypniewski W, Kalkura N, Moore K, DeLucas L, Stoeva S, Mikhailov A, Voelter W, Betzel C (2002) Crystallisation under microgravity of mistletoe lectin I from Viscum album with adenine monophosphate and the crystal structure at 1.9 A resolution. Acta Cryst D Biol Crystallogr 58:1704–1707
Kubota K (2011) Recent advances and limitations of surgical treatment for pancreatic cancer. World J Clin Oncol 2:225–228
Kuttan G, Menon LG, Antony S, Kuttan R (1997) Anticarcinogenic and antimetastatic activity of Iscador. Anticancer Drugs 8:15–16
Kwok R (2011) The real issues in vaccine safety. Nature 473:436–438
Kwon SB, Kim MJ, Yang JM, Lee HP, Hong JT, Jeong HS, Kim ES, Yoon DY (2016) Cudrania tricuspidata stem extract induces apoptosis via the extrinsic pathway in SiHa cervical cancer cells. PLoS One 11:e0150235
Landgrebe D, Haake C, Hopfner T, Beutel S, Hitzmann B, Scheper T, Rhiel M, Reardon KF (2010) On-line infrared spectroscopy for bioprocess monitoring. Appl Microbiol Biotechnol 88:11–22
Lannoo N, Van Damme EJ (2014) Lectin domains at the frontiers of plant defense. Front Plant Sci 5:397
Lannoo N, Vervecken W, Proost P, Rouge P, Van Damme EJ (2007) Expression of the nucleocytoplasmic tobacco lectin in the yeast Pichia pastoris. Protein Expr Purif 53:275–282
Larsen JS, Curtis WR (2008) Reducing batch-to-batch variability of agrobacterium-mediated transient protein expression in plant tissue culture. AIChE, Philadelphia, PA
Laukel M, Rogge P, Dudziak G (2011) Disposable downstream processing for clinical manufacturing. BioProcess Int 9:14–21
Lee RT, Gabius HJ, Lee YC (1994) The sugar-combining area of the galactose-specific toxic lectin of mistletoe extends beyond the terminal sugar residue: comparison with a homologous toxic lectin, ricin. Carbohydr Res 254:269–276
Li CH, Narhi LO, Wen J, Dimitrova M, Wen ZQ, Li J, Pollastrini J, Nguyen X, Tsuruda T, Jiang YJ (2012) Effect of pH, temperature, and salt on the stability of Escherichia coli- and Chinese hamster ovary cell-derived IgG1 Fc. Biochemistry 51:10056–10065
Li F, Vijayasankaran N, Shen A, Kiss R, Amanullah A (2010) Cell culture processes for monoclonal antibody production. MAbs 2:466–479
Li YF (2011) Self-cleaving fusion tags for recombinant protein production. Biotechnol Lett 33:869–881
Lightfoot EN, Moscariello JS (2004) Bioseparations. Biotechnol Bioeng 87:259–273
Lightfoot EN, Root TW, O'Dell JL (2008) Emergence of ideal membrane cascades for downstream processing. Biotechnol Prog 24:599–605
Linden W, Girgis A (2012) Psychological treatment outcomes for cancer patients: what do meta-analyses tell us about distress reduction? Psychooncology 21:343–350
Lolas A (2013) Microbial control strategies in bioprocessing falling short of assuring product quality and satisfying regulatory expectations. CompareNetworks, Fishers, IN
Lorence A, Verpoorte R (2004) Gene transfer and expression in plants. Methods Mol Biol 267:329–350
Ma JK, Chikwamba R, Sparrow P, Fischer R, Mahoney R, Twyman RM (2005) Plant-derived pharmaceuticals--the road forward. Trends Plant Sci 10:580–585
Ma JK, Drossard J, Lewis D, Altmann F, Boyle J, Christou P, Cole T, Dale P, van Dolleweerd CJ, Isitt V, Katinger D, Lobedan M, Mertens H, Paul MJ, Rademacher T, Sack M, Hundleby PA, Stiegler G, Stoger E, Twyman RM, Vcelar B, Fischer R (2015) Regulatory approval and a first-in-human phase I clinical trial of a monoclonal antibody produced in transgenic tobacco plants. Plant Biotechnol J 13:1106–1120

Manegold C (2014) Treatment algorithm in 2014 for advanced non-small cell lung cancer: therapy selection by tumour histology and molecular biology. Adv Med Sci 59:308–313

Markley N, Nykiforuk C, Boothe J, Moloney M (2006) Producing proteins using transgenic oilbody-oleosin technology. BioPharm Int 19:34–57

Massa S, Franconi R, Brandi R, Muller A, Mett V, Yusibov V, Venuti A (2007) Anti-cancer activity of plant-produced HPV16 E7 vaccine. Vaccine 25:3018–3021

Mauch P, Constine L, Greenberger J, Knospe W, Sullivan J, Liesveld JL, Deeg HJ (1995) Hematopoietic stem cell compartment: acute and late effects of radiation therapy and chemotherapy. Int J Radiat Oncol Biol Phys 31:1319–1339

McAllister SD, Soroceanu L, Desprez PY (2015) The antitumor activity of plant-derived non-psychoactive cannabinoids. J Neuroimmune Pharmacol 10:255–267

McCormack PL (2014) Quadrivalent human papillomavirus (types 6, 11, 16, 18) recombinant vaccine (Gardasil((R))): a review of its use in the prevention of premalignant anogenital lesions, cervical and anal cancers, and genital warts. Drugs 74:1253–1283

McCormick AA, Kumagai MH, Hanley K, Turpen TH, Hakim I, Grill LK, Tuse D, Levy S, Levy R (1999) Rapid production of specific vaccines for lymphoma by expression of the tumor-derived single-chain Fv epitopes in tobacco plants. Proc Natl Acad Sci U S A 96:703–708

McGuire S (2016) World cancer report 2014. Geneva, Switzerland: World Health Organization, International Agency for Research on Cancer, WHO Press, 2015. Adv Nutr 7:418–419

McLean MD, Chen RJ, Yu DQ, Mah KZ, Teat J, Wang HF, Zaplachinski S, Boothe J, Hall JC (2012) Purification of the therapeutic antibody trastuzumab from genetically modified plants using safflower Protein A-oleosin oilbody technology. Transgenic Res 21:1291–1301

Mellstedt H (2003) Monoclonal antibodies in human cancer. Drugs Today (Barc) 39:1–16

Menkhaus TJ, Bai Y, Zhang C, Nikolov ZL, Glatz CE (2004) Considerations for the recovery of recombinant proteins from plants. Biotechnol Prog 20:1001–1014

Menkhaus TJ, Glatz CE (2005) Antibody capture from corn endosperm extracts by packed bed and expanded bed adsorption. Biotechnol Prog 21:473–485

Menkhaus TJ, Roseland J (2008) Recovery of proteins from corn and soybean extracts by membrane adsorption. Biotechnol Prog 24:1075–1084

Menzel S, Holland T, Boes A, Spiegel H, Bolzenius J, Fischer R, Buyel JF (2016) Optimized blanching reduces the host cell protein content and substantially enhances the recovery and stability of two plant-derived malaria vaccine candidates. Front Plant Sci 7:1–15

Merrett ND (2014) Multimodality treatment of potentially curative gastric cancer: geographical variations and future prospects. World J Gastroenterol 20:12892–12899

Millner LM, Strotman LN (2016) The future of precision medicine in oncology. Clin Lab Med 36:557–573

Minchinton AI, Tannock IF (2006) Drug penetration in solid tumours. Nat Rev Cancer 6:583–592

Mor TS (2015) Molecular pharming's foot in the FDA's door: protalix's trailblazing story. Biotechnol Lett 37:2147–2150

Mukherjee AK, Basu S, Sarkar N, Ghosh AC (2001) Advances in cancer therapy with plant based natural products. Curr Med Chem 8:1467–1486

Murphy KP, Travers P, Walport M, Janeway C (2008) Janeway's immunobiology. Garland Science, New York, NY

Muthing J, Meisen I, Bulau P, Langer M, Witthohn K, Lentzen H, Neumann U, Peter-Katalinic J (2004) Mistletoe lectin I is a sialic acid-specific lectin with strict preference to gangliosides and glycoproteins with terminal Neu5Ac alpha 2-6Gal beta 1-4GlcNAc residues. Biochemistry 43:2996–3007

Napier JA, Stobart AK, Shewry PR (1996) The structure and biogenesis of plant oil bodies: the role of the ER membrane and the oleosin class of proteins. Plant Mol Biol 31:945–956

Nardin EH, Oliveira GA, Calvo-Calle JM, Wetzel K, Maier C, Birkett AJ, Sarpotdar P, Corado ML, Thornton GB, Schmidt A (2004) Phase I testing of a malaria vaccine composed of hepatitis B virus core particles expressing Plasmodium falciparum circumsporozoite epitopes. Infect Immun 72:6519–6527

Neidle S, Thurston DE (2005) Chemical approaches to the discovery and development of cancer therapies. Nat Rev Cancer 5:285–296
Newell CA (2000) Plant transformation technology - developments and applications. Mol Biotechnol 16:53–65
Nfor BK, Zuluaga DS, Verheijen PJT, Verhaert PDEM, van der Wielen LAM, Ottens M (2011) Model-based rational strategy for chromatographic resin selection. Biotechnol Prog 27:1629–1643
Niwa H, Tonevitsky AG, Agapov II, Saward S, Pfuller U, Palmer RA (2003) Crystal structure at 3 A of mistletoe lectin I, a dimeric type-II ribosome-inactivating protein, complexed with galactose. Eur J Biochem 270:2739–2749
Nuzzo R (2014) Scientific method: statistical errors. Nature 506:150–152
O'Brien TP, Brown LA, Battersby DG, Rudolph AS, Raman LP (2012) Large-scale, single-use depth filtration systems for mammalian cell culture clarification. BioProcess Int 10:50–57
O'Keefe BR, Vojdani F, Buffa V, Shattock RJ, Montefiori DC, Bakke J, Mirsalis J, d'Andrea AL, Hume SD, Bratcher B, Saucedo CJ, McMahon JB, Pogue GP, Palmer KE (2009) Scaleable manufacture of HIV-1 entry inhibitor griffithsin and validation of its safety and efficacy as a topical microbicide component. Proc Natl Acad Sci U S A 106:6099–6104
O'Neill KM, Larsen JS, Curtis WR (2008) Scale-up of Agrobacterium-mediated transient protein expression in bioreactor-grown Nicotiana glutinosa plant cell suspension culture. Biotechnol Prog 24:372–376
Oliveira C, Felix W, Moreira RA, Teixeira JA, Domingues L (2008) Expression of frutalin, an alpha-D-galactose-binding jacalin-related lectin, in the yeast Pichia pastoris. Protein Expr Purif 60:188–193
Olsnes S, Stirpe F, Sandvig K, Pihl A (1982) Isolation and characterization of viscumin, a toxic lectin from Viscum album L. (mistletoe). J Biol Chem 257:13263–13270
Orr V, Zhong LY, Moo-Young M, Chou CP (2013) Recent advances in bioprocessing application of membrane chromatography. Biotechnol Adv 31:450–465
Ouattara A, Laurens MB (2015) Vaccines against malaria. Clin Infect Dis 60:930–936
Pastores GM, Petakov M, Giraldo P, Rosenbaum H, Szer J, Deegan PB, Amato DJ, Mengel E, Tan ES, Chertkoff R, Brill-Almon E, Zimran A (2014) A Phase 3, multicenter, open-label, switchover trial to assess the safety and efficacy of taliglucerase alfa, a plant cell-expressed recombinant human glucocerebrosidase, in adult and pediatric patients with Gaucher disease previously treated with imiglucerase. Blood Cells Mol Dis 53:253–260
Paul M, Ma JK (2011) Plant-made pharmaceuticals: leading products and production platforms. Biotechnol Appl Biochem 58:58–67
Paul MJ, Teh AYH, Twyman RM, Ma JKC (2013) Target product selection - where can molecular pharming make the difference? Curr Pharm Des 19:5478–5485
Pereira DM, Valentao P, Correia-da-Silva G, Teixeira N, Andrade PB (2012) Plant secondary metabolites in cancer chemotherapy: where are we? Curr Pharm Biotechnol 13:632–650
Peters C, Brown S (2015) Antibody-drug conjugates as novel anti-cancer chemotherapeutics. Biosci Rep 35:1–20
PhRMA (2014) Pharmaceutical Industry Profile - PhRMA annual membership survey. Pharmaceutical Research and Manufacturers of America, Washington, DC
Polakis P (2016) Antibody drug conjugates for cancer therapy. Pharmacol Rev 68:3–19
Polireddy K, Chen Q (2016) Cancer of the pancreas: molecular pathways and current advancement in treatment. J Cancer 7:1497–1514
Poljak M (2012) Prophylactic human papillomavirus vaccination and primary prevention of cervical cancer: issues and challenges. Clin Microbiol Infect 18:64–69
Pritzkuleit R, Beske F, Katalinic A (2010) Demographic change and cancer. Onkologie 33:19–24
Puchta H, Fauser F (2014) Synthetic nucleases for genome engineering in plants: prospects for a bright future. Plant J 78:727–741
Raees MA, Hussain H, Al-Rawahi A, Csuk R, Muhammad SA, Khan HY, Rehman NU, Abbas G, Al-Broumi MA, Green IR, Elyassi A, Mahmood T, Al-Harrasi A (2016) Anti-proliferative and computational studies of two new pregnane glycosides from Desmidorchis flava. Bioorg Chem 67:95–104

Rathore AS (2009) Roadmap for implementation of quality by design (QbD) for biotechnology products. Trends Biotechnol 27:546–553
Rathore AS, Bhambure R, Ghare V (2010) Process analytical technology (PAT) for biopharmaceutical products. Anal Bioanal Chem 398:137–154
Rathore AS, Yu M, Yeboah S, Sharma A (2008) Case study and application of process analytical technology (PAT) towards bioprocessing: use of on-line high-performance liquid chromatography (HPLC) for making real-time pooling decisions for process chromatography. Biotechnol Bioeng 100:306–316
Read EK, Park JT, Shah RB, Riley BS, Brorson KA, Rathore AS (2010a) Process analytical technology (PAT) for biopharmaceutical products: Part I. Concepts and applications. Biotechnol Bioeng 105:276–284
Read EK, Shah RB, Riley BS, Park JT, Brorson KA, Rathore AS (2010b) Process analytical technology (PAT) for biopharmaceutical products: Part II. Concepts and applications. Biotechnol Bioeng 105:285–295
Research B (2013) Biologic therapeutic drugs: technologies and global markets. BCC Research LLC, Wellesley, MA
Reuter LJ, Bailey MJ, Joensuu JJ, Ritala A (2014) Scale-up of hydrophobin-assisted recombinant protein production in tobacco BY-2 suspension cells. Plant Biotechnol J 12:402–410
Rivera AL, Gomez-Lim M, Fernandez F, Loske AM (2012) Physical methods for genetic plant transformation. Phys Life Rev 9:308–345
Rosenberg Y, Sack M, Montefiori D, Forthal D, Mao LJ, Hernandez-Abanto S, Urban L, Landucci G, Fischer R, Jiang XM (2013) Rapid high-level production of functional HIV broadly neutralizing monoclonal antibodies in transient plant expression systems. PLoS One 8:e58724
Rottenberg Y, Barchana M, Liphshitz I, Peretz T (2010) The changing face of cancer in the elderly: only a demographic change? Arch Gerontol Geriatr 50:e59–e62
Sainz B, Antolin J, Lopez-Coronado M, de Castro C (2013) A novel low-cost sensor prototype for monitoring temperature during wine fermentation in tanks. Sensors (Basel) 13:2848–2861
Salama JK, Chmura SJ (2014) The role of surgery and ablative radiotherapy in oligometastatic breast cancer. Semin Oncol 41:790–797
Sankaranarayanan R (2014) Screening for cancer in low- and middle-income countries. Ann Glob Health 80:412–417
Santos RB, Abranches R, Fischer R, Sack M, Holland T (2016) Putting the spotlight back on plant suspension cultures. Front Plant Sci 7:297
Schillberg S, Raven N, Fischer R, Twyman RM, Schiermeyer A (2013) Molecular farming of pharmaceutical proteins using plant suspension cell and tissue cultures. Curr Pharm Des 19:5531–5542
Schilsky RL (2010) Personalized medicine in oncology: the future is now. Nat Rev Drug Discov 9:363–366
Schnekenburger M, Dicato M, Diederich M (2014) Plant-derived epigenetic modulators for cancer treatment and prevention. Biotechnol Adv 32:1123–1132
Schork NJ (2015) Personalized medicine: time for one-person trials. Nature 520:609–611
Schuster M, Nechansky A, Kircheis R (2006) Cancer immunotherapy. Biotechnol J 1:138–147
Scott AM, Wolchok JD, Old LJ (2012) Antibody therapy of cancer. Nat Rev Cancer 12:278–287
Scotti N, Rybicki EP (2013) Virus-like particles produced in plants as potential vaccines. Expert Rev Vaccines 12:211–224
Selvi BR, Pradhan SK, Shandilya J, Das C, Sailaja BS, Shankar GN, Gadad SS, Reddy A, Dasgupta D, Kundu TK (2009) Sanguinarine interacts with chromatin, modulates epigenetic modifications, and transcription in the context of chromatin. Chem Biol 16:203–216
Senter PD (2009) Potent antibody drug conjugates for cancer therapy. Curr Opin Chem Biol 13:235–244
Sfikakis PP, Gourgoulis GM, Moulopoulos LA, Kouvatseas G, Theofilopoulos AN, Dimopoulos MA (2005) Age-related thymic activity in adults following chemotherapy-induced lymphopenia. Eur J Clin Invest 35:380–387

Shahid S (2016) Review of hematological indices of cancer patients receiving combined chemotherapy & radiotherapy or receiving radiotherapy alone. Crit Rev Oncol/Hematol 105:145–155

Shoji Y, Farrance CE, Bautista J, Bi H, Musiychuk K, Horsey A, Park H, Jaje J, Green BJ, Shamloul M, Sharma S, Chichester JA, Mett V, Yusibov V (2012) A plant-based system for rapid production of influenza vaccine antigens. Influenza Other Respi Viruses 6:204–210

Shukla AA, Gottschalk U (2013) Single-use disposable technologies for biopharmaceutical manufacturing. Trends Biotechnol 31:147–154

Shukla AA, Thommes J (2010) Recent advances in large-scale production of monoclonal antibodies and related proteins. Trends Biotechnol 28:253–261

Siddiqui M, Rajkumar SV (2012) The high cost of cancer drugs and what we can do about it. Mayo Clin Proc 87:935–943

Silverstein A, Silverstein VB, Nunn LS (2006) Cancer: conquering a deadly disease. Twenty-first century books, Minneapolis, MN

Small E, Catling PM (1999) Canadian medicinal crops. NRC Research Press, Ottawa, ON

Sonnenschein C, Soto AM (2013) The aging of the 2000 and 2011 Hallmarks of cancer reviews: a critique. J Biosci 38:651–663

Souza MA, Carvalho FC, Ruas LP, Ricci-Azevedo R, Roque-Barreira MC (2013) The immunomodulatory effect of plant lectins: a review with emphasis on ArtinM properties. Glycoconj J 30:641–657

Sparrow PAC, Irwin JA, Dale PJ, Twyman RM, Ma JKC (2007) Pharma-planta: road testing the developing regulatory guidelines for plant-made pharmaceuticals. Transgenic Res 16:147–161

Spiegel H, Boes A, Voepel N, Beiss V, Edgue G, Rademacher T, Sack M, Schillberg S, Reimann A, Fischer R (2015) Application of a scalable plant transient gene expression platform for malaria vaccine development. Front Plant Sci 6:1169

Spiegel H, Stöger E, Twyman RM, Buyel JF (2016) Current status and perspectives of the molecular farming landscape. In: Kermode AR (ed) Molecular pharming: applications, challenges and emerging areas. Wiley-VCH, Weinheim

Spok A (2007) Molecular farming on the rise--GMO regulators still walking a tightrope. Trends Biotechnol 25:74–82

Stoger E, Fischer R, Moloney M, Ma JKC (2014) Plant molecular pharming for the treatment of chronic and infectious diseases. Annu Rev Plant Biol 65:743–768

Stoger E, Sack M, Perrin Y, Vaquero C, Torres E, Twyman RM, Christou P, Fischer R (2002) Practical considerations for pharmaceutical antibody production in different crop systems. Mol Breed 9:149–158

Strasser R (2013) Engineering of human-type O-glycosylation in Nicotiana benthamiana plants. Bioengineered 4:191–196

Strasser R (2016) Plant protein glycosylation. Glycobiology 26(9):926–939

Strasser R, Stadlmann J, Schahs M, Stiegler G, Quendler H, Mach L, Glossl J, Weterings K, Pabst M, Steinkellner H (2008) Generation of glyco-engineered Nicotiana benthamiana for the production of monoclonal antibodies with a homogeneous human-like N-glycan structure. Plant Biotechnol J 6:392–402

Sudhakar A (2009) History of cancer, ancient and modern treatment methods. J Cancer Sci Ther 1:1–4

Tabata H (2004) Paclitaxel production by plant-cell-culture technology. Adv Biochem Eng Biotechnol 87:1–23

The CMC Biotech Working Group (2009) A-Mab: a case study in bioprocess development. CASSS - An International Separation Science Society, Emeryville, CA

Thierry-Carstensen B, Dalby T, Stevner MA, Robbins JB, Schneerson R, Trollfors B (2013) Experience with monocomponent acellular pertussis combination vaccines for infants, children, adolescents and adults-A review of safety, immunogenicity, efficacy and effectiveness studies and 15 years of field experience. Vaccine 31:5178–5191

Tian L, Sun SSM (2011) A cost-effective ELP-intein coupling system for recombinant protein purification from plant production platform. PLoS One 6:e24183

Torrent M, Llompart B, Lasserre-Ramassamy S, Llop-Tous I, Bastida M, Marzabal P, Westerholm-Parvinen A, Saloheimo M, Heifetz PB, Ludevid MD (2009) Eukaryotic protein production in designed storage organelles. BMC Biol 7:5

Tremblay R, Diao H, Huner N, Jevnikar AM, Ma SW (2011) The development, characterization, and demonstration of a novel strategy for purification of recombinant proteins expressed in plants. Transgenic Res 20:1357–1366

Trueb RM (2010) Chemotherapy-induced hair loss. Skin Therapy Lett 15:5–7

Twyman RM, Stoger E, Schillberg S, Christou P, Fischer R (2003) Molecular farming in plants: host systems and expression technology. Trends Biotechnol 21:570–578

UniProt-Consortium (2012) Reorganizing the protein space at the Universal Protein Resource (UniProt). Nucleic Acids Res 40:D71–D75

Urry D (1988) Entropic elastic processes in protein mechanisms. I. Elastic structure due to an inverse temperature transition and elasticity due to internal chain dynamics. J Protein Chem 7:1–34

Valentine L (2003) Agrobacterium tumefaciens and the plant: the David and Goliath of modern genetics. Plant Physiol 133:948–955

Van Damme EJ (2014) History of plant lectin research. Methods Mol Biol 1200:3–13

Vandenborre G, Smagghe G, Van Damme EJ (2011) Plant lectins as defense proteins against phytophagous insects. Phytochemistry 72:1538–1550

Venuti A, Massa S, Mett V, Vedova LD, Paolini F, Franconi R, Yusibov V (2009) An E7-based therapeutic vaccine protects mice against HPV16 associated cancer. Vaccine 27:3395–3397

Von Hoff DD, Ervin T, Arena FP, Chiorean EG, Infante J, Moore M, Seay T, Tjulandin SA, Ma WW, Saleh MN, Harris M, Reni M, Dowden S, Laheru D, Bahary N, Ramanathan RK, Tabernero J, Hidalgo M, Goldstein D, Van Cutsem E, Wei X, Iglesias J, Renschler MF (2013) Increased survival in pancreatic cancer with nab-paclitaxel plus gemcitabine. N Engl J Med 369:1691–1703

Wahl O, Oswald M, Tretzel L, Herres E, Arend J, Efferth T (2011) Inhibition of tumor angiogenesis by antibodies, synthetic small molecules and natural products. Curr Med Chem 18:3136–3155

Walsh MJ, Dodd JE, Hautbergue GM (2013) Ribosome-inactivating proteins: potent poisons and molecular tools. Virulence 4:774–784

Wang JW, Roden RBS (2013) Virus-like particles for the prevention of human papillomavirus-associated malignancies. Expert Rev Vaccines 12:129–141

Wani MC, Horwitz SB (2014) Nature as a remarkable chemist: a personal story of the discovery and development of Taxol. Anticancer Drugs 25:482–487

Weinberg RA (2006) The biology of cancer. Garland Science, Milton Park

Weiner LM, Murray JC, Shuptrine CW (2012) Antibody-based immunotherapy of cancer. Cell 148:1081–1084

Whitford WG (2010) Single-use systems as principal components in bioproduction. BioProcess Int 8:34–44

Wilken LR, Nikolov ZL (2012) Recovery and purification of plant-made recombinant proteins. Biotechnol Adv 30:419–433

Wilson SA, Roberts SC (2012) Recent advances towards development and commercialization of plant cell culture processes for the synthesis of biomolecules. Plant Biotechnol J 10:249–268

Winkelnkemper T, Schembecker G (2010) Purification performance index and separation cost indicator for experimentally based systematic downstream process development. Sep Purif Technol 72:34–39

Wirz H, Sauer-Budge AF, Briggs J, Sharpe A, Shu SD, Sharon A (2012) Automated production of plant-based vaccines and pharmaceuticals. J Lab Autom 17:449–457

Wurm FM (2004) Production of recombinant protein therapeutics in cultivated mammalian cells. Nat Biotechnol 22:1393–1398

Yabroff KR, Lund J, Kepka D, Mariotto A (2011) Economic burden of cancer in the United States: estimates, projections, and future research. Cancer Epidemiol Biomarkers Prev 20:2006–2014

Zarkovic N, Kalisnik T, Loncaric I, Borovic S, Mang S, Kissel D, Konitzer M, Jurin M, Grainza S (1998) Comparison of the effects of Viscum album lectin ML-1 and fresh plant extract (Isorel) on the cell growth in vitro and tumorigenicity of melanoma B16F10. Cancer Biother Radiopharm 13:121–131

Zhong JJ (2002) Plant cell culture for production of paclitaxel and other taxanes. J Biosci Bioeng 94:591–599

Zimran A, Brill-Almon E, Chertkoff R, Petakov M, Blanco-Favela F, Munoz ET, Solorio-Meza SE, Amato D, Duran G, Giona F, Heitner R, Rosenbaum H, Giraldo P, Mehta A, Park G, Phillips M, Elstein D, Altarescu G, Szleifer M, Hashmueli S, Aviezer D (2011) Pivotal trial with plant cell-expressed recombinant glucocerebrosidase, taliglucerase alfa, a novel enzyme replacement therapy for Gaucher disease. Blood 118:5767–5773

Zischewski J, Sack M, Fischer R (2015) Overcoming low yields of plant-made antibodies by a protein engineering approach. Biotechnol J 11:107–116

Zlotogorski A, Dayan A, Dayan D, Chaushu G, Salo T, Vered M (2013) Nutraceuticals as new treatment approaches for oral cancer: II. Green tea extracts and resveratrol. Oral Oncol 49:502–506

Zwierzina H, Bergmann L, Fiebig H, Aamdal S, Schoffski P, Witthohn K, Lentzen H (2011) The preclinical and clinical activity of aviscumine: a potential anticancer drug. Eur J Cancer 47:1450–1457

Chapter 3
Cancer and Biotechnology: A Matchup that Should Never Slowdown

Ala'a Al-Hrout, Badriya Baig, Ali Hilal-Alnaqbi, and Amr Amin

Abstract Plant-based treatments propose a very attractive approach for cancer prevention and therapy due to their minimal toxicity and lower-to-nonassociated side effects. About 40% of FDA-approved therapeutic agents are natural-based components or their derivatives. Plant-based components have been reported to have anticancer properties in vivo and in vitro through the modulation of many cancer hallmarks and niche elements, including self-renewal properties of cancer stem cells. Screening for anticancer agents, synthetic or natural-based, requires a reliable disease model. Efforts to mimic in vivo conditions have led to the development of three-dimensional culture system, a biotechnology that allows cells to grow in three dimensions. Nanotechnology is also fast growing into a quite powerful tool to improve quality of life, particularly the life of cancer patients. Clearing away most hurdles that conventional/herbal medicine is often challenged with, nanotechnology can potentially help delivering anticancer herbal drugs more specifically and efficiently. Given the rather gloomy reality of cancer, nanomedicine becomes the major silver lining. A number of successful conjugates of nanoparticle with herbal products are discussed in this chapter.

Keywords Plant-derived nanoparticles • Cancer stem cells • Nanomedicine • 3D-culture system • Molecular docking

A. Al-Hrout • B. Baig
Biology Department, College of Science, UAE University, PO Box 15551, Al-Ain, UAE

A. Hilal-Alnaqbi
Department of Mechanical Engineering, College of Engineering, UAE University, PO Box 15551, Al-Ain, UAE

A. Amin (✉)
Biology Department, College of Science, UAE University, PO Box 15551, Al-Ain, UAE

Zoology Department, Faculty of Science, Cairo University, Giza, Egypt
e-mail: a.amin@uaeu.ac.ae

S. Malik (ed.), *Biotechnology and Production of Anti-Cancer Compounds*,
DOI 10.1007/978-3-319-53880-8_3

Introduction

Semantically, biotechnology is defined according to its prefix "bio" that means life and the core "technology" that refers to knowledge about targeted and cost-effective transformation process of natural resource into useful goods (Koltuniewicz 2014). Often, the development and utilization of biological processes, forms, and systems for obtaining maximum benefits to man and other forms of life is broadly accepted as a definition of biotechnology (Dubey 2014). Biotechnology contributes immensely to different fields of medicine, pharmaceutics, agriculture, marine, forensic, environment, industry, and many others. Medical biotechnology and how it relates to cancer research is the focus of this chapter.

The riddle of cancer remains unsolved and continues to stand out as a major global health burden. In fact, as recently reported, cancer is a leading cause of death worldwide (Stewart and Wild 2014). Cancer in its simplified definition is uncontrolled growth and division of genetically unstable cells. Cancer hold 11 biological capabilities that can be recognized as hallmarks of cancer which are genomic instability, sustaining proliferative signaling, evading growth suppressors, resisting cell death, enabling replicative immortality, inducing angiogenesis, activating invasion and metastasis, dysregulated metabolism, tumor promoting inflammation, and microenvironment (Hanahan and Weinberg 2011; Vogelstein et al. 2013; Block et al. 2015). The success rate of surviving cancer largely depends on the early detection (WHO 2003). Sadly, that is usually not the case where cancer is often diagnosed at advanced stages.

Responses to conventional cancer therapies needs further understanding; while some patients show complete cancer regression others show no significant changes. In fact, most conventional cancer therapies seem to settle for just attempting to prolong patients' life and palliate the disease rather than curing it (Lokich 2012). Main conventional cancer therapies include surgery, chemotherapy, and radiotherapy (Alibek et al. 2012). However, each of those has its own setback. Surgery, which once was considered the safest option for cancer treatment, is not risk-free to say the least. Surgery is intended to excise with high precision the bulk of a solid tumor. That however often disturbs the protective barrier, which normally keeps/holds the tumor isolated. Every incision made into cancerous tumor may help introducing cancer cells into bloodstream or inserting it into surrounding normal tissues. Therefore, disrupting tumor integrity through surgical process may facilitate cancer metastasis and induce local angiogenesis where cancer recurrence becomes certain (MacAdam 2003; van der Bij et al. 2009: Chen et al. 2011: Neeman and Ben-Eliyahu 2013). Chemotherapy is an approach that aims to stop cell division by targeting rapidly dividing cells. Unfortunately, it may target normal cells as well (Baudino 2015). Not to mention, that they are also considered to be "neocarcinogenic" as they tend to allow the development of new cancers that did not exist before the administration of chemotherapeutic agents (MacAdam 2003). Radiotherapy, on the other hand, often has temporary effects on survival rates. Its main downside stems from the fact that it is seemingly useless when it comes to cancer metastasis

(Jessy 2011). In addition, both chemotherapy and radiotherapy are known to be highly immunosuppressive. Thus, patients receiving chemoradiotherapy will not develop immunostimulation on the onset of an infection. Yet, worse, supplementing patients with antibiotics further deprive the patients from any response from their own immune systems (MacAdam 2003; Baden et al. 2012).

Out of frustrations, patients are exceedingly turning their back to conventional therapies. Interestingly, around early 1980s reports have emerged documenting cases of cancer patients switching from conventional treatment to complementary and alternative medicine, especially herbal medicine (Cassileth 1984). In this chapter, we will discuss herbal medicine and some of its associated cutting-edge technologies that assess the anticancer activities of herbal bioactive compounds.

Enormous efforts have been put forward to unravel the cellular and molecular mechanisms of cancer, optimistically, to overpower it by exploiting weaknesses. Emerging insights into the concept of Cancer Stem Cells (CSCs) may provide an effective way to defeat cancer metastasis and relapses (Grimes et al. 2012; Liao et al. 2014).

In this chapter, different advanced biotechnological approaches for cancer therapy will be addressed with emphasis on herbal medicine. Three-dimensional tumor spheroid models will be introduced and the concept of cancer stem cells and its different pathways and niches involvement will be discussed. In addition, nanotechnology will be investigated as an effective approach for CSCs targeting. We will finally introduce molecular docking commonly utilized for virtual screening in drug discovery.

Cancer Stem Cells

Despite the advances made in cancer diagnostics and chemotherapeutics, cancer remains an Everest to climb. The success of many chemotherapeutic agents is limited to shrinking the bulk of the tumor, which often fails to prevent the recurrence of a more aggressive tumor that is resistant to therapy. The heterogeneity nature of tumor is a major limiting factor when it comes to designing new therapeutic approaches to tackle issues like recurrence and resistance. Accumulated evidence referred to the unique population of cells in a tumor mass that behaves similarly to stem cells, as cancer stem cells (CSCs). Like "normal" stem cells, CSCs are characterized by their ongoing ability of self-renewal and differentiation, and they have been identified in many human cancers (Yang et al. 2008a; Schatton et al. 2008; Li et al. 2007; O'Brien et al. 2007; Singh et al. 2003; Al-Hajj et al. 2003; Bonnet and Dick 1997).

Thanks to the heterogeneity of cancer cells population, current therapeutic strategies, such as chemotherapy and radiotherapy, are only effective in eliminating "normal" cancer cells (Reya et al. 2001). Resistance to therapy by CSCs may be attributed to their unique proliferation machinery; CSCs are more quiescent and exhibit a slower proliferative rate in comparison to other highly proliferative cancer cells (Moore and Lyle 2011; Anjomshoaa et al. 2009; Roesch et al. 2010; Pece et al. 2010). Combinational therapy that targets the general population of cancer cells and CSCs is, therefore, desperately needed. Targeting CSCs can be achieved via modulating their self-renewal properties, metabolism, niche, and/or microenvironment (Dragu et al. 2015).

In recent years, plant-based treatments became the Holy Grail approach for cancer prevention and therapy thanks to their nontoxicity and minimal side effects. About 40% of FDA-approved therapeutic agents are natural-based components or their derivatives (Newman and Cragg 2016). To achieve more selectivity and specificity of natural-based therapeutics, natural components/derivatives may be structurally modified, which in turn enhances their pharmacological activity (Chen et al. 2015). Plant-based components have been reported to have anticancer properties in vivo and in vitro through the modulation of many cancer hallmarks and niche elements, including self-renewal properties of CSCs (Pistollato et al. 2014).

CSCs Self-Renewal Pathways

The self-renewal property of CSCs contributes to the tumor mass through continuous proliferation and differentiation (Reya et al. 2001; Liu et al. 2005). Various attempts have been put forward to refrain CSCs proliferative ability through forcing CSCs to differentiate and reduce its stemness-like properties (Persano et al. 2012; Paldino et al. 2014; You et al. 2014). Several signaling pathways that are regulated in stem cells have also been implicated in the regulation of CSCs self-renewal properties, including Wnt/β-catenin, Notch, and Hedgehog pathways (Klaus and Birchmeier 2008; Liu et al. 2006; Dontu et al. 2004). Many plant-based components have been shown to directly or indirectly inhibit self-renewal signaling pathways in CSCs, through regulating those specific pathways (Li et al. 2011). Targeting those signaling pathways by plant-based components may provide a more efficient sole and/or adjuvant treatment through disabling CSCs which would, otherwise, escape conventional cancer therapeutic approaches.

Wnt/β-Catenin Pathway

Several cellular processes such as proliferation, differentiation, motility, and self-renewal are regulated via Wnt/β-catenin pathway (Turashvili et al. 2006; Polakis 2012; Yamaguchi 2001; Akiyama 2000). Wnt/β-catenin pathway is also manifested throughout adulthood by maintaining tissue homeostasis via regulating adult stem cells and their microenvironment. Nonetheless, abnormal activation of Wnt pathway has been implicated in many diseases, including cancer. In fact, Wnt/β-catenin is crucial for the maintenance of CSCs in many cancer types, including leukemia, colon, liver, breast, lung, and melanoma (Kawaguchi-Ihara et al. 2008; Khan et al. 2007; Ysebaert et al. 2006; Schulenburg et al. 2007; Yang et al. 2008a, b; Woodward et al. 2007; Teng et al. 2010; Chien et al. 2009).

Curcumin (diferuloylmethane) is a yellow polyphenol found in turmeric (*Curcuma longa*) of the ginger family (*Zingiberaceae*). It has been part of folk medicine for centuries due to possessing many health benefits, including being potent analgesic, antibacterial, antiviral, antiparasitic, antioxidant, and anti-inflammatory agent. Curcumin has been the focus of many chemoprevention studies. Thus, it may be able to target several hallmarks of cancer. Jaiswal et al. (2002) demonstrated a dose-dependent inhibitory effect of curcumin on colon cancer cell lines. They suggested that apoptosis and cell-cycle arrest of HCT-116 cells after curcumin treatment is mediated by caspase-3-induced cleavage of β-catenin. In addition, decreasing transactivation of β-catenin/Tcf-Lef was found to lead to the inactivation of Wnt/β--catenin pathway. Moreover, treatment of HCT-116 cell line with curcumin decreased levels of c-Myc protein, a key mediator of gene regulation and differentiation that maintains self-renewal properties of embryonic stem cells. Alternatively, Ryu et al. (2008) offered another mechanism of curcumin inhibitory effect on Wnt/β-catenin pathway where curcumin derivatives were able to downregulate p300, a transcriptional coactivator of Wnt/β-catenin pathway. In addition, Ryu and his group reported the suppressive effect of curcumin derivatives on β-catenin transcription that was activated by Wnt3A conditioned-medium.

Saffron is a spice derived from *Crocus sativus* flower, stigmas to be more specific. Saffron's extract is rich in active components including oil-soluble carotenoid crocetin. Festuccia et al. (2014) investigated the ability of crocetin to inhibit the growth of aggressive prostate carcinoma cells lines (PC3 and 22rv1) when xenografted in nude mice. Crocetin was able to revert the Epithelial-Mesenchymal Transdifferentiation (EMT) by upregulating E-cadherin and downregulating N-cadherin and beta-catenin. It is important to mention that EMT is induced by Wnt/β-catenin signaling pathway activation. Moreover, crocetin demonstrated inhibitory effects on the PC3 and 22rv1 cell lines ability to invade and migrate by downregulating metalloproteinase, a family of extracellular matrix remodeling proteins.

Notch Pathway

Notch signaling pathway is evolutionarily conserved pathway that plays a key role in cell-fate determination during the development of multicellular organisms. It also controls key cellular processes such as proliferation and apoptosis that are involved in the development and homeostasis of multiple organs. Such regulation is executed via mediation of contact-dependent signaling that affects both signal-receiving and signal-sending cells. Additionally, Notch signaling is involved in a bidirectional cross talk with several other pathways, which adds another layer of complexity to this pathway. Deregulation of such a complex network can have drastic consequences and may initiate different diseases. Indeed, a great body of evidence from preclinical and clinical studies have demonstrated the role of Notch signaling pathway in driving growth of different solid tumors by promoting proliferation, survival, angiogenesis, migration, and self-renewal of CSCs (Kopan and Ilagan 2009; Fortini 2009; Wang et al. 2009; Wilson and Radtke 2006). Not surprisingly, Notch signaling pathway has attracted a great deal of interest as a potential therapeutic target. Hence, several inhibitors of Notch signaling have been proposed as candidate cancer treatments, such as gamma-secretase inhibitors (Yuan et al. 2015).

Withaferin-A (WA) is a bioactive component extracted from the root of *Withania somnifera*. WA has been utilized for many years for its medicinal properties, including being anti-inflammatory (Hamza et al. 2008). Koduru et al. (2010) reported the inhibitory effect of WA on several colon cancer cell lines. At doses of 4–5 μM, WA was able to reduce expression of Notch-1 protein, which is otherwise overexpressed during progression of colon carcinoma. WA treatment also downregulated Hes-1 and Hey-1, downstream targets of Notch-1, in addition to inhibiting the activation of Akt, a signaling cascade activated by Notch signaling. They also proposed another mechanism by which WA may exert its anticancer effects against colon cancer in vitro where WA may potentially interrupt the cross talk between Notch signaling and other signaling pathways, including mTOR and NF-kB. This, in turn, may lead to an activated JNK-mediated apoptosis. Similar findings were reported by Chen and his group (2014) where treatment with WA at doses 10–30 μM led to transcriptional inactivation of Notch-1 signaling pathway in osteosarcoma cell lines. That was evident where mRNA levels of Notch-1 and its downstream targets Hes-1, Hey-1 and Hey-2 were all reportedly decreased. Similarly, levels of protein expression of Notch-1 and its downstream targets Hes-1 and cyclin D1 were all downregulated (Chen et al. 2014).

Hedgehog Pathway

Hedgehog signaling pathway is a critical mediator in embryogenesis and is highly active in mammalian development. It is involved in many cellular processes, including protein trafficking, protein–protein interactions, and posttranslational modifications. Hedgehog pathway also regulates cellular proliferation and cell-fate determination in addition to modulating motility and adhesion to rearrange cellular structures (Teglund and Toftgård 2010). Hedgehog pathway is silenced in most adult tissues; hence, aberrant activation of this pathway is often implicated in the development of different types of cancer. Hedgehog signaling is also involved in maintaining self-renewing properties of adult stem cells and in the maintenance of CSCs. Aberrant activation of Hedgehog pathway may reduce the effectiveness and success of cancer treatments by promoting the progression of aggressive therapy-resistant cancer cells.

Crocetin, a carotenoid of saffron, has been purified to acquire crocetinic acid, a novel product that is 50 times more potent than any other carotenoid. Rangarajan et al. (2015) investigated the ability of crocetinic acid to inhibit the growth of pancreatic ductal adenocarcinoma in vitro and when xenografted. Crocetinic acid exerts its anticancer activity by targeting CSCs population through inhibiting Hedgehog pathway. Rangarajan et al. (2015) demonstrated the inhibitory effect of crocetinic acid treatment on molecules involved in hedgehog signaling pathway, including sonic hedgehog and its receptor Patched, and downstream targets smoothened, cyclin D1 and c-Myc. Crocetinic acid treatment at a dose of 0.5 mg/kg of body weight suppressed the growth of xenografted pancreatic ductal adenocarcinoma cells in mice.

CSCs Niche

CSCs are localized at distinct regions of the tumor microenvironment called niches. These regions are discriminated by being acidic and hypoxic in addition of being associated with exceeding levels of oxidative stress and inflammation. CSCs niche assures the maintenance of the stemness phenotype and tumor growth by the secretion of certain factors, including VEGF and TGF-β and other chemoattractants. Many plant-based components have been investigated for their ability to modulate CSCs niche through targeting tumor microenvironmental elements (Pistollato et al. 2014).

Hypoxia

During progression and growth of the tumor, some regions become hypoxic due to the limited access to vasculature, which initiates a cascade of processes to guarantee the survival of the tumor. Hypoxia-Inducible Factor 1 (HIF-1) is a key transcriptional factor that plays a critical role in hypoxia-related signaling pathway, which activates an array of genes involved in metastasis, angiogenesis, invasion, and other tumor microenvironmental elements (Liu et al. 2012; Warfel and El-Deiry 2014). Hypoxia also activates other signaling pathways that are important in promoting CSCs survival, including reactive oxygen species (ROS)-activated stress response pathways (Liu et al. 2008).

Mulberry (*Morus alba L*) leaf is heavily used in Asia for its medicinal properties, including antioxidant, anti-inflammatory, anticancer properties. Mulberry leaves extract has been reported to inhibit neuroblastoma stemness phenotype. Park et al. (2013) studied the effect of mulberry leaf extract on the invasiveness of neuroblastoma cells and on the expression of HIF-1α and its downstream targets. At a concentration of 40 μg/mL, mulberry leaves extract was able to decrease the expression of HIF-1α and MMP-2, under hypoxic conditions. In addition, neuroblastoma cells were treated with the same dose of mulberry extract prior to induction of hypoxia to examine the effect of the extract on hypoxia-regulated genes. Treatment with mulberry leaf extract also downregulated the mRNA levels of HIF-1α downstream targets VEGF and GLUT-1.

Inflammation

An escalating body of evidence continues to support a long-standing association between inflammation and tumor progression. Infection and chronic inflammation are hallmarks of many types of cancer, where the inflammatory response promotes tissue repair and regeneration via different signaling molecules. In fact, an inflammatory microenvironment is characteristic of hepatocellular carcinoma, cervical carcinoma, and Kaposi's sarcoma, where chronic infection of hepatitis virus (B and C), human papilloma virus, and human herpes virus 8, respectively, are involved in the initiation of those cancers. Similarly, colon cancer may be the result of inflammatory microenvironment accumulated in cases such as chronic inflammatory bowel disease. To fully grasp the notion of the association between cancer progression and inflammation, one should acknowledge the role of the entire spectrum of released signaling molecules and activated inflammation-related pathways in a typical inflammatory response. In return, cancer cells release their own chemical signals to recruit immune cells that secrete a range of cytokines to promote angiogenesis and other mechanisms needed for survival. It has been suggested that factors associated with chronic inflammation may facilitate the development and maintenance of CSCs. These factors include oxidative stress resulting from increased levels of ROS, in addition to inflammatory cytokines such as TNF-α, IL-6, and IL-8 (Tanno and Matsui 2011).

Crocin, among the few naturally occurring (saffron-based) carotenoids easily soluble in water, has recently gained great attention as a chemopreventive agent. Its multiple medicinal benefits are mainly attributed to its potent antioxidant, antidepressant, and anti-inflammation properties. Our laboratory has demonstrated the anti-inflammatory, antioxidant, antiproliferative, pro-apoptotic effects of saffron crude extract on liver cancer in vivo and in vitro (Amin et al. 2011). Most recently, we have also reported the preventive effect of crocin on early lesions of liver cancer (Amin et al. 2016) and on colorectal cancer cells (Amin et al. 2015). Using a chemically induce liver cancer animal model, we were able to demonstrate that the anticancer effects of crocin were mediated through its anti-inflammatory effects. Treatment with crocin downregulated many inflammation mediators, including NF-κB, a key regulator of proliferation and inflammation. Crocin treatment also restored hepatic MPO levels, a marker of neutrophil infiltration, in liver cancer bearing animals, to almost control levels. In addition, crocin decreased the number of hepatic ED1 and ED2 macrophages, and also decreased the expression of p-TNF-R1 and the content of TNF-α. A dramatic decrease in levels of COX-2 and iNOS, two key players in inflammation and promotion of cancer growth, was evident after crocin treatment. Crocin also restored ROS levels and eliminated oxidative stress in liver cancer bearing animals. Taken together, our findings suggested that the mechanism by which crocin exerts its protective effects against liver cancer is mainly through the inhibition of inflammation (Fig. 3.1).

Three-Dimensional Tumor Spheroid Models

Screening for anticancer agents, synthetic or natural-based, requires a reliable disease model. Prior introduction to human patients, any given potential agent has to go through preclinical trials, which includes testing on animal models and on human

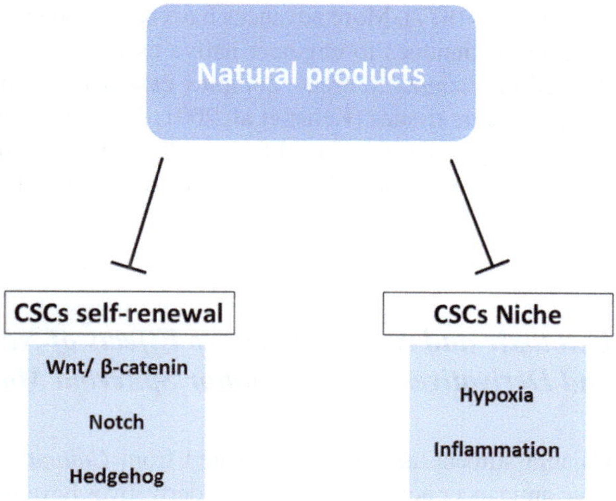

Fig. 3.1 Natural products disable CSCs through targeting their self-renewal ability and their supportive niche

cells grown in culture. However, multiple limitations to these trials stand in the way, including unsuccessful translation to human patients from animal experimentation data and unreliability of current cell culturing methods to reflect the complexity of a living organism.

In the past several years, other cell culturing methods have been trending widely. Concerted efforts have been doubling down on developing a system to mimic in vivo conditions using cell culture. That has led to the development of three-dimensional (3D) cultures that allows cells to grow into three-dimensional structures rather than flattened structures resulting from conventional two-dimensional (2D) culturing methods. In addition to distorting cellular architecture, conventional 2D culture disrupts cell–cell and cell–matrix interactions (Unger et al. 2014), and increase drug sensitivity (Breslin and O'Driscoll 2013).

Tumor spheroids are one of the most widely used 3D culture models. It is highly versatile and easily manipulated model that can be developed by promoting single cell in a suspension to form a sphere cluster of cells. What makes this model particularly attractive is the presence of three distinct regions in a spheroid: a highly proliferative outer region, a middle quiescent region, and a hypoxic core region. Such stratification creates a diffusional gradient of oxygen, nutrients, and drugs exposure among all three regions of the spheroid, which is characteristic of most solid tumors. Spheroids can be generated using several methods, the earliest of which is the "hanging drop method" that depends on the fact that cells adhere to each other. Thus, cells in a drop of a cell suspension tend to aggregate together at the bottom of a drop hanging from a plate (Harrison 1907). Spheroid generation has advanced significantly since then where several other methods are currently available to generate spheroids, including the static liquid overlay. This technique utilizes culturing substrates that are pre-coated with ultra-low attachment coating that prevents cells from adhering to the culturing substrate, but rather to each other to form aggregates

(Yuhas et al. 1977). More advances have been made through implementation of 3D culturing techniques to engineer native tissue. For example, spheroids have been utilized in attempts to create complex environments of cardiac muscle, cartilage, and pancreatic tissues (Kehat et al. 2001; Jukes et al. 2008; Lumelsky et al. 2001). Additionally, spheroids have been employed in the formation of organs, including optic cup, layered cortical tissue, and pituitary gland (Eiraku et al. 2011, 2008; Suga et al. 2011).

Cytotoxic and Antiproliferative Effects of Natural Products and Derivatives Against Tumor Spheroid Models

Cajanin stilbene acid (CSA), isolated from *Cajanus cajan*, has been reported to have anticancer activities. Two of its derivatives have recently been investigated for their cytotoxic effects against breast cancer spheroids. Seo et al. (2015) assessed the toxicity of CSA derivatives (CSA 6, CSA19) on MCF-7 mammospheres, where IC_{50} value was reached after treatment with 4.98 ± 0.64 μM and 7.51 ± 1.83 μM of CSA 6 and CSA 19, respectively.

Oplopanax horridus, commonly known as Devil's club (DC), is an understory shrub heavily used for its medicinal properties against a wide range of diseases, including different cancer types. Tai et al. (2014) studied the antiproliferative activity of DC against pancreatic cancer in 3D culture. The investigators assessed the effects of DC extract on proliferation of pancreatic cancer PANC-1 spheroids, alone or in combination with chemotherapeutic agents "gemcitabine," "cisplatin," and "paclitaxel." PANC-1 spheroids demonstrated a significant resistance to DC extract and two of the three tested chemotherapeutic agents. However, when combined with cisplatin and gemcitabine, antiproliferative effects on PANC-1 spheroids were significantly enhanced.

Natural Products Inhibit Formation and Invasion of Tumor Spheroid Models

Rangarajan and his colleagues (2015) demonstrated an inhibitory effect of crocetinic acid on formation of pancreatic cancer MiaPaCa-2 spheroids. Treatment with crocetinic acid inhibited pancosphere formation and decreased the number and size of pancospheres in a dose (1–20 μM) dependent manner.

Dolfini et al. (2007) have investigated the effect of resveratrol, a natural phenol present in many fruits, on breast cancer MDA-MB-231 spheroid formation. Incubation with resveratrol at a dose of 64 μM led to the formation of significantly smaller MDA-MB-231 spheroids that were unable to form colonies after reseeding. In contrast, treatment with resveratrol at the same dose had no effect on the already formed MDA-MB-231 spheroids. In addition to resveratrol, other polyphenols including baicalein, epicatechin, epigallocatechin, and polyphenon 60 have been investigated for their effects on tumor spheroids shedding, and cell invasion. Günther et al. (2007) incubated spheroids from 4T1 cells, a murine mammary cell line, with resveratrol, baicalein, epicatechin, epigallocatechin, and polyphenon 60. Those treatments caused tumor growth arrest and abolished formation of subspheroids and cell shedding. In addition, treatment with such polyphenols downregulated expression of MMP-9, reduced ROS generation, and significantly inhibited cell invasion of 4T1 cells into embryonic stem cell-derived tissues.

Nanotechnology in Nanomedicine

With the fast paced growing technology, possibilities seem limitless; it is only a matter of how to utilize technology to serve and advance the human race. Nanotechnology is an emerging field of technology that is fast progressing worldwide and is attracting the attention of scientists and clinicians alike. Although the concept of nanotechnology is based on manipulating and manufacturing matters into nanoscales, the dimensions are typically less than 100 nm (Sanchez and Sobolev 2010). These nanoscale particles, nanoparticles (NP), are often used for a wide range of applications. Most of those applications make use of the fact that NP's high surface area to mass ratio provides a large functional surface which is able to bind, adsorb, and carry other compounds. NPs are commonly classified based on their origins as the source material can be biological (such as phospholipids, lipids, lactic acid, dextran, chitosan) or chemical (such as carbon, silica, gold, silver, and other metals) (De Jong and Borm 2008).

Nanomedicine is a major field of nanotechnology that involves nanotechnological applications utilized for diagnosis, monitoring, biological systems, and therapeutics (Ansari et al. 2012; Moghimi 2005). The term nanomedicine itself appeared by the beginning of this century; however, the essence of nanomedicine has surfaced around late 1970s, particularly in respect of nanoparticles for biomedical use (Astruc 2015).

Herbal Medicine

Without a doubt and for centuries herbal medicine has been well recognized and *sometimes* most popular natural treatment for different human diseases. Although modern medical practices have gradually replaced the herbal medicine, herbal medicine still hold a lot of ground with patients at different parts of the world. Exhausting methods of extraction, standardization, molecular characterization, and possible modification of biological properties of natural biomolecules have increasingly promoted herbal medicine among the scientific community in the last decade (Ansari et al. 2012; Silva et al. 2013; Verma and Singh 2008).

The therapeutic manifestation of herbal medicine may be attained through the active component/s or the whole plant extract. The impressive biological results of some medicinal plants and/or natural products have raised the bar for other synthetic drugs that may be toxic and cause serious side effects. Concerns about solubility, low bioavailability, low absorption, degradation, and toxicity continue to be a bottleneck toward global appreciation of herbal medicine (Ansari et al. 2012; De Jong and Borm 2008; Yadav et al. 2011).

Implication of Nanotechnology Within Herbal Medicine: Drug Delivery System

Obstacles hindering wider acceptance the herbal medicine can potentially be minimized through nanotechnology. Nanotechnology is well equipped to provide a remarkable system for drug delivery mediated through nanocarriers/NPs. Interestingly in some cases, the drug itself can work as its own carrier as it can be formulated into nanoscale dimensions (Somwanshi et al. 2013). Such assembly system between the nanocarrier and the drug tend to maximize the benefits over risk ratio. Furthermore, delivering drugs through NPs do enhance the solubility and hence enhance the bioavailability of the drug. It also reduces the rate of drug degradation rate within the body and control the drug release, thus allowing to more sustained release over specific time interval. In addition, drug delivery via NPs increases specificity as they deliver a specific cargo drug to an exact location without compromising the drug dose concentration through crossing the body natural barriers. NP-mediated drug delivery also reduces toxicity (and thus side effects) as much lower doses of drugs are now utilized (Ansari et al. 2012; De Jong and Borm 2008; Astruc 2015).

Features (chemical and physical) are quite crucial to evaluate nanoparticles and account for their particle size. Chemical properties include total chemical composition, mixing state (internal/external), surface composition, electrochemistry, and oxidation state while number and mass concentration, size, surface area, total mass, morphology, and optical properties are essential physical properties (Thompson 2010). Particle size measurements helps determining the biodistribution and the retention of nanoparticle within target tissues (Cho et al. 2013) and alteration of particle size inversely affects the surface area and hence drug carrying capacity (Merkus 2009). Surface charge (zeta potential) is the measurement of the electrostatic potential of nanoparticle at the diffuse layer surrounding the nanoparticle in solution. This property is associated with the nanoparticle suspension stability and the particle surface morphology, both of which are crucial for evaluating the NP's stability and surface adsorption (Delgado et al. 2007). As rule of thumb, NPs are neutral when their zeta potential ranges between −10 mV and +10 mV. However, NPs are strongly cationic and strongly anionic when the zeta potential is greater than +30 mV or less than −30 mV, respectively (McNeil 2011).

Drug release analysis is the evaluation of the release of the cargo drug over a specific time interval. Such analysis would eventually determine the drug bioavailability within target sites and the concentration sufficient to fulfill the proposed therapeutic effect (Lu et al. 2011). To develop a successful drug delivery system, stability of the NPs is an essence. Typically, NPs should be stable and capable of retaining the drug in the bloodstream until they properly release the drug in target tissues. Instability of NPs is often reflected by aggregation, degradation, or unintended drug release. Techniques such as determination of critical aggregation/micelle concentration, determination of low critical solution temperature, gel permeation chromatography, and forster resonance energy transfer technique are often adopted to investigate nanoparticle's stability in vitro before ultimately testing it out in vivo (Cho et al. 2013). It is quite essential that these developed nanoparticles be very well characterized in a reproducible manner.

Following the assurance that both physical and chemical properties of NPs are in check, biological competence of those NPs must be evaluated. Two cell culture models are widely applied: two-dimensional (2D) monolayer cell culture and three-dimensional (3D) cell culture system. 2D cell culture is the one most frequently used in most laboratories; it assesses cellular uptake, intracellular trafficking, and bioactivity of the NP-delivered drug (Cho et al. 2013).

The 3D cell culture system is known to recapitulate the whole in vivo experience where the NPs have to overcome various barriers unlike in 2D systems where NP-associated drugs are directly exposed to their target cells with no loss of drug concentration (Holback and Yeo 2011). Many 3D cell culture systems have been developed over the past few years to determine NPs efficiency. Those systems include cells encapsulated in scaffolds, multicellular spheroids (Khademhosseini et al. 2006) and a combination of both scaffolds and spheroids (Ho et al. 2010). 3D system models with excised tissues or tissue components have also been developed and are reported to provide more physiologically relevant results (Astashkina et al. 2012).

In order to provide a proof of concept of developed drug-coated NPs with proven therapeutic potential, safety and efficiency of those structures must be examined in vivo. In vivo analyses are mostly carried out in mice with allograft (murine tumor transplant) or xenograft (human tumor transplant) (Cho et al. 2013; Voskoglou-Nomikos et al. 2003). Clinical trials with different phases are still needed to acquire Food and Drug Administration (FDA) final approval for a wider utilization of such "magic bullets," a term that has been first introduced by Paul Ehrlich (1960) (De Jong, 2008; Astruc 2015).

Nanoparticles for Anticancer Drug Delivery

Despite a number of sound successes in our combat against cancer, it remains one of the deadliest diseases ever known to mankind, where, in many cases, conventional chemo/radio/therapy offer more harm than good. Herbal medicine provides less toxic and cheaper novel therapeutic and adjuvant alternative (Prakash et al.

2013; Block et al. 2015). Developing NPs loaded with herbal drugs, their delivery and efficiency have been extensively investigated and few astonishing examples have surfaced.

Several studies have shown curcumin's health-promoting benefits against cancer such as prevention of tumor initiation, promotion, metastasis, and angiogenesis in numerous cancer types with staggering amount of data that spans bench- to bed-side (Bar-Sela et al. 2010; Maheshwari et al. 2006; Jurenka 2009). Interestingly, a phase I clinical trials published in 2004 conducted by Dr. Sharma and his colleagues on advanced colon cancer patients concluded curcumin's poor bioavailability. A daily 3.6 g oral curcumin administrated for approximately 4 months resulted in 10 nM concentration of curcumin in the peripheral blood. Hence, to overcome this challenge, nanotechnology was employed. Through the, then, new technology, the particle size was reduced and the surface area was increased resulting in a significant increase of the bioavailability and a dramatic solubility improvement. Dr. Bisht and his colleagues (2010) have engineered a polymeric nanoparticle encapsulated curcumin formulation named "NanoCurc." Pharmacokinetic analysis showed a remarkable increase in the bioavailability of curcumin as NanoCurc in plasma and tissues when compared to the free curcumin. In addition, NanoCurc was administrated to pancreatic cancer xenograft model and exhibited significant primary tumor growth inhibition. NanoCurc was also administrated along with gemcitabine to study its effect as therapeutic additive which surely did enhance the tumor growth inhibition more effectively compared to either NanoCurc alone or gemcitabine alone. Several other studies were also conducted on NanoCurc indicating its neuroprotective abilities, and breast cancer chemoprevention activities (Ray et al. 2011; Lim et al. 2011; Chun et al. 2012).

Saffron and its biomolecules possess a unique therapeutic potential, including anticancer, antioxidant, anti-inflammatory, among many other properties (Das et al. 2010; Bhandari 2015; Amin et al. 2011). Out of over 160 unique biomolecules that saffron is made of, safranal, crocin, and crocetin represent the major three active molecules (Giaccio 2004). Crocin and crocetin were found to have significant anticancer activity among breast, lung, pancreatic, and leukemic cells (Samarghandian et al. 2013). Safranal was also reported to have anticancer properties against neuroblastoma cells, human prostate cancer cells, human cervical cancer cells, and breast cancer cell line (Farahzad et al. 2014; Shabestari and Samarghandian 2013; Malaekeh-Nikouei et al. 2013). Our lab has long interest in saffron and its derivative's anticancer activities. Most recently, a therapeutic model of early liver cancer using crocin-coated magnetite nanoparticle has been developed in our lab (El-kharrag et al. 2017). In an attempt to increase the therapeutic index of crocin, magnetite nanoparticles were chosen with dextran as coating agent crosslinked with crocin. These developed NPs were then tested out both in vitro and in vivo where mice were injected with carcinogenic agent to induce liver cancer. The developed crocin-coated NPs exhibited improved anti-tumorigenic activities compared to free crocin in both in vitro and in vivo systems.

Crocin is known to have a low stability and to suffer a loss of functionality if exposed to conditions like heat, oxygen, light, acidic environment, and the existence of additives which usually are applied during its processing (Kanakis et al. 2007; Maggi et al. 2009). To improve crocin's stability, Rahaiee and his group (2015) have encapsulated crocin into chitosan—sodium alginate nanoparticles. The zeta potential measurement for the produced NPs loaded with crocin was found to be −33.52 mV, indicating that the synthesized nanoparticle is relatively stable under such, otherwise, unfavorable environmental conditions.

Implication of Nanomedicine and Cancer Stem Cells Therapies

As discussed in cancer stem cells section, conventional therapies might actually provoke the enrichment of CSCs instead of eliminating them. Therefore, adopting a new technology capable of an effective CSCs eradication is necessary. Nanomedicine can be one way to specifically tackle CSCs. Nanomedicine is uniquely set up to deliver anti-CSC agents. Applying NPs-based delivery will immediately boast the bioavailability and the biological activity. NPs may accumulate either passively due to enhanced permeation retention (EPR) by exploiting the abnormal vascular nature of tumors or positively due specific targeting where NPs can be decorated with antibodies (or other molecules with high affinity toward receptors or CSC markers) against CSCs (Gao et al. 2012; Xia 2014). Developing NPs to target CSCs can effectively reach the CSCs niches and eradicate them, providing the perfect opportunity to focus on nanotechnological approach for CSCs treatment.

The research in the field of nanomedicine has grown tremendously in the last decade with undeniably a wide spectrum of NPs and nanocarries being developed with one purpose of becoming a successful drug delivery system. However, going through multiple stages of characterization, investigation of different biological properties, and then proceeding through serious stages of clinical trials, continues to be a major challenge. This long list of prerequisites allows only a few candidate NPs to proceed through FDA for final approval (Fig. 3.2).

Molecular Docking in Drug Discovery

Many advances have been made to elucidate the structure of a countless number of proteins and protein–ligand complexes through data provided by crystallography, nuclear magnetic resonance spectroscopy, and high-throughput protein purification (Meng et al. 2011). At this age of technology, this feed of information can best be utilized through computational approaches to facilitate the process of drug discovery, through means of virtual screening (VS) of potential therapeutic agents (Jorgensen 2004; Bajorath 2002; Walters et al. 1998; Langer and Hoffmann 2001) (Fig. 3.3).

VS is a computational technique to automatically search, and evaluate large libraries of compounds in their ability to bind to targets of interest (Rollinger et al.

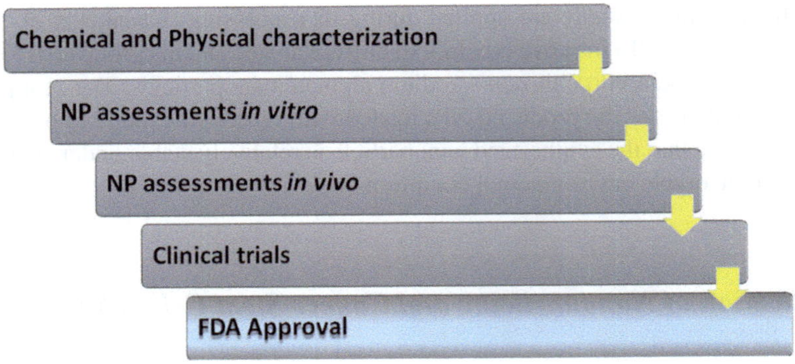

Fig. 3.2 FDA approval of nanoparticles is a multistep process

Fig. 3.3 Natural products-based anticancer therapeutics have become more efficient and practical with advances made available by biotechnology

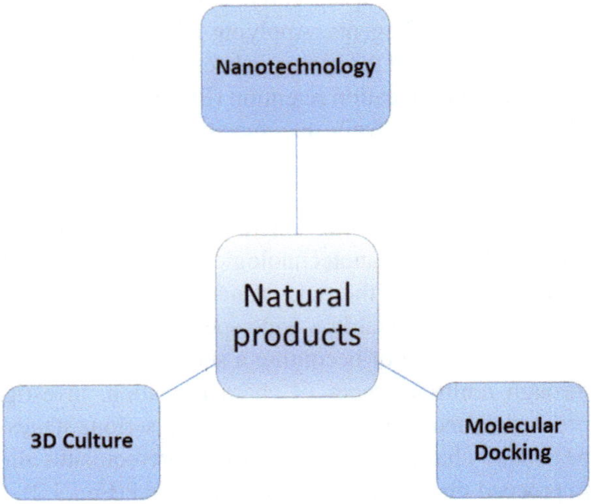

2008; Walters et al. 1998). With the enormous number of compounds to be screened, and equally large number of potential drug targets, the possibilities are countless; this is where VS comes to play. The feed of information is filtered according to specific set of criteria through computational approaches until a reasonable number of candidate compounds can be eventually synthesized and tested (Bohacek et al. 1996). In addition, VS has been shown to provide more effective screening at much lower cost than traditional experimental high-throughput screening (Moitessier et al. 2008; Shoichet et al. 2002; Bailey and Brown 2001). Molecular docking is a structure-based virtual screening technique that predicts the ability of candidate components to bind to protein targets with high affinity; those with the highest affinity according to a scoring function will be put through for experimental validation (Kroemer 2007; Cavasotto and Orry 2007).

Many synthetic therapeutic agents, including those used in cancer treatments, have been heavily studied, and their pharmacokinetics and pharmacodynamics are well known. However, this is not the case with plant-derived products, where their active components are still largely unknown (Usha et al. 2014). In contrast to the numerous number of synthetic compounds databases, there are only a handful of natural products databases, including ZINC and SPECS. Nonetheless, screening of natural-based compounds is considerably limited due to the time-consuming processes of isolation and purification of natural products. Recent attempts have been made to identify bioactive phytochemicals and predict their potential using molecular docking. An exciting progress have been made toward phytochemicals virtual screening by Wang et al. (2016), by creating a tool for target-based identification of bioactive phytochemicals, called Herbalog. Through molecular docking, Herbalog is able to identify potential herb candidates against a particular therapeutic target, from a database containing 5112 phytochemicals from 179 common herbs. These candidate herbs identified by Herbalog contain the large number of potential activators or inhibitors against certain targets. Such advances facilitate the progress of screening and identifying candidate natural products with anticancer potential.

The technique of molecular docking facilitated the identification of natural compounds with exciting potential to be used as anticancer agents through predicting their binding affinity to therapeutic targets. Earlier this year, Thiyagarajan et al. (2016) identified novel S6K1 and FAK dual inhibitors from natural compounds through docking 3D structures of S6K1 and FAK obtained from the Protein Data Bank (PDB) with 60 natural compounds. S6K1 and FAK have been shown to play a role in cancer progression. Once activated, S6K1 mediates various cellular processes regulated by PIK3/mTOR signaling pathway, including cell proliferation, protein synthesis, and mRNA processing. Hyperactivation of S6K1 has been implicated in cancer, including breast cancer and brain tumors (Pérez-Tenorio et al. 2011; Yamnik et al. 2009; Ismail 2012). Additionally, FAK is involved in the regulation of many cellular processes, including survival, proliferation, and migration, in response to extracellular signals. When overexpressed, FAK promotes cancer cell invasion, metastasis, and angiogenesis (Golubovskaya and Cance 2007). Indeed, overexpression of FAK has been implicated in invasive liver, colon, prostate, thyroid, breast, brain, and head and neck tumors (Sood et al. 2004; Cance et al. 2000; Thiyagarajan et al. 2013).

Considering their roles in cancer progression and survival, S6K1 and FAK present an attractive therapeutic target for which inhibitors are identified. Thiyagarajan and colleagues were able to identify 30 molecules with high scores, out of which, further analysis revealed three potential inhibitors out of a set of candidate compounds, neferine B, neferine A, and antroquinonol D. The dual inhibitory effect of these three candidates on S6K1 and FAK was then investigated in vitro on C6 glioma cell line. Treatment with neferine B, neferine A, and antroquinonol D showed an IC_{50} value at doses of 10 μM, 12 μM, and 16 μM, respectively. When treated with the three candidates for 24 h, C6 glioma cells exhibited a significant decrease in the expression of p-S6K1 and p-FAK. Findings reported by Thiyagarajan and colleagues pave the way for the development of neferine B, neferine A, and antroquinonol D as a conjugate to target cancer progression and metastasis.

Apigenin has been shown to induce G1 cell-cycle arrest and p53-dependent apoptosis. It has also been shown to decrease expression of Bcl-2, and to inhibit the growth of resistant breast cancer cells through neutralizing protein kinase-mediated growth factor signaling (Zheng et al. 2005; Long et al. 2008). Saeed et al. (2015) investigated the way apigenin binds to ATP-binding cassette (ABC) transporters P-glycoprotein and its close relative ABCB5 through molecular docking approaches. They also investigated the activity of apigenin in drug-sensitive and -resistant cell lines specifically overexpressing ATP-binding cassette (ABC) transporters BCRP, P-glycoprotein, and ABCB5, in addition to investigating the synergistic effect of apigenin in combination with established anticancer drugs. After generation of human P-glycoprotein and ABCB5 models based on homology with the crystal structure of murine P-glycoprotein, Saeed and colleagues performed molecular docking of apigenin against the nucleotide-binding domains (NBDs) of P-glycoprotein and ABCB5. Their findings revealed that apigenin may compete with ATP for binding to the NBDs, interfering with ATP binding and cleavage. These findings were supported with experimental data demonstrating that multidrug-resistant tumor cells are not resistant against apigenin treatment and that P-glycoprotein and BCRP were inhibited by increasing cellular uptake of doxorubicin. Apigenin treatment also increased the synergistic inhibition of multidrug-resistant tumor cells viability when combined with doxorubicin or docetaxel, both of which are established anticancer drugs.

Acknowledgment This work was supported by ZCHS Grant 31R050.

References

Akiyama T (2000) Wnt/beta-catenin signaling. Cytokine Growth Factor Rev 11:273–282

Al-Hajj M, Wicha MS, Benito-Hernandez A et al (2003) Prospective identification of tumorigenic breast cancer cells. Proc Natl Acad Sci U S A 100:3983–3988. doi:10.1073/pnas.0530291100

Alibek K, Bekmurzayeva A, Mussabekova A, Sultankulov B (2012) Using antimicrobial adjuvant therapy in cancer treatment: a review. Infect Agents Cancer 7(1):33

Amin A, Bajbouj K, Koch A et al (2015) Defective autophagosome formation in p53-null colorectal cancer reinforces crocin-induced apoptosis. Int J Mol Sci 16:1544–1561

Amin A, Hamza AA, Bajbouj K et al (2011) Saffron: a potential candidate for a novel anticancer drug against hepatocellular carcinoma. Hepatology 54:857–867. doi:10.1002/hep.24433

Amin A, Hamza A, Daoud S et al (2016) Saffron-based crocin prevents early lesions of liver cancer: in vivo, in vitro and network analyses. Recent Pat Anticancer Drug Discov 11:121–133

Anjomshoaa A, Nasri S, Humar B et al (2009) Slow proliferation as a biological feature of colorectal cancer metastasis. Br J Cancer 101:822–828. doi:10.1038/sj.bjc.6605229

Ansari S, Sameem M, Islam F (2012) Influence of nanotechnology on herbal drugs: a review. J Adv Pharm Technol Res 3(3):142

Astashkina A, Mann B, Prestwich G, Grainger D (2012) A 3-D organoid kidney culture model engineered for high-throughput nephrotoxicity assays. Biomaterials 33(18):4700–4711

Astruc D (2015) Intro Nanomed Mol 21(1):4

Baden L, Bensinger W, Angarone M, Casper C, Dubberke E, Freifeld A et al (2012) Prevention and treatment of cancer-related infections. J Natl Compr Canc Netw 10:1412–1445

Bailey D, Brown D (2001) High through-put chemistry and structure-based design. Drug Discov Today 6:57–59

Bajorath J (2002) Integration of virtual and high-throughput screening. Nat Rev Drug Discov 1:882–894. doi:10.1038/nrd941

Bar-Sela G, Epelbaum R, Schaffer M (2010) Curcumin as an anti-cancer agent: review of the gap between basic and clinical applications. Curr Med Chem 17(3):190–197

Baudino T (2015) Targeted cancer therapy: the next generation of cancer treatment. Curr Drug Discov Technol 12(1):3–20

Bhandari P (2015) Crocus sativus L. (saffron) for cancer chemoprevention: a mini review. J Trad Complement Med 5(2):81–87

Bisht S, Mizuma M, Feldmann G, Ottenhof N, Hong S, Pramanik D et al (2010) Systemic administration of polymeric nanoparticle-encapsulated curcumin (NanoCurc) blocks tumor growth and metastases in preclinical models of pancreatic cancer. Mol Cancer Ther 9(8):2255–2264

Block K, Gyllenhaal C, Lowe L, Amedei A, Amin A, Amin A et al (2015) Designing a broad-spectrum integrative approach for cancer prevention and treatment. Semin Cancer Biol 35:S276–S304

Bohacek RS, McMartin C, Guida WC (1996) The art and practice of structure-based drug design: a molecular modeling perspective. Med Res Rev 16:3–50

Bonnet D, Dick JE (1997) Human acute myeloid leukemia is organized as a hierarchy that originates from a primitive hematopoietic cell. Nat Med 3:730–737. doi:10.1038/nm0797-730

Breslin S, O'Driscoll L (2013) Three-dimensional cell culture: the missing link in drug discovery. Drug Discov Today 18:240–249. doi:10.1016/j.drudis.2012.10.003

Cance WG, Harris JE, Iacocca MV et al (2000) Immunohistochemical analyses of focal adhesion kinase expression in benign and malignant human breast and colon tissues: correlation with preinvasive and invasive phenotypes. Clin Cancer Res 6:2417–2423. doi:10.1073/pnas.89.11.5192

Cassileth B (1984) Contemporary unorthodox treatments in cancer medicine. Ann Intern Med 101(1):105

Cavasotto CN, Orry AJW (2007) Ligand docking and structure-based virtual screening in drug discovery. Curr Top Med Chem 7:1006–1014. doi:10.2174/156802607780906753

Chen J, Li W, Yao H, Xu J (2015) Insights into drug discovery from natural products through structural modification. Fitoterapia 103:231–241. doi:10.1016/j.fitote.2015.04.012

Chen Y, Han X-Z, Wang W et al (2014) Withaferin A inhibits osteosarcoma cells through inactivation of Notch-1 signalling. Bangladesh J Pharmacol 9:364–370

Chen X, Zhang H, Fang X (2011) Surgery-induced immunomodulation in breast cancer. J Surg Oncol 103(2):197–197

Chien AJ, Moore EC, Lonsdorf AS et al (2009) Activated Wnt/beta-catenin signaling in melanoma is associated with decreased proliferation in patient tumors and a murine melanoma model. Proc Natl Acad Sci U S A 106:1193–1198. doi:10.1073/pnas.0811902106

Cho E, Holback H, Liu K, Abouelmagd S, Park J, Yeo Y (2013) Nanoparticle characterization: state of the art, challenges, and emerging technologies. Mol Pharm 10(6):2093–2110

Chun Y, Bisht S, Chenna V, Pramanik D, Yoshida T, Hong S et al (2012) Intraductal administration of a polymeric nanoparticle formulation of curcumin (NanoCurc) significantly attenuates incidence of mammary tumors in a rodent chemical carcinogenesis model: implications for breast cancer chemoprevention in at-risk populations. Carcinogenesis 33(11):2242–2249

Das I, Das S, Saha T (2010) Saffron suppresses oxidative stress in DMBA-induced skin carcinoma: a histopathological study. Acta Histochem 112(4):317–327

De Jong W, Borm P (2008) Drug delivery and nanoparticles: applications and hazards. Int J Nanomedicine 3(2):133–149

Delgado A, González-Caballero F, Hunter R, Koopal L, Lyklema J (2007) Measurement and interpretation of electrokinetic phenomena. J Colloid Interface Sci 309(2):194–224

Dolfini E, Roncoroni L, Dogliotti E et al (2007) Resveratrol impairs the formation of MDA-MB-231 multicellular tumor spheroids concomitant with ceramide accumulation. Cancer Lett 249:143–147. doi:10.1016/j.canlet.2006.08.013

Dontu G, Jackson KW, McNicholas E et al (2004) Role of Notch signaling in cell-fate determination of human mammary stem/progenitor cells. Breast Cancer Res 6:R605–R615. doi:10.1186/bcr920

Dragu DL, Necula LG, Bleotu C et al (2015) Therapies targeting cancer stem cells: current trends and future challenges. World J Stem Cells 7:1185–1201. doi:10.4252/wjsc.v7.i9.1185

Dubey R (2014) Advanced biotechnology. S Chand, New Delhi

Ehrlich P (1960) Experimental researches on specific therapy: on immunity with special reference to the relationship between distribution and action of antigens*: first Harben lecture. Collect Papers Paul Ehrlich 1960:106–117

Eiraku M, Takata N, Ishibashi H et al (2011) Self-organizing optic-cup morphogenesis in three-dimensional culture. Nature 472:51–56. doi:10.1038/nature09941

Eiraku M, Watanabe K, Matsuo-Takasaki M et al (2008) Self-organized formation of polarized cortical tissues from ESCs and its active manipulation by extrinsic? Signals. Cell Stem Cell 3:519–532. doi:10.1016/j.stem.2008.09.002

El-Kharrag R, Amin A, Hisaindee S, Greish Y, Karam SM (2017) Development of a therapeutic model of precancerous liver using crocin-coated magnetite nanoparticles. Int J Oncol 50(1):212–222. doi:10.3892/ijo.2016.3769

Farahzad J, Samarghandian S, Shoshtari M, Sargolzaei J, Hossinimoghadam H (2014) Anti-tumor activity of safranal against neuroblastoma cells. Pharmacogn Mag 10(38):419

Festuccia C, Mancini A, Gravina GL et al (2014) Antitumor effects of saffron-derived carotenoids in prostate cancer cell models. Biomed Res Int. doi:10.1155/2014/135048

Fortini ME (2009) Notch signaling: the core pathway and its posttranslational regulation. Dev Cell 16:633–647

Günther S, Ruhe C, Derikito MG et al (2007) Polyphenols prevent cell shedding from mouse mammary cancer spheroids and inhibit cancer cell invasion in confrontation cultures derived from embryonic stem cells. Cancer Lett 250:25–35. doi:10.1016/j.canlet.2006.09.014

Gao Z, Zhang L, Sun Y (2012) Nanotechnology applied to overcome tumor drug resistance. J Control Release 162(1):45–55

Giaccio M (2004) Crocetin from saffron: an active component of an ancient spice. Crit Rev Food Sci Nutr 44(3):155–172

Golubovskaya VM, Cance WG (2007) Focal adhesion kinase and p53 signaling in cancer cells. Int Rev Cytol 263:103–153

Grimes C, Margolin DA, Li L (2012) Are cancer stem cells responsible for cancer recurrence? Cell Biol Res Ther 1(1)

Hamza A, Amin A, Daoud S (2008) The protective effect of a purified extract of Withania somnifera against doxorubicin-induced cardiac toxicity in rats. Cell Biol Toxicol 24:63–73. doi:10.1007/s10565-007-9016-z

Hanahan D, Weinberg R (2011) Hallmarks of cancer: the next generation. Cell 144(5):646–674

Harrison RG (1907) Observations on the living developing nerve fiber. Proc Soc Exp Biol Med 4:140–143. doi:10.1002/ar.1090010503

Ho W, Pham E, Kim J, Ng C, Kim J, Kamei D, Wu B (2010) Incorporation of multicellular spheroids into 3-D polymeric scaffolds provides an improved tumor model for screening anticancer drugs. Cancer Sci 101(12):2637–2643

Holback H, Yeo Y (2011) Intratumoral drug delivery with nanoparticulate carriers. Pharm Res 28(8):1819–1830

Ismail HMS (2012) Overexpression of S6 kinase 1 in brain tumours is associated with induction of hypoxia-responsive genes and predicts patients' survival. J Oncol. doi:10.1155/2012/416927

Jaiswal AS, Marlow BP, Gupta N, Narayan S (2002) Beta-catenin-mediated transactivation and cell-cell adhesion pathways are important in curcumin (diferuylmethane)-induced growth arrest and apoptosis in colon cancer cells. Oncogene 21:8414–8427. doi:10.1038/sj.onc.1205947

Jessy T (2011) Immunity over inability: the spontaneous regression of cancer. J Nat Sci Biol Med 2(1):43

Jorgensen WL (2004) The many roles of computation in drug discovery. Science 303:1813–1818. doi:10.1126/science.1096361

Jukes JM, Both SK, Leusink A et al (2008) Endochondral bone tissue engineering using embryonic stem cells. Proc Natl Acad Sci U S A 105:6840–6845. doi:10.1073/pnas.0711662105

Jurenka J (2009) Anti-inflammatory properties of curcumin, a major constituent of Curcuma longa: a review of preclinical and clinical research. Altern Med Rev 14(2):141–153

Kanakis C, Tarantilis P, Tajmir-Riahi H, Polissiou M (2007) Crocetin, dimethylcrocetin, and safranal bind human serum albumin: stability and antioxidative properties. J Agric Food Chem 55(3):970–977

Kawaguchi-Ihara N, Murohashi I, Nara N, Tohda S (2008) Promotion of the self-renewal capacity of human acute leukemia cells by Wnt3A. Anticancer Res 28:2701–2704

Kehat I, Kenyagin-Karsenti D, Snir M et al (2001) Human embryonic stem cells can differentiate into myocytes with structural and functional properties of cardiomyocytes. J Clin Invest 108:407–414. doi:10.1172/JCI12131

Khademhosseini A, Eng G, Yeh J, Fukuda J, Blumling J, Langer R, Burdick J (2006) Micromolding of photocrosslinkable hyaluronic acid for cell encapsulation and entrapment. J Biomed Mater Res A 79A(3):522–532

Khan NI, Bradstock KF, Bendall LJ (2007) Activation of Wnt/beta-catenin pathway mediates growth and survival in B-cell progenitor acute lymphoblastic leukaemia. Br J Haematol 138:338–348. doi:10.1111/j.1365-2141.2007.06667.x

Klaus A, Birchmeier W (2008) Wnt signalling and its impact on development and cancer. Nat Rev Cancer 8:387–398. doi:10.1038/nrc2389

Koduru S, Kumar R, Srinivasan S et al (2010) Notch-1 inhibition by Withaferin-A: a therapeutic target against colon carcinogenesis. Mol Cancer Ther 9:202–210. doi:10.1158/1535-7163.MCT-09-0771

Koltuniewicz A (2014) Sustainable process engineering. De Gruyter, Berlin

Kopan R, Ilagan MXG (2009) The canonical notch signaling pathway: unfolding the activation mechanism. Cell 137:216–233

Kroemer RT (2007) Structure-based drug design: docking and scoring. Curr Protein Pept Sci 8:312–328. doi:10.2174/138920307781369382

Langer T, Hoffmann RDD (2001) Virtual screening: an effective tool for lead structure discovery? Curr Pharm Des 7:509–527. doi:10.2174/1381612013397861

Li C, Heidt DG, Dalerba P et al (2007) Identification of pancreatic cancer stem cells. Cancer Res 67:1030–1037. doi:10.1158/0008-5472.CAN-06-2030

Li Y, Wicha MS, Schwartz SJ, Sun D (2011) Implications of cancer stem cell theory for cancer chemoprevention by natural dietary compounds. J Nutr Biochem 22:799–806

Liao W-T, Ye Y-P, Deng Y-J, Bian X-W, Ding Y-Q (2014) Metastatic cancer stem cells: from the concept to therapeutics. Am J Stem Cells 3(2):46–62

Lim K, Bisht S, Bar E, Maitra A, Eberhart C (2011) A polymeric nanoparticle formulation of curcumin inhibits growth, clonogenicity and stem-like fraction in malignant brain tumors. Cancer Biol Ther 11(5):464–473

Liu L, Wise DR, Diehl JA, Simon MC (2008) Hypoxic reactive oxygen species regulate the integrated stress response and cell survival. J Biol Chem 283:31153–31162. doi:10.1074/jbc.M805056200

Liu S, Dontu G, Mantle ID et al (2006) Hedgehog signaling and Bmi-1 regulate self-renewal of normal and malignant human mammary stem cells. Cancer Res 66:6063–6071. doi:10.1158/0008-5472.CAN-06-0054

Liu S, Dontu G, Wicha MS (2005) Mammary stem cells, self-renewal pathways, and carcinogenesis. Breast Cancer Res 7:86–95. doi:10.1186/bcr1021

Liu W, Shen SM, Zhao XY, Chen GQ (2012) Targeted genes and interacting proteins of hypoxia inducible factor-1. Int J Biochem Mol Biol 3:165–178

Lokich J (2012) Clinical cancer medicine: treatment tactics. Springer, Boston, MA, p 15

Long X, Fan M, Bigsby RM, Nephew KP (2008) Apigenin inhibits antiestrogen-resistant breast cancer cell growth through estrogen receptor-alpha-dependent and estrogen receptor-alpha-independent mechanisms. Mol Cancer Ther 7:2096–2108. doi:10.1158/1535-7163.MCT-07-2350

Lu X, Wu D, Li Z, Chen G (2011) Polymer nanoparticles. Prog Mol Biol Transl Sci 104:299–323

Lumelsky N, Blondel O, Laeng P et al (2001) Differentiation of embryonic stem cells to insulin-secreting structures similar to pancreatic islets. Science 292:1389–1394. doi:10.1126/science.1058866

Moore N, Lyle S (2011) Quiescent, slow-cycling stem cell populations in cancer: a review of the evidence and discussion of significance. J Oncol 2011:396076

MacAdam D (2003) Spontaneous regression: cancer and the immune system. Xlibris Corp, Philadelphia, PA

Maggi L, Carmona M, Zalacain A, Tomé M, Murcia M, Alonso G (2009) Parabens as agents for improving crocetin esters' shelf-life in aqueous saffron extracts. Molecules 14(3):1160–1170

Maheshwari R, Singh A, Gaddipati J, Srimal R (2006) Multiple biological activities of curcumin: a short review. Life Sci 78(18):2081–2087

Malaekeh-Nikouei B, Mousavi S, Shahsavand S, Mehri S, Nassirli H, Moallem S (2013) Assessment of cytotoxic properties of safranal and nanoliposomal safranal in various cancer cell lines. Phytother Res 27(12):1868–1873

McNeil S (2011) Characterization of nanoparticles intended for drug delivery. Humana Press, New York, NY

Meng X-Y, Zhang H-X, Mezei M, Cui M (2011) Molecular docking: a powerful approach for structure-based drug discovery. Curr Comput Aided Drug Des 7:146–157. doi:10.2174/157340911795677602

Merkus H (2009) Particle size measurements. Springer, Dordrecht

Moghimi S (2005) Nanomedicine: current status and future prospects. FASEB J 19(3):311–330

Moitessier N, Englebienne P, Lee D et al (2008) Towards the development of universal, fast and highly accurate docking/scoring methods: a long way to go. Br J Pharmacol 153(Suppl):S7–26. doi:10.1038/sj.bjp.0707515

Neeman E, Ben-Eliyahu S (2013) Surgery and stress promote cancer metastasis: new outlooks on perioperative mediating mechanisms and immune involvement. Brain Behav Immun 30:S32–S40

Newman DJ, Cragg GM (2016) Natural Products as sources of new drugs from 1981 to 2014. J Nat Prod 79(3):629–662. doi:10.1021/acs.jnatprod.5b01055

O'Brien CA, Pollett A, Gallinger S, Dick JE (2007) A human colon cancer cell capable of initiating tumour growth in immunodeficient mice. Nature 445:106–110. doi:10.1038/nature05372

Paldino E, Tesori V, Casalbore P et al (2014) Tumor initiating cells and chemoresistance: which is the best strategy to target colon cancer stem cells? Biomed Res Int 2014:859871

Park S, Kim Y-S, Lee H-A et al (2013) Mulberry leaf extract inhibits invasive potential and down-regulates hypoxia-inducible factor-1α (HIF-1α) in SK-N-BE2C neuroblastoma cells. Biosci Biotechnol Biochem 77:722–728. doi:10.1271/bbb.120763

Pece S, Tosoni D, Confalonieri S et al (2010) Biological and molecular heterogeneity of breast cancers correlates with their cancer stem cell content. Cell 140:62–73. doi:10.1016/j.cell.2009.12.007

Pérez-Tenorio G, Karlsson E, Waltersson MA et al (2011) Clinical potential of the mTOR targets S6K1 and S6K2 in breast cancer. Breast Cancer Res Treat 128:713–723. doi:10.1007/s10549-010-1058-x

Persano L, Pistollato F, Rampazzo E et al (2012) BMP2 sensitizes glioblastoma stem-like cells to Temozolomide by affecting HIF-1α stability and MGMT expression. Cell Death Dis 3:e412. doi:10.1038/cddis.2012.153

Pistollato F, Giampieri F, Battino M (2014) The use of plant-derived bioactive compounds to target cancer stem cells and modulate tumor microenvironment. Food Chem Toxicol 75C:58–70. doi:10.1016/j.fct.2014.11.004

Polakis P (2012) Wnt signaling in cancer. Cold Spring Harb Perspect Biol 4:9. doi:10.1101/cshperspect.a008052

Prakash O, Kumar A, Kumar P, Ajeet A (2013) Anticancer potential of plants and natural products: a review. Am J Pharmacol Sci 1(6):104–115

Rahaiee S, Shojaosadati S, Hashemi M, Moini S, Razavi S (2015) Improvement of crocin stability by biodegradeble nanoparticles of chitosan-alginate. Int J Biol Macromol 79:423–432

Rangarajan P, Subramaniam D, Paul S, et al (2015) Crocetinic acid inhibits hedgehog signaling to inhibit pancreatic cancer stem cells. Oncotarget; 6(29):27661-27673. doi: 10.18632/oncotarget.4871

Ray B, Bisht S, Maitra A, Maitra A, Lahiri D (2011) Neuroprotective and neurorescue effects of a novel polymeric nanoparticle formulation of curcumin (NanoCurc™) in the neuronal cell culture and animal model: implications for Alzheimer's disease. J Alzheimers Dis 23(1):61–77

Reya T, Morrison SJ, Clarke MF, Weissman IL (2001) Stem cells, cancer, and cancer stem cells. Nature 414:105–111. doi:10.1038/35102167

Roesch A, Fukunaga-Kalabis M, Schmidt EC et al (2010) A temporarily distinct subpopulation of slow-cycling melanoma cells is required for continuous tumor growth. Cell 141:583–594. doi:10.1016/j.cell.2010.04.020

Rollinger JM, Stuppner H, Langer T (2008) Virtual screening for the discovery of bioactive natural products. Prog Drug Res 65(211):213–249. doi:10.1007/978-3-7643-8117-2_6

Ryu MJ, Cho M, Song JY et al (2008) Natural derivatives of curcumin attenuate the Wnt/β-catenin pathway through down-regulation of the transcriptional coactivator p300. Biochem Biophys Res Commun 377:1304–1308. doi:10.1016/j.bbrc.2008.10.171

Saeed M, Kadioglu O, Khalid H et al (2015) Activity of the dietary flavonoid, apigenin, against multidrug-resistant tumor cells as determined by pharmacogenomics and molecular docking. J Nutr Biochem 26:44–56. doi:10.1016/j.jnutbio.2014.09.008

Samarghandian S, Borji A, Farahmand S, Afshari R, Davoodi S (2013) Crocus sativusL. (Saffron) stigma aqueous extract induces apoptosis in alveolar human lung cancer cells through caspase-dependent pathways activation. Biomed Res Int 2013:1–12

Sanchez F, Sobolev K (2010) Nanotechnology in concrete – a review. Construct Build Mater 24(11):2060–2071

Schatton T, Murphy GF, Frank NY et al (2008) Identification of cells initiating human melanomas. Nature 451:345–349. doi:10.1038/nature06489

Schulenburg A, Cech P, Herbacek I et al (2007) CD44-positive colorectal adenoma cells express the potential stem cell markers musashi antigen (msiI) and ephrin B2 receptor (EphB2). J Pathol 213:152–160. doi:10.1002/path.2220

Seo EJ, Wiench B, Hamm R et al (2015) Cytotoxicity of natural products and derivatives toward MCF-7 cell monolayers and cancer stem-like mammospheres. Phytomedicine 22:438–443. doi:10.1016/j.phymed.2015.01.012

Shabestari M, Samarghandian S (2013) DNA fragmentation and apoptosis induced by safranal in human prostate cancer cell line. Indian J Urol 29(3):177

Sharma R, Euden S, Platton S, Cooke D, Shafayat A, Hewitt H et al (2004) Phase I clinical trial of oral curcumin: biomarkers of systemic activity and compliance. Clin Cancer Res 10(20): 6847–6854

Shoichet B, Mcgovern S, Wei B, Irwin J (2002) Hits, leads and artifacts from virtual and high-throughput screening. In: Hicks MG, Kettner C (eds) Molecular informatics: confronting complexity. Proceedings of the Beilstein-Institut Workshop, May 13th–16th 2002, Bozen, Italy. http://www.beilstein-institut.de/bozen2002/proceedings/Shoichet/Shoichet.pdf

Silva P, Bonifácio B, Ramos M, Negri K, Maria Bauab T, Chorilli M (2013) Nanotechnology-based drug delivery systems and herbal medicines: a review. Int J Nanomedicine 9:1–15

Singh SK, Clarke ID, Terasaki M et al (2003) Identification of a cancer stem cell in human brain tumors. Cancer Res 63:5821–5828. doi:10.1038/nature03128

Somwanshi S, Dola R, Siddheshwa S, Mereka A, Godg R, Patta S (2013) Nanomedicine drug delivery system. Asian J Biomed Pharmaceut Sci 3(22):9–15

Sood AK, Coffin JE, Schneider GB et al (2004) Biological significance of focal adhesion kinase in ovarian cancer: role in migration and invasion. Am J Pathol 165:1087–1095. doi:10.1016/S0002-9440(10)63370-6

Stewart B, Wild C (2014) World cancer report 2014. IARC Press, Lyon
Suga H, Kadoshima T, Minaguchi M et al (2011) Self-formation of functional adeno- hypophysis in three-dimensional culture. Nature. doi:10.1038/nature10637
Tai J, Cheung S, Ou D et al (2014) Antiproliferation activity of Devil's club (Oplopanax horridus) and anticancer agents on human pancreatic cancer multicellular spheroids. Phytomedicine 21:506–514
Tanno T, Matsui W (2011) Development and maintenance of cancer stem cells under chronic inflammation. J Nippon Med Sch 78:138–145
Teglund S, Toftgård R (2010) Hedgehog beyond medulloblastoma and basal cell carcinoma. Biochim Biophys Acta 1805:181–208
Teng Y, Wang X, Wang Y, Ma D (2010) Wnt/beta-catenin signaling regulates cancer stem cells in lung cancer A549 cells. Biochem Biophys Res Commun 392:373–379. doi:10.1016/j.bbrc.2010.01.028
Thiyagarajan V, Lin S, Chang Y, Weng C (2016) Identification of novel FAK and S6K1 dual inhibitors from natural compounds via ADMET screening and molecular docking. Biomed Pharmacother 80:52–62
Thiyagarajan V, Lin SH, Chia YC, Weng CF (2013) A novel inhibitor, 16-hydroxy-cleroda-3,13-dien-16,15-olide, blocks the autophosphorylation site of focal adhesion kinase (Y397) by molecular docking. Biochim Biophys Acta 1830:4091–4101. doi:10.1016/j.bbagen.2013.04.027
Thompson M (2010) The characterisation of nanoparticles. Royal Society of Chemistry, London. Retrieved 28 August 2016, from http://www.rsc.org/images/characterisation-nanoparticles-technical-brief-48_tcm18-214815.pdf
Turashvili G, Bouchal J, Burkadze G, Kolar Z (2006) Wnt signaling pathway in mammary gland development and carcinogenesis. Pathobiology 73:213–223. doi:10.1159/000098207
Unger C, Kramer N, Walzl A et al (2014) Modeling human carcinomas: physiological relevant 3D models to improve anti-cancer drug development. Adv Drug Deliv Rev 79–80:50–67. doi:10.1016/j.addr.2014.10.015
Usha T, Middha SK, Goyal AK et al (2014) Molecular docking studies of anti-cancerous candidates in Hippophae rhamnoides and Hippophae salicifolia. J Biomed Res 28:406–415. doi:10.7555/JBR.28.20130110
van der Bij G, Oosterling S, Beelen R, Meijer S, Coffey J, van Egmond M (2009) The perioperative period is an underutilized window of therapeutic opportunity in patients with colorectal cancer. Ann Surg 249(5):727–734
Verma S, Singh S (2008) Current and future status of herbal medicines. Vet World 2(2):347
Vogelstein B, Papadopoulos N, Velculescu V, Zhou S, Diaz L, Kinzler K (2013) Cancer genome landscapes. Science 339(6127):1546–1558
Voskoglou-Nomikos T, Pater J, Seymour L (2003) Clinical predictive value of the in vitro cell line, human xenograft, and mouse allograft preclinical cancer models. Clin Cancer Res 9(11):4227–4239
Walters WP, Stahl MT, Murcko MA (1998) Virtual screening—an overview. Drug Discov Today 3:160–178. doi:10.1016/S1359-6446(97)01163-X
Wang Y, Hu J, Lin H et al (2016) Herbalog: a tool for target-based identification of herbal drug efficacy through molecular docking. Phytomedicine 23(12):1469–1474. doi:10.1016/j.phymed.2016.08.008
Wang Z, Li Y, Banerjee S, Sarkar FH (2009) Emerging role of Notch in stem cells and cancer. Cancer Lett 279:8–12
Warfel NA, El-Deiry WS (2014) HIF-1 signaling in drug resistance to chemotherapy. Curr Med Chem 21:3021–3028
WHO (2003) Global action against cancer. World Health Organization, Geneva, pp 1–25
Wilson A, Radtke F (2006) Multiple functions of Notch signaling in self-renewing organs and cancer. FEBS Lett 580:2860–2868
Woodward WA, Chen MS, Behbod F et al (2007) WNT/beta-catenin mediates radiation resistance of mouse mammary progenitor cells. Proc Natl Acad Sci U S A 104:618–623. doi:10.1073/pnas.0606599104

Xia P (2014) Surface markers of cancer stem cells in solid tumors. Curr Stem Cell Res Ther 9(2):102–111

Yadav D, Suri S, Choudhary A, Sikender M, Hemant, Beg M et al (2011) A novel approach: herbal remedies and natural products in pharmaceutical science as nano drug delivery systems. Int J Pharm Technol 3(3):3092–3116

Yamaguchi TP (2001) Heads or tails: wnts and anterior-posterior patterning. Curr Biol 11(17): R713–R724

Yamnik RL, Digilova A, Davis DC et al (2009) S6 kinase 1 regulates estrogen receptor alpha in control of breast cancer cell proliferation. J Biol Chem 284:6361–6369. doi:10.1074/jbc.M807532200

Yang W, Yan HX, Chen L et al (2008a) Wnt/β-catenin signaling contributes to activation of normal and tumorigenic liver progenitor cells. Cancer Res 68:4287–4295. doi:10.1158/0008-5472.CAN-07-6691

Yang ZF, Ho DW, Ng MN et al (2008b) Significance of CD90+ cancer stem cells in human liver cancer. Cancer Cell 13:153–166. doi:10.1016/j.ccr.2008.01.013

You N, Zheng L, Liu W et al (2014) Proliferation inhibition and differentiation induction of hepatic cancer stem cells by knockdown of BC047440: a potential therapeutic target of stem cell treatment for hepatocellular carcinoma. Oncol Rep 31:1911–1920. doi:10.3892/or.2014.3043

Ysebaert L, Chicanne G, Demur C et al (2006) Expression of beta-catenin by acute myeloid leukemia cells predicts enhanced clonogenic capacities and poor prognosis. Leukemia 20:1211–1216. doi:10.1038/sj.leu.2404239

Yuan X, Wu H, Xu H et al (2015) Notch signaling: an emerging therapeutic target for cancer treatment. Cancer Lett 369:20–27

Yuhas JM, Li AP, Martinez AO, Ladman AJ (1977) A simplified method for production and growth of multicellular tumor spheroids. Cancer Res 37:3639–3643

Zheng P-W, Chiang L-C, Lin C-C (2005) Apigenin induced apoptosis through p53-dependent pathway in human cervical carcinoma cells. Life Sci 76:1367–1379. doi:10.1016/j.lfs.2004.08.023

Chapter 4
Plant-Derived Compounds with Anticancer Properties: From Folklore to Practice

Tripti Tewari, Ruchi Singh, Vartika Pant, Ajit Kumar, and Preeti Chaturvedi

Abstract Autotrophs have had an important role in the folklore of ancient cultures. In addition to the use as food and spices, autotrophs have also been utilized as medicines for over 5000 years. It is estimated that 70–95% of the population in developing countries continues to use traditional medicines even today. A new trend, that involved the isolation of plant active compounds, led to the discovery of different active compounds that are derived from plants. Radio- and chemotherapy is routinely used for cancer treatment. In the last decades, plant-derived compounds have been subscribed as medicines with anticancer activity. Cancer is the second leading cause of death worldwide. The number of new cases is expected to rise by about 70% over the next two decades. Thus, there is a real need of plant-derived compounds as a promising source for new efficient anticancer drugs with reduced side effects. Here we focus on three plants exhibiting anti-cancereous activity, their mode of action, and bioavailability. These include betulinic acid, berberine, and palmatin. Also discussed are commercial considerations and future prospects for development of plant-derived substances with anticancer activity.

Keywords Anticancer compounds • Chemoprevention • Plants • Medicinal plants

Introduction

Plants play an important role in traditional world. Two important traditions, the traditional Chinese medicine (TCM) and the traditional Indian medicine (TIM), Ayurveda, have implemented the accustomed knowledge pertinent to traditional important plants. It is predicted that traditional medicines have been used by 70–95% of the population in various countries. Today, medicinal plants are defined as plants

T. Tewari (✉) • R. Singh • V. Pant • P. Chaturvedi
Department of Biological Sciences, College of Basic Sciences and Humanities, US Nagar, Pantnagar, Uttarakhand 263145, India
e-mail: tripti.tewari2015@gmail.com

A. Kumar
Department of Horticulture, College of Agriculture, G.B. Pant University of Agriculture and Technology, US Nagar, Pantnagar, Uttarakhand 263145, India

S. Malik (ed.), *Biotechnology and Production of Anti-Cancer Compounds*,
DOI 10.1007/978-3-319-53880-8_4

that contain valuable substances with therapeutic effect in healing and prevention of multifarious ailments in various human organisms. Due to the competence needed to identify plant-derived compounds, different such compounds were disclosed during last 200 years. Initially, botanist or ethnobotanist pinpoint a particular plant. Next, by biological screening assays plant extracts are supervened which are later executed by a phytochemist to determine the conceivable ameliorative activity after isolation of the effective compound. Certainly, molecular biology studies are enforced to acknowledge the mode of action and significant molecular targets.

Cancer is the most paramount cause of despair and fatality worldwide. Epidemiological data determined an escalation in the cancer prevalence and lethality. According to the GLOBOCAN 2008 assessments, there have comparatively been 12.7 million new cancer cases pronounced and 7.6 million deaths omnipresent in 2008 (Ferlay et al. 2010).

Moreover, it has also been predicted that cancer will eclipse heart diseases as the preeminent matter of death in the world, encompassing deliberate commutable and economic emanation (Jemal et al. 2010). The most predominant challenge to oncology is the eloquent reinforcement of advanced surgical techniques, chemo-, and proposed therapy, misstep in tumor treatment (Jemal et al. 2011). Chemotherapy is customarily used for cancer prescription. Since cancer cells deplete manifold regulatory provinces present in normal cells, they persevere to divide which natural cells do not. This feature makes cancer cells vulnerable to chemotherapeutic drugs.

Due to the number of displeasing side effects that consistently occur during chemotherapy, chemoprevention by innumerable plant-derived secondary metabolites epitomizes a disparate perspective in the battle against cancer. A peculiar advent to the cancer treatment has acknowledged the key constituents of altered signaling pathways in neoplastic cells without affecting noncancerous cells. Commonly occurring substances have been used for combating human diseases for several years which play a flourishing role in drug discovery and augmentation.

Cell cycle of an organism is regulated by cyclin-dependent kinases (CDKs) which is a catalytic subunit; these enzymes control the transition between the different states of the cell cycle. All these catalytic units depend on regulatory subunits known as cyclins for their activities. Cell cycle possesses different classes of cyclins and CDKs (Taiz and Zeiger 2006):

G_1/S cyclin: CDK-2 with cyclin E, activated in late G1.

S cyclin: CDK-2 with cyclin A, which is activated at the beginning of the S phase.

M cyclin: CDK-1 with cyclin B, activated just prior to the mitotic phase.

G_1 cyclin: CDK-4 and CDK-6 with cyclin D which is activated in the early G_1 phase.

Cell Cycle and Checkpoints

Checkpoints are mechanisms that stop the progression of cell cycle, if there is any damage in the chromosomal DNA, presence of unreplicated DNA, or misalignment of chromosome during M phase on mitotic spindle assembly. When there is

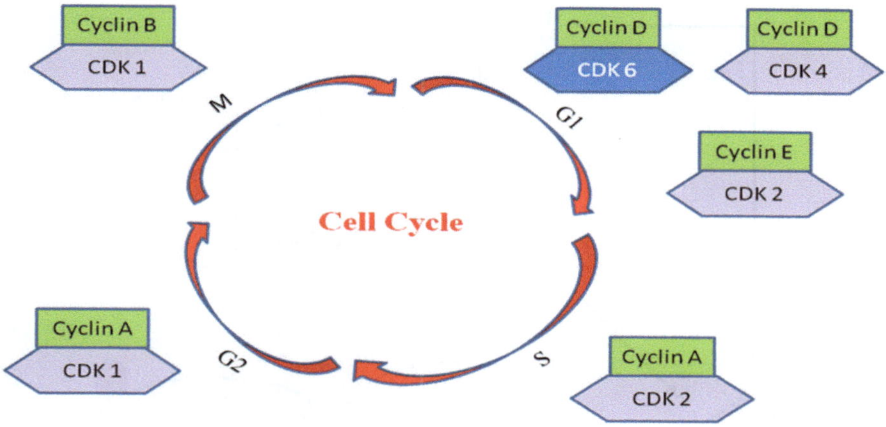

Fig. 4.1 Cell cycle regulation by cyclin-dependent kinase

any DNA damage, the cell cycle is arrested at the G_1, S, and G_2 checkpoints (Fig. 4.1). Major checkpoint of cell cycle is G_2 arrest. First of all sensor proteins are bound to the DNA damage that lead to the activation of signaling pathway causing cell cycle arrest and activate the DNA repair mechanism and in some cases even cell death. Immediate target of sensor proteins are ATM/ATR (ataxia-telangiectasia-mutated and ataxia telangiectasia- and Rad3-related protein). ATM senses ds DNA break whereas ATR senses ss DNA break. Mutation in the gene of ATM/ATR causes ataxia which is an inherited recessive disorder resulting in nervous and immune system defect making the organism highly susceptible for cancer. ATM activates chk-2 protein which phosphorylate p53 protein and this p53 protein induces expression of the p21 gene lead to the production of p21 protein form a complex with CDK-4/CDK-6 and cyclin D and block the cycle at G_1 phase. ATR gene activates the chk-1 which in turn activate cdc25 phosphates which inhibit the activity of CDK by removing activatory phosphate and thus cyclin B and CDK-1 become inactive which block the cell cycle and it remain in G_2 phase unless DNA is repaired (Fig. 4.2).

Natural products have also been allied their ability against tumor action to trigger cell death pathways and their associated approaches. Apoptosis or programmed cell death plays a pivotal role in maintaining tissue homeostasis that is highly conserved among different species (Evan and Vousden 2001). A hallmark of human cancers is evasion of apoptosis (Hanahan and Weinberg 2000). Myriad of many plant products exist that have shown very promising anticancer properties, but have yet to be evaluated in humans. Further study is required to determine the efficacy of these plant products in treating cancers in humans. This chapter focuses on the three plant-derived compounds derived from *Tinospora cordifolia*, *Ocimum sanctum*, and *Bauhinia variegata* that have, in recent years, shown promise as anticancer agents and outlines their potential mechanism of action.

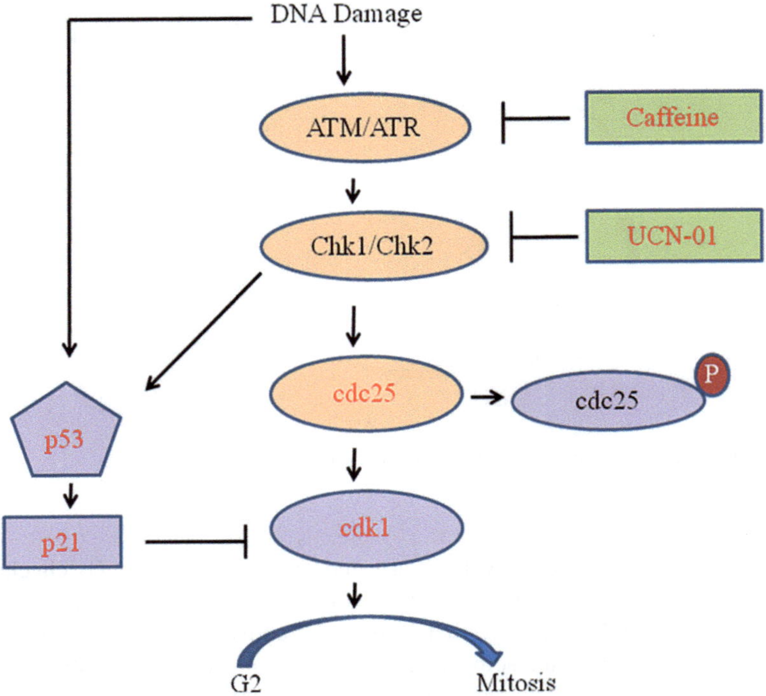

Fig. 4.2 DNA damage mechanism

Medicinal Plants and Cancer

Tinospora cordifolia

Tinospora cordifolia (Willd.) Miers ex Hook. F. & Thoms commonly known as Guduchi belonging to family Menispermaceae is a traditional herb used in the various ailments. It is native of India but also found in Burma and Sri Lanka. It thrives well in the tropical region up to 1200 m above sea level (Reddy and Reddy 2015; Kavya et al. 2015).

Plant Profile

Tinospora cordifolia is a large, deciduous climbing shrub; leaves are simple, alternate, and exstipulate with long petiole (15 cm); and lamina is broadly ovate or ovate cordate, 10–20 cm long or 8–15 cm broad, seven nerved, and deeply cordate at base; and flowers are unisexual, where male flowers are clustered and female usually solitary. Six sepals arrange in two whorls where the outer ones are smaller than the inner one. Petals are smaller than sepals, obovate, and membranous. Fruit is aggregate of 1–3 and red in color (Veeraiah and Reddy 2012; Sinha et al. 2004).

Anticancerous Compounds

The anticancerous compounds present in the *Tinospora cordifolia* are berberine and palmatine. Berberine is present in the stem whereas palmatine is present in both root and stem (Mittal et al. 2014). The berberine and palmatine both are protoberberine alkaloids that show anticancerous activity. Palmatine is a close structural analogue of berberine that both bear the same tetracyclic structure (7,8,13,13-tetrahydro 9,10 dimethoxy berberinium) but differ in the nature of substitutes at position 2,3 on the benzo ring, where it is dimethoxy for palmatine and methylene dioxy for berberine (Bhadra et al. 2007).

Berberine: Berberine is a naturally occurring isoquinoline alkaloid, which is present in the roots, rhizome, and stem bark of a number of important medicinal plants such as *Berberis* species, *Xanthorhiza simplicissima, Coptis chinensis,* and *Tinospora cordifolia* (Mantena et al. 2006).

Biosynthesis Pathway of Berberine

Tyrosine is the precursor for the biosynthesis of berberine which involves 13 different enzymatic reactions. In berberine biosynthesis, the first committed step is catalysis by the berberine bridge enzyme (BBE) which converts *N*-methyl group of reticuline to berberine bridge carbon C-8 of scoulerine and this BBE is a major branch point enzyme in benzylisoquinoline alkaloid biosynthesis. The reaction involves the oxidation of the *N*-methyl group followed by ring closure. This reaction is unique in nature and cannot be achieved by using synthetic methods. All the enzymes involved in the biosynthesis of berberine are sterio-specific and (S) enantiomers are the only effective substrate.

Action Mechanism of Berberine in Cancer Treatment

Berberine is an isoquinoline alkaloid that possesses anticancerous, antioxidant, immune-enhancing, and anti-inflammatory properties. It exhibits the strong anticancer activity against the various cancers such as prostrate cancer, liver cancer, and leukemia due to its antioxidant activity which reduces the free radical or reactive oxygen species that cause the cancer or by arresting the cancer cell cycle in G1 phase and inducing apoptosis (Umadevi et al. 2013).

Antioxidant Activity of Berberine

Free radicals, reactive oxygen species (ROS), and radiations induce the DNA damage, including the oxidation, strand breakage, and ionization which may lead to defects in the genes involved in proliferation and cell signaling pathways that are crucial for tumor growth and cancer progression, so antioxidants play a protective role in preventing cellular damage due to oxidation. Lipoxygenase, xanthine oxidase, and cyclooxygenase (COX2) are the three important enzymes that are responsible for generating the reactive oxygen species in the cells and berberine has the ability to reduce the activity of xanthine oxidase and COX2, which finally decrease the level of reactive oxygen level in the cell. But sometime berberine increases the amount of reactive oxygen species to induce several apoptotic signaling pathways, including mitogen-activated protein kinase (MAPK), extracellular signal-regulated kinase 1/2 (ERK1/2), c-Jun N-terminal protein kinase (JNK) and Akt, as well as calcium-dependent pathways. So these different effects of berberine, whether it increases or decreases the cellular reactive oxygen species content, depend upon the cell conditions and types. For example, berberine induces reactive oxygen species production in prostate cancer cells, but not in the normal epithelial cells. Berberine induces apoptosis and inhibits cell proliferation in various cell lines derived from breast, lung, colon, and liver cancer by different pathways: Fas-dependent pathway and p53-dependent pathway as follows.

Fas-Dependent Apoptosis/Caspase-Dependent Pathway

Berberine induces apoptosis by decreasing the levels of the apoptosis inhibitors (IAP, XIAP, and Bcl-2), and increasing the amounts of apoptosis activators (caspase-3, caspase-8, caspase-9, Bax, Bid, and IAF). Berberine induces the expression of both Fas (death receptor present on tenth chromosome number in human) and FasL (Fas ligand) in cancer cell lines to induce caspase activity. First of all this FasL is attachment to the Fas receptor (FasR) and activates the caspase-8, followed by the activation of Bax and in turn Bcl-2 deactivation. And this activated Bax leads to the formation of pores in the mitochondrial membrane; releases the cytochrome *c*, Smac/Diablo, and Apaf-1 from the mitochondria; and leads to the subsequent activation of a sequential caspase cascade, which activates the caspase-activated DNase (CAD) which is normally inhibited by ICAD, and finally this activated CAD induces the DNA fragmentation (Kaboli et al. 2014).

Caspase-Independent Pathway

In this pathway, berberine directly activates the generation of reactive oxygen species. This increased level of reactive oxygen species can alter the mitochondrial membrane potential, and lead to mitochondrial collapse. This reactive oxygen species production leads to the activation of apoptosis-inducing factor (AIF), a protein that is involved in caspase-independent apoptosis and occurs to induce apoptosis.

Apoptosis Pathway Mediated by Berberine in SCC-4 Human Tongue Cancer Cells

Berberine increases production of reactive oxygen species (ROS) and cytosolic Ca^{2+} levels, resulting in stimulation of Bax protein expression and reduction of Bcl-2 same as in above case, and reduces the mitochondrial membrane potential ($\Delta\Psi m$) and releases the cytochrome *c* which activates caspases-9- and -3-producing apoptosis in the SCC-4 human tongue squamous cell carcinoma cancer cell line.

p53-Dependent and -Independent Apoptosis Pathway

BRCA1 and BRCA2 function in DNA repair in breast cancer cells. When DNA damage occurs, BRCA1 is phosphorylated by ATM. Mutated BRCA1 and BRCA2 cause checkpoint dysfunction that in turn leads to development of cancerous breast cells. Mutated ATM also affects normal DNA repair processes and can inhibit cell cycle arrest in response to DNA damage. Berberine is a crucial anticancer compound that affects ATM such that apoptosis is triggered via p53-independent pathway.

Ocimum sanctum

Origin

Ocimum L., belonging to family Lamiaceae is the best known genus for its medicinal aromatic oils. This genus comprises a wide range of intra- and interspecific genetic diversity encompassing at least 65 to more than 150 species dispersed all over the world. Among these, *Ocimum sanctum* L. (*Ocimum tenuiflorum* L.) and *Ocimum basilicum* L. are the two important species used broadly for their medicinal and industrial importance. *O. sanctum*, known as "the holy basil," is native to Asian tropics, though *O. basilicum* L. or "the sweet basil" is construed to be of African origin as per the Germplasm Resources Information Network of United States Department of Agriculture. Both of the two *Ocimum* species are rich source of myriad phytochemicals, which are composed of essential phenylpropanoids and terpenoids with various anticancerous properties. Besides anticancerous activity, *O. sanctum* is also known to possess antibacterial, antihistaminic, wound-healing, radioprotective, antidiabetic, larvicidal, anti-genotoxic, neuroprotective, and cardio-possessive activity.

The leaves and stem of holy basil contain a variety of biologically active constituents like saponins, flavonoids, triterpenoids, as well as tannins. Of the 40 secondary metabolites that have medicinal value, the genes and enzymes censurable for the production of 16 metabolites have been mapped on the genome. Of the 14 metabolites that have been mapped, 8 have anticancer equity and the remaining 6 have antifungal, antioxidant, anti-inflammatory, antiseptic, and cardioprotective properties. Various compounds found in the *Ocimum* sp. are being appraised for their anticancer properties in various clinical trials. But there is a critical limitation—the final product is isolated from the tulsi leaves. Betulinic acid (BA) from *O. sanctum* L. is reported to have cardioprotective effect. It is a natural product that exhibits potent antitumor activities by triggering the mitochondrial path to apoptosis. Mitochondrion-targeted agents such as betulinic acid may open new prospects to overcome some forms of drug resistance (Galluzzi et al. 2006). Thus, the uses of *Ocimum* sp. for beneficial purposes in addition to their industrial importance for aromatic properties bolster the importance of ethno-botanical access as a potential source of bioactive substances.

Betulinic Acid, Anticancerous Phytochemical

Terpenes are a large group of comprehensive secondary metabolites of plants and are considered as probably useful in cancer pharmacotherapy, because of their selective cytotoxicity towards numerous human cancer cells. Triterpenes are one of terpene classes, are formed from six isoprene units, and occur as complex cyclic structures called triterpenoids. Betulinic acid (3β, hydroxy-lup-20(29)-en-28-oic acid) is a pentacyclic triterpenoid of plant origin that is widely distributed in the plant kingdom throughout the world (Cichewicz and Kouzi 2004a, b). The reduced congener of betulinic acid, betulin (3β-lup-20(29)-en-3,28-diol), was one of the first natural products identified and isolated from plants in 1788 (Alakurtti et al. 2006). Betulinic acid exerts a number of biological activities and has shown to have antitumor properties.

Anticancer Activity of Betulinic Acid

The antitumor activity of betulinic acid has been broadly studied in cancer cell lines in various organisms. Betulinic acid possesses anticancer activity which was subsequently also reported against other types of human cancers including neuroblastoma, glioblastoma, medulloblastoma, as well as several carcinomas, i.e., head and neck, colon, breast, hepatocellular, lung, prostate, renal cell, ovarian, or cervix carcinoma (Wick et al. 1999; Tan and Pezzuto 2003). Further, there is evidence that betulinic acid is conversely cytotoxic against metastatic over nonmetastatic melanoma cell lines. Moreover, betulinic acid collaborated with different cytotoxic stimuli to quench tumor growth, including ionizing radiation (Selzer et al. 2000), chemotherapeutic drugs (Fulda and Debatin 2004; Sawada et al. 2004), or the death receptor ligand TRAIL. Therefore, betulinic acid may be used as a sensitizer in combination regimens to enhance the efficacy of anticancer therapy. Besides its potent antitumor activity in vitro, betulinic acid also suppressed tumor growth in several animal models of human cancer. In a xenograft mouse model of ovarian cancer administration of betulinic acid significantly increased the survival time (Zuco et al. 2002). Also, betulinic acid suppressed tumor growth in a melanoma xenograft model (Pisha et al. 1995). Also in vivo, betulinic acid cooperated with chemotherapeutic agents such as vincristine to reduce lung metastasis in a metastatic melanoma model (Sawada et al. 2004).

Mechanisms of BE-Mediated Anticancer Activity

Apoptotic cell death is a prerequisite mechanism of anticancer agent activity (Brown and Attardi 2005; Elmore 2007) along with BE. Apoptosis is a type of programmed cell death. Two important pathways of apoptosis have been established, the extrinsic (death receptor related) as well as intrinsic (mitochondrion dependent). The extrinsic pathway is proposed by external signals, the imperative of ligands, such as Fas, TNF, or TRAIL, to their corresponding death receptors, limited in the cell surface.

The apoptosis pathway is triggered by stimuli, such as DNA damages, oxidative stress, radiation, as well as growth factor withdrawal (Green et al. 2004).

Apoptosis pathways can be proposed by intrinsic pathway at the level of the mitochondria by the release of apoptogenic factors such as cytochrome *c*, Smac, or AIF from the mitochondrial intermembrane space into the cytosol. Smac promotes by neutralizing "inhibitor of apoptosis protein" (IAP)-mediated restraint of caspase-3 and -9 (Saelens et al. 2004). In comparison, death receptor stimulation (extrinsic pathway) in turn points to receptor trimerization, contracting of adaptor molecules such as FADD, and incitement of the initiator caspase-8, which proliferates the death signal to effector caspases such as caspase-3 (Fulda and Debatin 2004). The BH3 domain protein Bid links the receptor to the mitochondrial pathway (Adams and Cory 2007). Bid is stimulated by caspase-8 and translocates to mitochondria to promote cytochrome *c* release. Apoptosis can be subdued at various levels, e.g., by FLIP, Bcl-2, or IAPs (Salvesen and Duckett 2002).

Provocation of the Intrinsic Pathway by Anticancer Therapeutics

The intrinsic pathway of apoptosis is prompted with chemotherapeutic agents as a result of a DNA damage. A pivotal step in the activation of the intrinsic pathway is the permeabilization of the outer mitochondrial membrane which in turn results in the release of soluble proteins from the mitochondrial interspace into the cytosol like cytochrome *c*, AIF, or Smac (Green and Kroemer 2004). A number of second messengers have been diagnosed that can regulate outer mitochondrial membrane permeabilization (Galluzzi et al. 2006). Therefore, factors that can directly urge mitochondrial outer membrane permeabilization can act as adequate cytotoxic agents.

Introduction of Mitochondrial Outer Membrane Permeabilization by Betulinic Acid

Betulinic acid induces apoptosis by inducing loss of mitochondrial membrane potential and was constrained by inhibitor of the permeability transition pore complex (Fulda et al. 1998a, b). Betulinic acid was shown to prompt cytochrome *c* in a permeability transition pore-dependent manner (Andre et al. 2002). Pertinence of mitochondria in betulinic acid-induced apoptosis leads to caspase activation and apoptotic DNA fragmentation. Cytochrome *c* from mitochondria undergoing betulinic acid-mediated permeability transition activated caspase-3 in a cell-free system. Separation of caspase-3 and -8 was anticipated by uprising of mitochondrial membrane potential. Overexpression of Bcl-2 and Bcl-XL negotiates contention to betulinic acid at the level of mitochondrial dysfunction, protease activation, and nuclear fragmentation which points out that these events eventuate downstream of the Bcl-2- or Bcl-XL-restrained checkpoint of apoptosis. This proposes that caspase-8 is stimulated downstream of mitochondria during betulinic acid-induced apoptosis. Activation of the caspase cascade was requisite for betulinic acid to bring about apoptosis, as broad-spectrum peptide inhibitors of caspases completely invalidate betulinic acid-triggered apoptosis. Interestingly, neuroblastoma cells opposing to doxorubicin-mediated apoptosis were still conscious to treatment with betulinic acid (Fulda et al. 1998a, b). This proposes that betulinic acid may affect some forms of drug resistance. Formation of reactive oxygen species (ROS) with betulinic acid has been recorded in mitochondrial membrane permeabilization. To this end, ROS generation was disclosing in cancer cell lines and was treated with betulinic acid (Wick et al. 1999). Administration of betulinic acid prior to antioxidants rescued cells from apoptosis suggesting that ROS production was involved in resolving cell death. Also, ROS generation was linked to activation of pro-apoptotic p38 and SAP/JNK kinases with no change in the phosphorylation of ERK which reveals that ROS act upstream of the MAPKs in the signaling pathway of betulinic acid (Tan and Pezzuto 2003).

Modulation of Betulinic Acid-Induced Apoptosis by Bcl-2 Family Proteins

Bcl-2 family proteins are the signal transduction proteins that regulate outer membrane permeabilization on mitochondria. Bcl-2 family proteins encompass both anti-apoptotic members, viz., Bcl-2, Bcl-XL, and Mcl-1, as well as pro-apoptotic molecules like Bax, Bak, Bad, and BH3 domains (Adams and Cory 2007). Inequality in the proportion of anti-apoptotic versus pro-apoptotic Bcl-2 proteins may favor tumor cell survival instead of cell death (Adams and Cory 2007). Betulinic acid has been recorded to modulate expression levels of different Bcl-2 family proteins. For example, betulinic acid emerges in upregulation of the pro-apoptotic Bcl-2 family protein Bax in neuroblastoma, glioblastoma, as well as melanoma cells, though Bcl-XS was found at inflated levels in betulinic acid-treated neuroblastoma cells (Selzer et al. 2000). Elucidation levels of pro-apoptotic proteins Bak and Bad were not reformed in feedback to betulinic acid in melanoma cells (Selzer et al. 2002). While expression levels of anti-apoptotic Bcl-2 remained unchanged upon incubation with betulinic acid in neuroblastoma and squamous cell carcinoma cells, an enhancement in Bcl-2 protein levels was recorded in glioblastoma cells (Thurnher et al. 2003). This suggests that betulinic acid regulates Bcl-2 family proteins in a condition-dependent manner. Furthermore, betulinic acid has been described to induce apoptosis in a p53- and CD95-independent manner. Moreover, apoptosis upon analysis with betulinic acid was not in league with accumulation of wild-type p53 protein. Also, betulinic acid similarly induced apoptosis in p53 mutant and p53 wild-type cell lines and was also active in p53-deficient melanoma cells (Zuco et al. 2002). Moreover, betulinic acid provokes apoptosis independent of CD95-ligand/receptor interaction.

Inflection of NF-κB Activity by Betulinic Acid

Betulinic acid has also been reputed to inflect the activity of the transcription nuclear factor-κB (NF-κB), a key regulator of stress-induced transcriptional incitement. Betulinic acid was pinpointed as a potent activator of NF-κB in a number of cancer cell lines (Kasperczyk et al. 2005). Betulinic acid-induced NF-κB activation elaborates increased IKK activity, phosphorylation of IκBα at serine 32/36 supervened by deterioration of IκBα, and nuclear translocation of the NF-κB subunit p65. Assays entrenched that NF-κB that was triggered by betulinic acid is transcriptionally active. Interestingly, suppression of betulinic acid-induced NF-κB activation by disparate chemical inhibitors (proteasome inhibitor, antioxidant, IKK inhibitor) also harmed betulinic acid-convinced apoptosis.

Eminently, specific NF-κB prohibition by transient or stable elucidation of IκBα super-repressor subdued betulinic acid-induced apoptosis in neuroblastoma cells, although transient expression of IκBα super-repressor had no consequences on betulinic acid-induced apoptosis in other cell lines. These recommendations reveal that incitement of NF-κB by betulinic acid bolsters betulinic acid-induced apoptosis. By resemblance, betulinic acid was shown to intrude with NF-κB activation and NF-κB-adapted gene expression provoked by carcinogens (Takada and Aggarwal 2003). These conclusions may contribute a molecular basis for the capacity of betulinic acid to subdue inflammation and attune the immune response. Together, these recommendations point to a context-dependant function of NF-κB in the direction of betulinic acid-mediated apoptosis.

Conceivable Application in Therapy

No exemplary clinical trials have been reported using betulinic acid for the prescription of human cancer so far (Laszczyk 2009). BE has been exposed to elicit anticancer properties by constraining cancer cell growth. Copious studies over the last years intended at exemplifying the mechanisms of betulinic acid-mediated antitumor activity. One distinctive feature of betulinic acid's cytotoxicity is its capacity to prompt the mitochondrial pathway of apoptosis in cancer cells.

Bauhinia variegata L.

Bauhinia variegata L. is commonly known as mountain ebony, orchid tree, or poor man's orchid (Orwa et al. 2009), and belongs to the family Leguminosae. It is common in warm temperate and subtropical regions. It is native to Southeast Asia and grows in tropical and subtropical climate (Kanak et al. 2012).

Plant Profile

Bauhinia variegata L. is a small- to medium-sized deciduous tree with a short bole and spreading crown, attaining a height of up to 15 m and diameter of 50 cm. The bark is light brownish, hairy, and angled, becoming brownish grey. Leaves have 1–2 mm sized stipules, early caducous; petiole puberulous to glabrous, 3–4 cm;

lamina broadly ovate to circular, often broader than long, 6–16 cm diameter; 11–13 nerved; tips of lobes broadly rounded, base cordate; upper surface glabrous, lower glaucous but glabrous when fully grown. Flower clusters (racemes) are unbranched at the ends of twigs. The few flowers have short, stout stalks and a stalk like, green, narrow basal tube (hypanthium). The light green, fairly hairy calyx forms a pointed five-angled bud and splits open on one side, remaining attached; petals, slightly unequal, wavy margined and narrowed to the base; five curved stamens; very slender, stalked, curved pistil, with narrow, green, one-celled ovary, style and dotlike stigma. Pods are strap-shaped, obliquely striate, 20–30 by 2–25 cm; long, hard, flat with 10–15 seeds in each; seeds brown, flat, nearly circular with coriaceus testa (Orwa et al. 2009).

Biosynthesis Pathway

Kaempferol (flavonoids) has a diphenylpropane structure (C6–C3–C6) and is synthesized by condensation of 4-coumaroyl-CoA (C6–C3) with three molecules of malonyl-CoA (C6). This reaction, catalyzed by the enzyme chalcone synthase (EC 2.3.1.74), results in the formation of the flavonoid naringenin chalcone (C6–C3–C6). This chalcone is transformed into the flavanone naringenin by the enzyme chalcone isomerase (EC 5.5.1.6), which catalyzes the closure of the C3 ring. The enzyme flavanone 3-dioxygenase (EC 1.14.11.9) introduces a hydroxyl group in naringenin at C3 to form dihydrokaempferol. Finally, the enzyme flavonol synthase (EC 1.14.11.23) introduces a double bond in dihydrokaempferol at C2–C3 to produce kaempferol (Fig. 4.1). Because the enzymes involved in the biosynthesis of kaempferol are relatively common in the plant kingdom, it is not surprising that this flavonoid is widely distributed in plants (Calderon-Montano et al. 2011).

Anticancer Activity

The ethanolic extract of *B. variegate* has antitumor effect on Dalton's ascitic lymphomas (RajKapoor et al. 2003) and protection of liver from diethyl nitrosamine which has cytotoxic effect on human cell line and showed cytotoxic against human breast cancer (HBL-100) cells (RajKapoor et al. 2006). Protocol of methanolic extract of stem bark of *B. variegate* also exerted anticancer effects in skin papilloma cells against 7,12-dimethylbenz(a)anthracene and croton oil-induced skin carcinogenesis in mice. Various studies suggest that the consumption of kaempferol-rich foods may reduce the risk of developing different types of cancer, including lung cancer, gastric cancer, pancreatic cancer, and ovarian cancer (Calderon-Montano et al. 2011). Malignant tumors are known to activate angiogenesis, which is necessary for the formation of solid tumors, and kaempferol has been shown to inhibit angiogenesis in vitro (Ahn et al. 2009). It is recognized that inflammatory diseases increase the risk of developing different types of cancer, including bladder, cervical, gastric, esophageal, ovarian, prostate, as well as thyroid cancer. By P-glycoprotein-mediated efflux, kaempferol has found to reduce cellular level of carcinogens (Phang et al. 1993). Numerous reports have shown that kaempferol as well as kaempferol glycosides induce cell death in various cancer cells from different tissues, including lung, breast, colon, prostate, pancreas, blood, skin, esophagus, brain uterus, ovary, thyroid, and bone (Calderon-Montano et al. 2011). Caspases, family of cysteine proteases involved in the initiation of apoptosis, and kaempferol have been found to induce the activation of caspase-3 (Bestwick et al. 2007), caspase-7 (Kang et al. 2009), and caspase-9 (Zhang et al. 2009). The antiproliferative effects of kaempferol in cancer cells may also inhibit MAPK/ERK pathway (Jeong et al. 2009). Although low concentrations of kaempferol can reduce the cellular levels of ROS and induce antioxidant effects, higher concentrations of this flavonoid are known to generate ROS (Marfe et al. 2009). Therefore, kaempferol could be used in combination with several anticancer drugs to improve their therapeutic effects. It has been reported that kaempferol can sensitize cancer cells to the cytotoxic effects of cisplatin (Luo et al. 2010), 5-fluorouracil (Zhang et al. 2008), cytarabine (Nadova et al. 2007), doxorubicin (Sharma et al. 2007), mitoxantrone, and the active metabolite of irinotecan (SN-38) (Imai et al. 2004). This flavonoid can also enhance the cytotoxic effects of tumor necrosis factor-related apoptosis-inducing ligand (TRAIL) (Yoshida et al. 2008). These investigations suggest that kaempferol may have clinical applications as adjuvant therapy in the treatment of some cancers.

Functions of Kaempferol

(a) (HO)-1

Kaempferol-mediated MAPK activation can help in preventing DNA damage leading to cell transformation. Kaempferol was found to increase heme oxygenase (HO)-1 gene expression, which increases antioxidant capacity of cells (Hong et al. 2009). Therefore, kaempferol treatment heightened cell viability in response to oxidative stress, which includes unstable radicals prone to harm DNA. In this way, kaempferol-induced MAPK initiation protects healthy cells from transforming into cancerous ones.

(b) RSK2

Along with activation of pathway, kaempferol also modifies different proteins involved in it. RSK2 is a key suppressor of apoptosis which downregulates the apoptosis-promoting protein BAD and upregulates Bcl-2 protein (Luo et al. 2011). Kaempferol binds directly to the RSK2 protein, specifically at the Val^{82} and the Lys^{100} sites (Cho et al. 2009). Thereby kaempferol paralyzes the RSK2 protein. As expected, treatment was reported to drop Bcl levels and boost concentrations of tumor-suppressor proteins BAD and p53 (Luo et al. 2011).

(c) Src

Kaempferol disrupts Src kinase activity (Lee et al. 2010). Src activates MAPK which turns on the COX-2 protein, which is a warning marker for skin tumors (Athar et al. 2001). UVB radiation is one of the major contributors to Src kinase activity. Kaempferol, on the other hand, is a potential inhibitor of Src, which requires ATP molecule in order to function. Kaempferol binds to Src at its ATP site, disturbing its skin cancer-promoting activity. Therefore, the MAPK/ERK pathway is altered at various locations by kaempferol (Fig. 4.2).

Other Uses of Bauhinia

It also helps to enrich soil fertility by nitrogen fixing. Leaves and buds are eaten as vegetable. It is also used for coloring furniture dying fiber. The flowers have laxative action (Figs. 4.3, 4.4, 4.5, 4.6, 4.7, 4.8, and 4.9).

Commercial Considerations and Future Prospects

The conventional assumption is committed on the presumption that human diseases have a assertion genetic basis across patient's populations; therefore a single drug is satisfactory to treat a disease. However, recent advances in genomics exhibit that different patient populations may require disparate drugs to their treatment, as illustrated by medicine. Moreover, the deficit in acknowledged new chemical existence together with a focus of the pharmaceutical industry and the rearrangement of companies from uncertain research to a more reliable businesses and revenues had spawn a crisis and loss of conviction of public presumption in the chemical industry. The practice of herbal medicines overtures a way to mitigate this crisis in drug advancement. There are two preeminent approaches for herbal medicine. First one is that promoting the herbal medicine knowledge may give rise to a reasonable and more prompt discovery of recent drugs, and second is that herbal remedies offer a comprehensive approach that accompanies the disease-targeted approach of chemicals. The main prejudice related to herbal medicines is the depletion of international standardization in terms of proceedings for assessing their composition, efficacy,

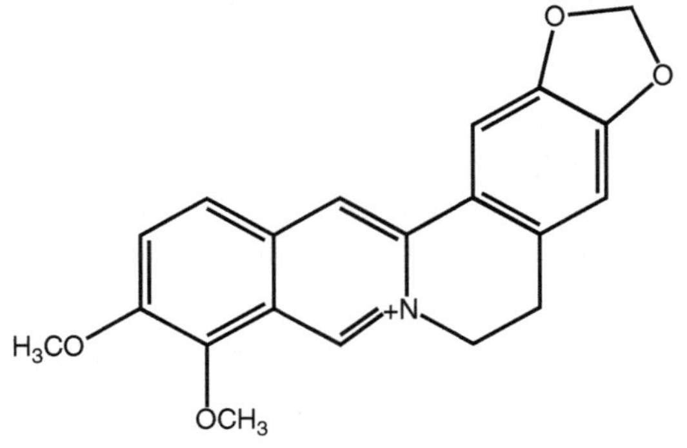

Fig. 4.3 Structure of berberine (Kaboli et al. 2014)

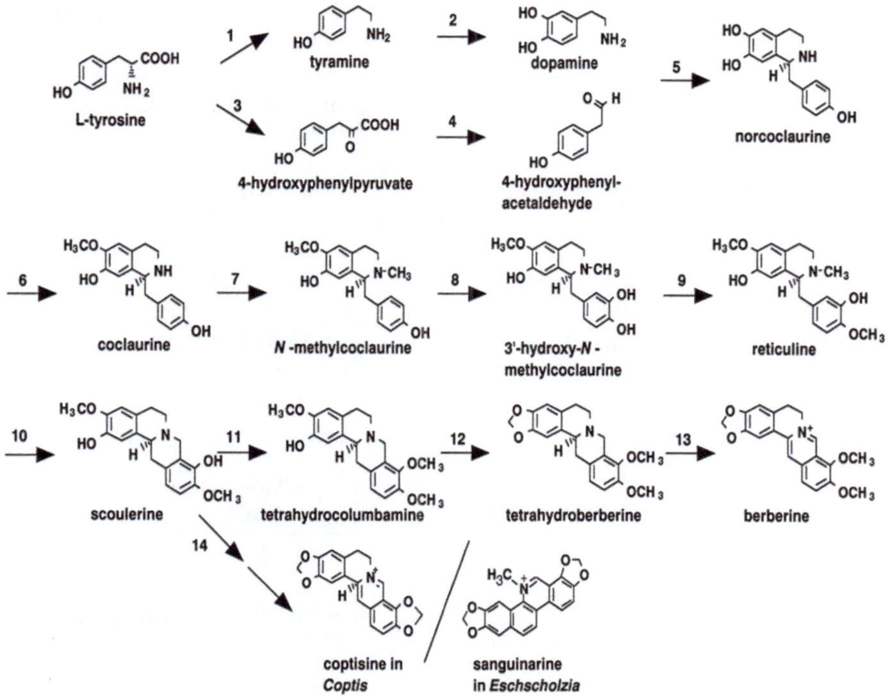

Fig. 4.4 Schematic biosynthetic pathway of berberine alkaloid. (*1*) L-Tyrosine decarboxylase; (*2*) phenolase; (*3*) L-tyrosine transaminase; (*4*) *p*-hydroxyphenylpyruvate decarboxylase; (*5*) (*S*)-norcoclaurine synthase; (*6*) (*S*)-adenosyl-L-methionine:norcoclaurine 6OMT; (*7*) (*S*)-adenosyl-L-methionine:coclaurine *N*-methyltransferase; (*8*) *N*-methylcoclaurine-4'-hydroxylase; (*9*) *S*-adenosyl-L-methionine:3'-hydroxy-*N*-methylcoclaurine 4'OMT; (*10*) berberine bridge enzyme; (*11*) *S*-adenosyl-L-methionine:scoulerine 9-*O*-methyltransferase; (*12*) (*S*)-canadine synthase; (*13*) tetrahydroprotoberberine oxidase; (*14*) cheilanthifolin synthase (Sato et al. 2001)

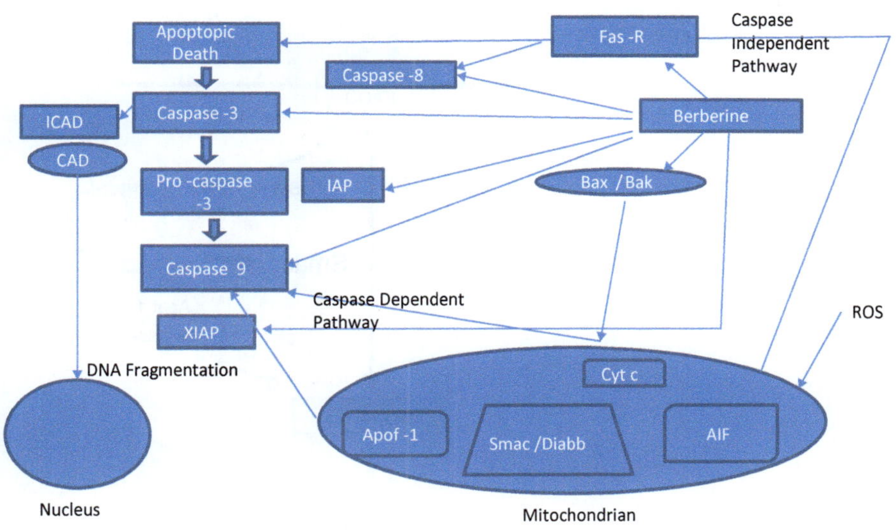

Fig. 4.5 Apoptosis pathway induced by berberine

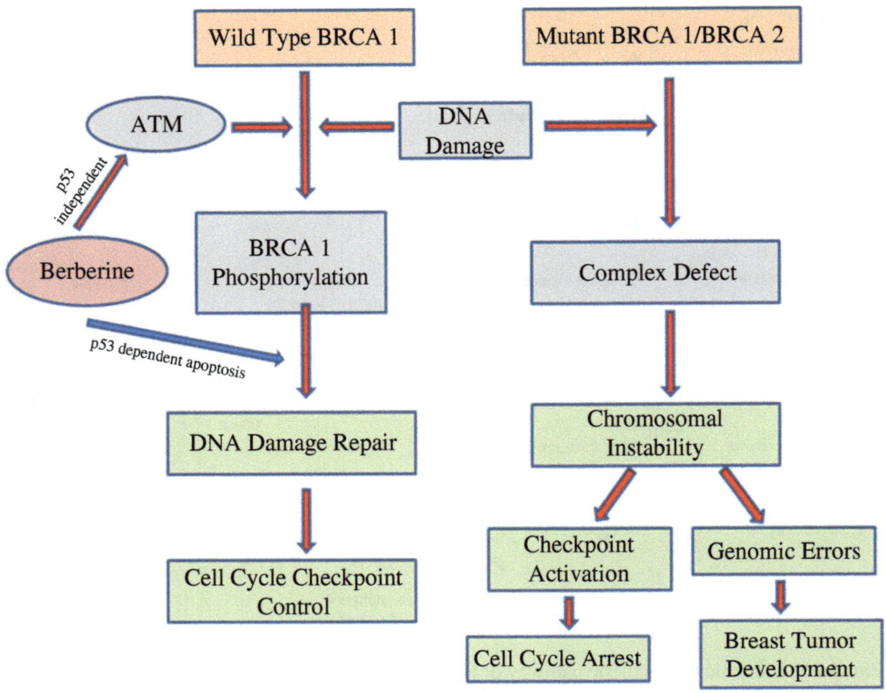

Fig. 4.6 Different checkpoints of berberine

Fig. 4.7 Apoptosis pathway at the level of mitochondria

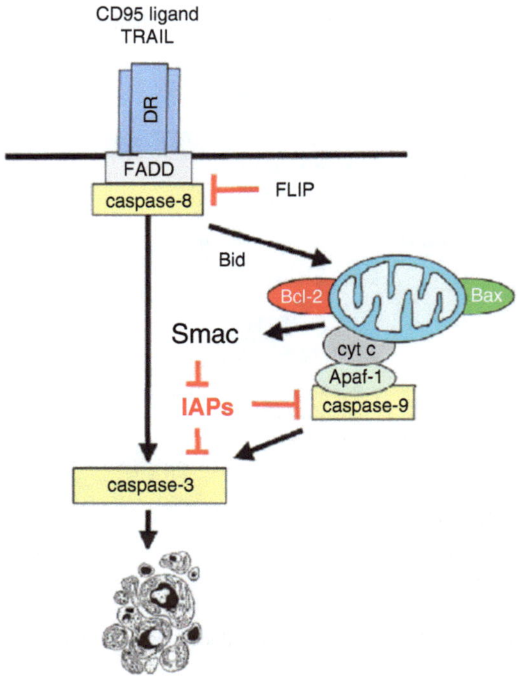

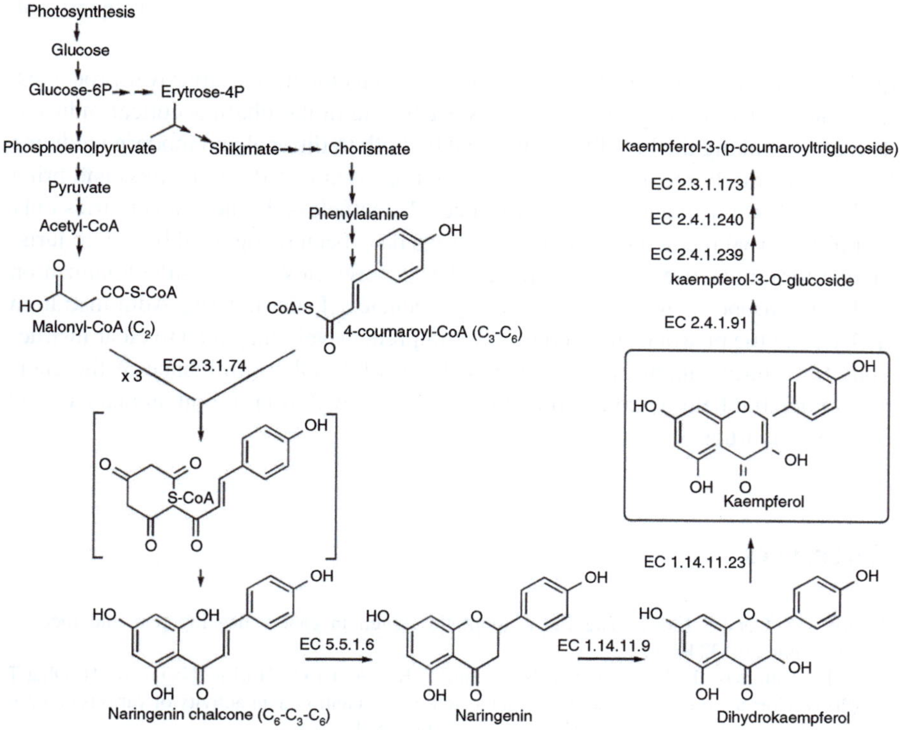

Fig. 4.8 Biosynthesis of kaempferol and some glycosides of kaempferol (see text for further details). EC 2.3.1.74: chalcone synthase; EC 5.5.1.6: chalcone isomerase; EC 1.14.11.9: flavanone 3-dioxygenase; EC 1.14.11.23: flavonol synthase; EC 2.4.1.91: flavonol 3-O-glucosyltransferase; EC 2.4.1.239: flavonol-3-O-glucoside glucosyltransferase; EC 2.4.1.240: flavonol-3-O-diglycoside glucosyltransferase; EC 2.3.1.173: flavonol-3-O-triglucoside *p*-coumaroyltransferase. (Calderon-Montano et al. 2011)

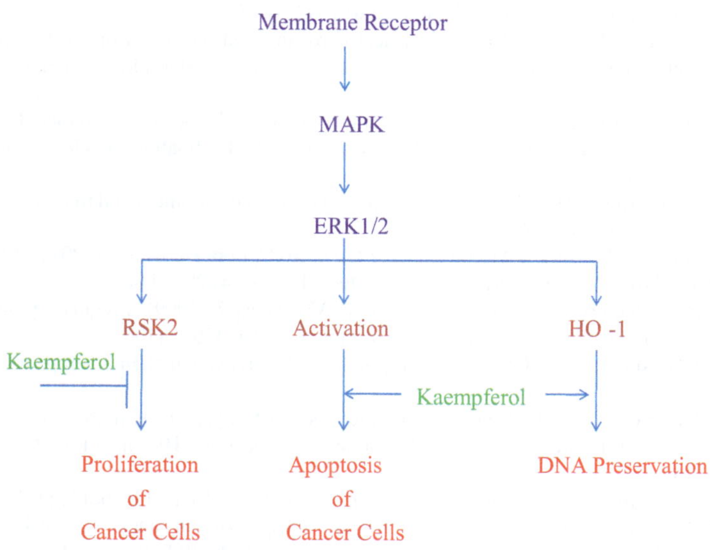

Fig. 4.9 The MAPK pathway plays an integral role in both the promotion and regulation of cell growth

quality, persistent manufacturing practices, management, and approval processes. Vast knowledge in drug development is accessible in the pharmaceutical industry. Therefore, linking the benefits contributed by both traditional and modern medicine has been formerly recommended as a rising approach in order to confess and bring to the market new plant-derived substances. Nonetheless, in the last centuries only manifold herbal medicines or botanical drugs have been recognized by health jurisdiction for human use. Collusion and gradation between World Health Organization (WHO), European and other regulatory agencies, Federal Drug Administration (FDA), and the pharmaceutical industry comprehensively may lead to clear instructions for reinforcement of herbal medications while taking influence of the enormous potential held by conventional medicine for evolution of both anticancer and other health drugs.

References

Adams JM, Cory S (2007) The Bcl-2 apoptotic switch in cancer development and therapy. Oncogene 26:1324–1337

Ahn MR, Kunimasa K, Kumazawa S, Nakayama T, Kaji K, Uto Y, Hori H, Nagasawa H, Ohta T (2009) Correlation between antiangiogenic activity and antioxidant activity of various components from propolis. Mol Nutr Food Res 53(5):643–651

Alakurtti S, Makela T, Koskimies S, Yli-Kauhaluoma J (2006) Pharmacological properties of the ubiquitous natural product betulin. Eur J Pharm Sci 29:1–13

Andre N, Carre M, Brasseur G, Pourroy B, Kovacic H, Briand C, Braguer D (2002) Paclitaxel targets mitochondria upstream of caspase activation in intact human neuroblastoma cells. FEBS Lett 532:256–260

Athar M, An KP, Morel KD, Kim AL, Aszterbaum M, Longley J, Epstein EH Jr, Bickers DR (2001) Ultraviolet B(UVB)-induced cox-2 expression in murine skin: an immunohistochemical study. Biochem Biophys Res Commun 280:1042–1047

Bestwick CS, Milne L, Duthie SJ (2007) Kaempferol induced inhibition of HL-60 cell growth results from a heterogeneous response, dominated by cell cycle alterations. Chem Biol Interact 170(2):76–85

Bhadra K, Maiti M, Kumar GS (2007) Molecular recognition of DNA by small molecules: AT base pair specific intercalative binding of cytotoxic plant alkaloid palmatine. Biochim Biophys Acta 1770(7):1071–1080

Brown M, Attardi LD (2005) The role of apoptosis in cancer development and treatment response. Nat Rev Cancer 5:231–237

Calderon-Montano JM, Burgos-Morón E, Pérez-Guerrero C, López-Lázaro M (2011) A review on the dietary flavonoid kaempferol. Mini-Rev Med Chem 11(4):298–344

Cho YY, Yao K, Pugliese A, Malakhova ML, Bode AM, Dong Z (2009) A regulatory mechanism for RSK2 NH2-terminal kinase activity. Cancer Res 69(10):4398–4406

Cichewicz RH, Kouzi SA (2004a) Chemistry, biological activity, and chemotherapeutic potential Acid. Neoplasia 7:162–170

Cichewicz RH, Kouzi SA (2004b) Chemistry, biological activity, and chemotherapeutic potential of betulinic acid for the prevention and treatment of cancer and HIV infection. Med Res Rev 24:90–114

Elmore S (2007) Apoptosis: a review of programmed cell death. Toxicol Pathol 35:495–516

Evan GI, Vousden KH (2001) Proliferation, cell cycle and apoptosis in cancer. Nature 411:342–348

Ferlay J, Shin HR, Bray F, Forman D, Mathers C, Parkin DM (2010) Estimates of worldwide burden of cancer in 2008: GLOBOCAN2008. Int J Cancer 127:2893–2917

Fulda S, Debatin KM (2004) Targeting apoptosis pathways in cancer therapy. Curr Cancer Drug Targets 4:569–576

Fulda S, Scaffidi C, Susin SA, Krammer PH, Kroemer G, Peter ME, Debatin KM (1998a) Activation of mitochondria and release of mitochondrial apoptogenic factors by betulinic acid. J Biol Chem 273:33942–33948

Fulda S, Susin SA, Kroemer G, Debatin KM (1998b) Molecular ordering of apoptosis induced by anticancer drugs in neuroblastoma cells. Cancer Res 58:4453–4460

Galluzzi L, Larochette N, Zamzami N, Kroemer G (2006) Mitochondria as therapeutic targets for cancer chemotherapy. Oncogene 25:4812–4830

Green DR, Kroemer G (2004) The pathophysiology of mitochondrial cell death. Science 305:626–629

Green DR, Knight RA, Melino GA, Finazzi-Agro S (2004) Ten years of publication in cell death. Cell Death Differ 11:2–3

Hanahan D, Weinberg RA (2000) The hallmarks of cancer. Cell 100:57–70

Hong JT, Yen JH, Wang L, Lo YH, Chen ZT, Wu MJ (2009) Regulation of heme oxygenase-1 expression and MAPK pathways in response to kaempferol and rhamnocitrin in PC12 cells. Toxicol Appl Pharmacol 237:59–68

Imai Y, Tsukahara S, Asada S, Sugimoto Y (2004) Phytoestrogens/flavonoids reverse breast cancer resistance protein/ABCG2-mediated multidrug resistance. Cancer Res 64(12):4346–4352

Jemal A, Center MM, De Santis C, Ward EM (2010) Global patterns of cancer incidence and mortality rates and trends. Cancer Epidemiol Biomark Prev 19(8):1893–1907

Jemal A, Bray F, Center MM, Ferlay J, Ward E, Forman D (2011) Global cancer statistics. CA Cancer J Clin 61:69–90

Jeong JC, Kim MS, Kim TH, Kim YK (2009) Kaempferol induces cell death through ERK and Akt-dependent down-regulation of XIAP and survivin in human glioma cells. Neurochem Res 34(5):991–1001

Kanak S, Verma Anita K (2012) Evaluation of antimicrobial and anticancer activities of methanol extract of in vivo and in vitro grown *Bauhinia variegata* L. International Research. J Biol Sci 1(6):26–30

Kang GY, Lee ER, Kim JH, Jung JW, Lim J, Kim SK, Cho SG, Kim KP (2009) Downregulation of PLK-1 expression in kaempferol-induced apoptosis of MCF-7 cells. Eur J Pharmacol 611(1):17–21

Kasperczyk H, La Ferla-Bruhl K, Westhoff MA, Behrend L, Zwacka RM, Debatin KM, Fulda S (2005) Betulinic acid as new activator of NF-kappaB: molecular mechanisms and implications for cancer therapy. Oncogene 24:6945–6956

Kavya B, Kavya N, Ramarao V, Venkatateshwar G (2015) Tinospora cordifolia (Willd.) Miers: nutritional, ethanomedical and therapeutic utility. Ind J Res Ayurveda Pharm 6(2):195–198

Laszczyk MN (2009) Pentacyclic triterpenes of the lupane, oleanane and ursane group as tools in cancer therapy. Planta Med 75:1549–1560

Lee KM, Lee KW, Jung SK, Lee EJ, Heo YS, Bode AM, Lubet RA, Lee HJ, Dong Z (2010) Kaempferol inhibits UVB-induced COX-2 expression by suppressing Src kinase activity. Biochem Pharmacol 80(12):2042–2049

Luo H, Daddysman MK, Rankin GO, Jiang BH, Chen YC (2010) Kaempferol enhances cisplatin's effect on ovarian cancer cells through promoting apoptosis caused by down regulation of cMyc. Cancer Cell Int 10(1):1

Luo H, Rankin GO, Li Z, Depriest L, Chen YC (2011) Kaempferol induces apoptosis in ovarian cancer cells through activating p53 in the intrinsic pathway. Food Chem 128(2):513–519

Mantena SK, Sharma SD, Katiyar SK (2006) Berberine inhibits growth, induces G1 arrest and apoptosis in human epidermoid carcinoma A431 cells by regulating Cdki–Cdk-cyclin cascade, disruption of mitochondrial membrane potential and cleavage of caspase 3 and PARP. Carcinogenesis 27(10):2018–2027

Marfe G, Tafani M, Indelicato M, Sinibaldi-Salimei P, Reali V, Pucci B, Fini M, Russo MA (2009) Kaempferol induces apoptosis in two different cell lines via Akt inactivation, Bax and SIRT3 activation, and mitochondrial dysfunction. J Cell Biochem 106(4):643–650

Mittal J, Sharma MM, Batra A (2014) *Tinospora cordifolia*: a multipurpose medicinal plant-A review. J Med Plant Stud 2(2):32–47

Nadova S, Miadokova E, Cipak L (2007) Flavonoids potentiate the efficacy of cytarabine through modulation of drug-induced apoptosis. Neoplasma 54(3):202

Orwa C, Mutua A, Kindt R, Jamnadass R, Anthony S (2009) Agroforestree database: a tree reference and selection guide version 4.0. http://www.worldagroforestry.org/sites/treedbs/treedatabases.asp

Phang JM, Poore CM, Lopaczynska J, Yeh GC (1993) Flavonol-stimulated efflux of 7,12-dimethylbenz (a) anthracene in multidrug-resistant breast cancer cells. Cancer Res 53(24):5977–5981

Pisha E, Chai H, Lee IS, Chagwedera TE, Farnsworth NR, Cordell GA, Beecher CW, Fong HH, Kinghorn AD, Brown DM (1995) Discovery of betulinic acid as a selective inhibitor of human melanoma that functions by induction of apoptosis. Nat Med 1:1046–1051

RajKapoor B, Jayakar B, Murugesh N (2003) Antitumor activity of *Bauhinia variegata* on Dalton's ascitic lymphoma. J Ethnopharmacol 89:107–109

RajKapoor B, Jayakar B, Murugesh N, Sakthisekaran D (2006) Chemoprevention and cytotoxic effect of *Bauhinia variegate* against N-nitrosodiethylamine induced liver tumors and human cancer cell lines. J Ethnopharmacol 104:407–409

Reddy NM, Reddy R (2015) Tinospora cordifolia chemical constituents and medicinal properties: a review. Sch Acad J Pharm 4(8):364–369

Saelens X, Festjens N, Vande Walle L, van Gurp M, van Loo G, Vandenabeele P (2004) Toxic proteins released from mitochondria in cell death. Oncogene 23:2861–2874

Salvesen GS, Duckett CS (2002) IAP proteins: blocking the road to death's door. Nat Rev Mol Cell Biol 3:401–410

Sato F, Hashimoto T, Hachiya A, Tamura KI, Choi KB, Morishige T, Yamada Y (2001) Metabolic engineering of plant alkaloid biosynthesis. Proc Natl Acad Sci 98(1):367–372

Sawada N, Kataoka K, Kondo K, Arimochi H, Fujino H, Takahashi Y, Miyoshi T, Kuwahara T, Monden Y, Ohnishi Y (2004) Betulinic acid augments the inhibitory effects of vincristine on growth and lung metastasis of B16F10 melanoma cells in mice. Br J Cancer 90:1672–1678

Selzer E, Pimentel E, Wacheck V, Schlegel W, Pehamberger H, Jansen B, Kodym R (2000) Effects of betulinic acid alone and in combination with irradiation in human melanoma cells. J Invest Dermatol 114:935–940

Selzer E, Thallinger C, Hoeller C, Oberkleiner P, Wacheck V, Pehamberger H, Jansen B (2002) Betulinic acid-induced Mcl-1 expression in human melanoma—mode of action and functional significance. Mol Med 8:877–884

Sharma V, Joseph C, Ghosh S, Agarwal A, Mishra MK, Sen E (2007) Kaempferol induces apoptosis in glioblastoma cells through oxidative stress. Mol Cancer Ther 6(9):2544–2553

Sinha K, Mishra NP, Singh J, Khanuja SPS (2004) Tinospora cordifolia (Guduchi), a reservoir plant for therapeutic applications: a review. Indian J Tradit Knowl 3(3):257–270

Taiz L, Zeiger E (2006) Plant Physiology. Sinauer Associated Inc, Massachusetts

Takada Y, Aggarwal BB (2003) Betulinic acid suppresses carcinogen-induced NF-kappa B activation through inhibition of I kappa B alpha kinase and p65 phosphorylation: abrogation of cyclooxygenase-2 and matrix metalloprotease-9. J Immunol 171:3278–3286

Tan Y, Yu R, Pezzuto JM (2003) Betulinic acid-induced programmed cell death in human melanoma cells involves mitogen-activated protein kinase activation. Clin Cancer Res 9:2866–2875

Thurnher D, Turhani D, Pelzmann M, Wannemacher B, Knerer B, Formanek M, Wacheck V, Selzer E (2003) Betulinic acid: a new cytotoxic compound against malignant head and neck cancer cells. Head Neck 25:732–740

Umadevi M, Kumar KS, Bhowmik D, Duraivel S (2013) Traditionally used anticancer herbs in India. J Med Plant Stud 1(3):56–74

Veeraiah S, Reddy KJ (2012) Current strategic approaches in ethnomedicinal plants of Tinospora cordifolia and Gloriosa superb—A review. Int J Pharm Bio Sci 3(2):320–326

Wick W, Grimmel C, Wagenknecht B, Dichgans J, Weller M (1999) Betulinic acid-induced apoptosis in glioma cells: a sequential requirement for new protein synthesis, formation of reactive oxygen species, and caspase processing. J Pharmacol Exp Ther 289:1306–1312

Yoshida T, Konishi M, Horinaka M, Yasuda T, Goda AE, Taniguchi H, Yano K, Wakada M, Sakai T (2008) Kaempferol sensitizes colon cancer cells to TRAIL-induced apoptosis. Biochem Biophys Res Commun 375(1):129–133

Zhang Y, Chen AY, Li M, Chen C, Yao Q (2008) *Ginkgo biloba* extract kaempferol inhibits cell proliferation and induces apoptosis in pancreatic cancer cells. J Surg Res 148(1):17–23

Zhang Q, Zhao XH, Wang ZJ (2009) Cytotoxicity of flavones and flavonols to a human esophageal squamous cell carcinoma cell line (KYSE-510) by induction of G 2/M arrest and apoptosis. Toxicol in Vitro 23(5):797–807

Zuco V, Supino R, Righetti SC, Cleris L, Marchesi E, Gambacorti-Passerini C, Formelli F (2002) Selective cytotoxicity of betulinic acid on tumor cell lines, but not on normal cells. Cancer Lett 175:17–25

Chapter 5
Anticancer Drugs from Plants

Denise Fernandes Coutinho Moraes, Ludmilla Santos Silva de Mesquita, Flavia Maria Mendonça do Amaral, Maria Nilce de Sousa Ribeiro, and Sonia Malik

Abstract Cancer is a major problem of public health and one of the main causes of death around the globe. According to World Health Organization, the prevalence of this disease is rising, however, more rapidly in Africa, Asia, and Central and South America that account for about 70% of cancer deaths in the world. The chemotherapy is one of the ways to treat this disease and the advances in anticancer drugs have improved patient care. Plants have been used to treat different diseases since ancient times. Among the anticancer drugs, about 50% come from natural products as isolated or semisynthetic or related synthetic compounds and plants represent important source of these substances. Taxol, vinca alkaloids, camptothecin, and podophyllotoxins, as well as their semisynthetic or synthetic derivatives, are the most important anticancer drugs obtained from plants. In this chapter, we review the importance of plants as source of drugs and describe these anticancer compounds. The continuing search for antitumor agents from plants is extremely necessary to find the possible ways to have safe and more effective treatment for this health problem.

Keywords Antitumoral compounds • Medicinal plants • Neoplasm • Taxol • Vinca alkaloids

D.F.C. Moraes (✉) • L.S.S. de Mesquita • F.M.M. do Amaral • M.N. de Sousa Ribeiro
Department of Pharmacy, Biological and Health Sciences Center, Federal University of Maranhão, São Luís, MA 65085-580, Brazil
e-mail: deniseufma2013@gmail.com; ludmilla.ssm@gmail.com; flavia.amaral@ufma.br; mnribeiro@ufma.br

S. Malik (✉)
Graduate Program in Health Sciences, Biological and Health Sciences Center, Federal University of Maranhão, São Luís, MA 65085-580, Brazil
e-mail: 777soniamalik@gmail.com

S. Malik (ed.), *Biotechnology and Production of Anti-Cancer Compounds*,
DOI 10.1007/978-3-319-53880-8_5

Introduction

Cancer is a disease that occurs when normal cells begin to grow in disordered and uncontrolled manner, forming tumor masses. It can affect people of different gender, social class, and age; however its risk increases with the age. Cancer is one of the main causes of death in the world (Fonseca and Farías 2016; Sena et al. 2016). According to the World Health Organization (WHO 2015), there were 14 million new cases of cancer worldwide in 2012, and among them, 8.2 million have ended to death. This organization also cited that the prevalence of this disease is still rising and the number of cases can rise about 70% in the next two decades. Although it is an important public health problem in both developed and developing countries, the recent studies have demonstrated that the rates have increased more in Africa, Asia, and Central and South America that account for 70% of cancer deaths in the world (Sterwart and Wild 2014).

There are several types of cancers, which are classified according to the type/origin of affected cell in the organism. Theoretically, all cells in the human body may show flaws in their division and initiate this disordered growth. Among men, the cancers with more prevalence are lung, prostate, colorectal, stomach, and liver, and among women, normally the breast, rectum, lung, cervix, and stomach cancer has been more commonly diagnosed (WHO 2015). If we consider the ability to lead to death, the types of cancer with the highest mortality rates are lung, stomach, liver, rectum, and breast (Sena et al. 2016).

This global pandemic is considered multifactorial and its main risk factors are age, genetic, and some issues linked to behavior. Smoking, obesity, lack of physical exercise, and high-fat diet are some of the triggering factors that can cause failure in the division of normal cells (de Sant'Ana et al. 2016). The genetic issue has been much studied recently and the discovery of mutations of specific genes important for oncogenesis has shown perspectives in the early treatment of this disease, increasing the cure rates or allowing procedures that prevent the tumor (Xue et al. 2014; Futreal et al. 2004). Some cancers are caused by virus like the cervix cancer that is associated with the presence of certain types of human papillomavirus (HPV) which is transmitted sexually. HPV is also linked with other types of cancer such as anus, vulva, vagina, penis, oropharynx, and rectum which have been growing lately (Rodrigues et al. 2016). The hepatitis B virus (HBV) is extremely linked with development of liver cancer that is the fifth most common cancer in the world (Du et al. 2012). The several types of skin cancer are caused mainly by the expose to UV radiation (McWhirter and Hoffman-Goetz 2016). Occupational factors are also important in development of tumor processes when it exposes workers to carcinogenic substances in their workplace. Lung and skin cancer are the most common occupational cancers (Petter 2015; Labreche et al. 2014).

The World Health Organization has suggested the governments of all countries to invest in campaigns for preventing cancer. For skin cancer, the campaigns aim to inform measures to protect against ultraviolet rays such as the use of sun block and clothes resistant against this irradiation. For breast cancer, early diagnosis is essential for the treatment of patients, so the campaigns guide the self-examination and inform the need to perform annual mammography after 40 years (de Sant'Ana et al. 2016). To prevent cervical cancer, which occurs from infection with human papillomavirus, it is extremely important to increase the rate of vaccination against this virus, in addition to normal campaigns to prevent sexually transmitted diseases. This type of cancer can also be prevented with annual Pap smear test of the cervix, known as Papanicolaou test to detect precancerous cells that can be treated before the occurrence of cancer (Rodrigues et al. 2016). Taking balanced diet, avoiding smoking, and doing physical exercise are also general procedures to avoid cancer (Wang et al. 2012).

There are different strategies for cancer treatment that depend on several factors including type, patient's age, and stage of disease. The treatment for cancer has improved recently due to advancements and the patients have more chances to be cured. These different approaches include surgery, use of antineoplastic drugs (chemotherapy), radiation, immunotherapy, stem cell transplant, laser technique, blood production transfusion, photodynamic therapy, and hyperthermia (Weeks et al. 2014; Volgin et al. 2014; Holgado et al. 2015).

Chemotherapy is the use of drugs to treat cancer. In fact, this term is also used when any drug is used in treatment of disease; however it is more often to refer to cancer. This procedure can cure, control, or just relieve some symptoms as a palliative. In this approach, one or several drugs may be used. The most important feature of anticancer drugs is not to affect the normal cells and should possess a wide range between safety and effectiveness of doses (Polovich et al. 2014).

Among the anticancer drugs used these days, about 50% come from the natural products as isolated or semisynthetic or related synthetic compounds and plants represent important source of these substances (Newman and Cragg 2012; Malik et al. 2014a). Besides the usage of plants for isolation of anticancer drugs, many vegetal species are used for cancer prevention. So, for both approaches, plants have played an important role in cancer therapy. However, it is important to check that the bioactive compounds obtained from plants have studied for their effectiveness against cancer cells without any harmful effect on other normal cells of the body. Use of plants without this kind of evaluation should be avoided due to the risk of toxicity.

Plants as Source of Drugs

Plants have been used to treat different diseases since the time immemorial. After the development of modern chemistry by the nineteenth century, scientists started identifying the compounds that have therapeutic properties in plants (Shankari and Gurunathan 2015).

Morphine, an alkaloid with analgesic activity from the opium resin (*Papaver somniferum*), was isolated by German Friedrich Serturner in 1804. It marked for the first isolation of natural products, and in the middle of 1820s, Heinrich Emanuel

Merck commercialized this compound as a pain reliever. The structure of this substance was identified in the beginning of nineteenth century and although having addictive property morphine has been used as a sedative for the past 200 years (Huxtable and Schwarz 2001).

After the discovery and the great success of morphine, many other scientists isolated several compounds from plants. *Salix alba* L., known as white willow, has been used for pain and fever relief since ancient times. Attempts to isolate the active substance from white willow stem bark extract began in the early nineteenth century, but only in 1828, the purification of salicin was possible by the chemist Buchner from Monique University. The pure form of these crystals was obtained by the pharmacist Henri Leroux in 1829, who demonstrated its therapeutic use in the treatment of rheumatism. Later in 1935, it has been discovered with acid hydrolysis by an Italian chemist, Raffaelle Pirin, that the salicin was comprised by a sugar and an aromatic alcohol, known as salicyl alcohol. The double oxidization of this compound leads to salicylic acid that has been used for analgesic and antipyretic purposes. Due to its capacity to promote gastric irritation, Felix Hoffmann, in 1897, converted the salicylic acid into acetylsalicylic acid by acetylation of its hydroxyl group. The salicylic acid was registered by Bayer Company as the commercial name aspirin in 1899 (Mahdi et al. 2006). According to literature, Hoffmann's father used the salicylic acid by oral administration to relieve the rheumatism; however he suffered with the stomach issues caused by this molecule and because of this the scientist decided to decrease the acidic property of this substance and discovered one of the most popular drugs used worldwide (Golberg 2009). Although the idea for aspirin came from natural product salicin, nowadays this important drug is obtained by synthesis.

Quinine is another example of natural product isolated from the bark of *Cinchona* species in the eighteenth century and was the first compound used to treat malaria caused by *Plasmodium falciparum*. This bark is indigenous from Peru in South America and it was introduced in Europe probably in the seventeenth century to treat malaria fever. In 1820, Pierre Joseph Pelletier and Joseph Bienaimé Caventou isolated this alkaloid as the main compound of the *Cinchona* bark. The scientists named it quinine after the Inka name of the *Cinchona* species quina (Hoogte and Pieters 2014; Achan et al. 2011; Kinsley-Scott and Norton 2003). Besides quinine, other cinchona alkaloids including quinidine, cinchonine, and cinchonidine have also been used as antimalarial drugs (Goss 2014). During the mid-nineteenth century, it became possible to synthesize quinine, which could serve as a starting point for the synthesis of other antimalarial drugs such as chloroquine and primaquine (de Sa 2011). Recently the Chinese researcher Youyou Tu won the Nobel prize in Medicine in 2015, for isolating drug Artemisinin from *Artemisia annua* that was introduced to treat cases of malaria where the quinine and their derivatives were resistant (Rao et al. 2015). Artemisinin and its derivatives are still being used for the treatment of malaria in many countries (O'Neill and Posner 2004).

Nowadays, the modern therapy all over the world has a great variety of compounds with different proven pharmacological activities and this therapeutic action could not be possible without the contribution of natural products, especially of plants. Among all modern drugs available, it has been estimated that 40% come from natural products, and out of this, 25% is directly or indirectly are plant derivatives (Giachetti and Monti 2005). A total of 547 natural products and their derivatives have been approved for use as therapeutics by the Food and Drug Administration, USA, in 2013, corresponding a 38% of the total of all new molecular entities approved. These products include the ones from mammalians, plants, bacteria, fungi, and marine organisms (Patridge et al., 2016). According to Veeresham (2012), among the drugs considered essential for World Health Organization (WHO), 11% are derivatives from plants.

Thus, the natural products including plants play a great role in therapy especially in the development of some classes of medicines such as antitumor ones. The aim of this chapter is to review the main natural compounds derived from plant species which have been used as anticancer drugs and approved by pharmaceutical regulatory agencies around the world.

Anticancer Drugs from Plant Sources

The natural products derived from plants have been used as active compounds to treat several diseases since a long time. The secondary or natural compounds derived from plants play an important role in the pharmacological activities, including the antineoplasic action. Taxol, vinca alkaloids, camptothecin, podophyllotoxins, as well as their semisynthetic or synthetic derivatives are the most important examples of this cooperation (Malik et al. 2014b; Malik et al. 2013; Cragg and Newman 2013; Newman and Cragg 2012).

Vinca *Alkaloids*

Catharanthus roseus (L.) G. Don, commonly known as periwinkle (family Apocynaceae) has gained pharmaceutical interest because of the presence of secondary metabolites, mainly alkaloids (Fig. 5.1). The most important alkaloids isolated from this plant species are vinblastine (**1**) and vincristine (**2**), two bisindole alkaloids, which were isolated and identified by Robert Noble and Charles Beer in 1950s and are known to possess strong anticancer activities (Yang et al. 2016). They inhibit the tubulin polymerization of tumor cells and also cause mitotic spindle destruction.

The extraction and purification of these alkaloids are quite laborious and expensive due to their extremely low concentration in *C. roseus* leaves (Noble 1990). However, these alkaloids are still isolated by semisynthetic routes using undesired and inactive alkaloids in the *C. roseus* leaves (Salim and de Luca 2013).

The molecular structures of vinblastine and vincristine are very similar, differing just for one moiety at nitrogen. Vinblastine has a methyl group at this nitrogen while

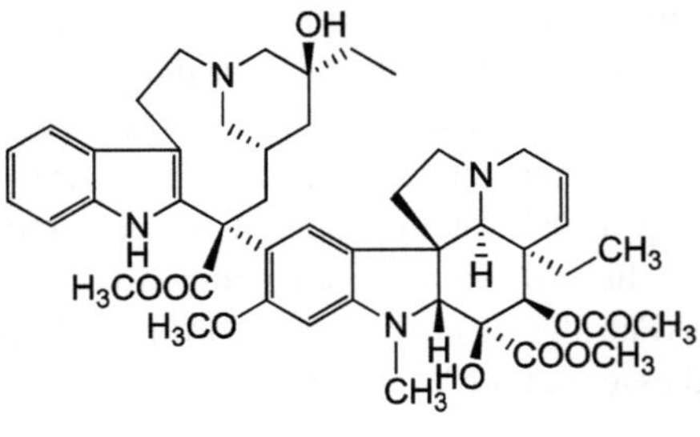

1 - vinblastin

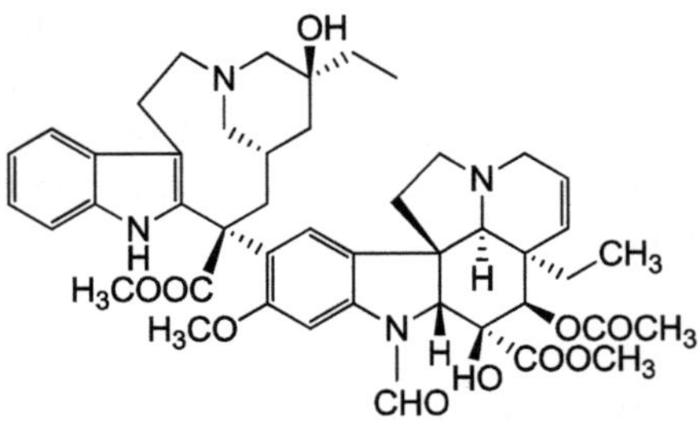

2 - vincristin

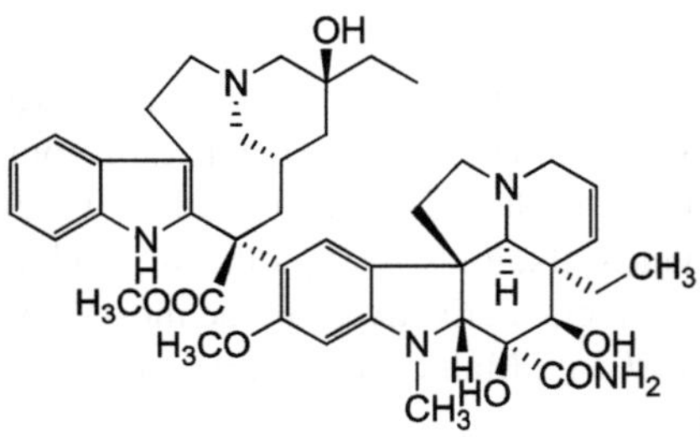

3 - vindesin

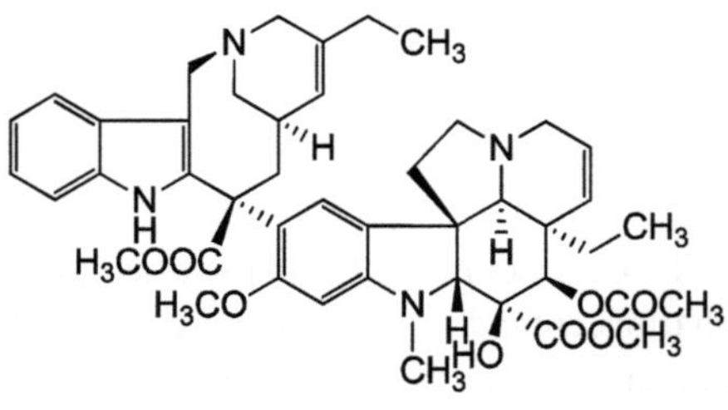

4 - vinolrebin

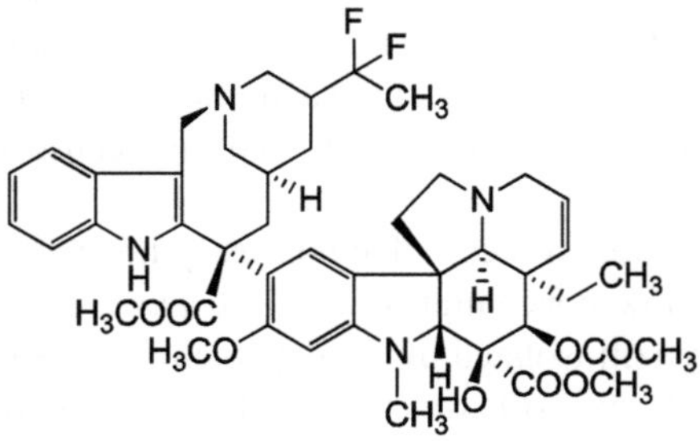

5 - vinfluvine

Fig. 5.1 Anticancer compounds derived from plant sources

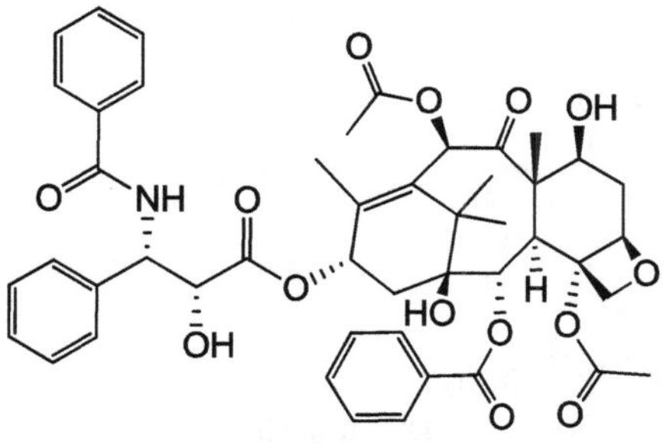

6 - taxol

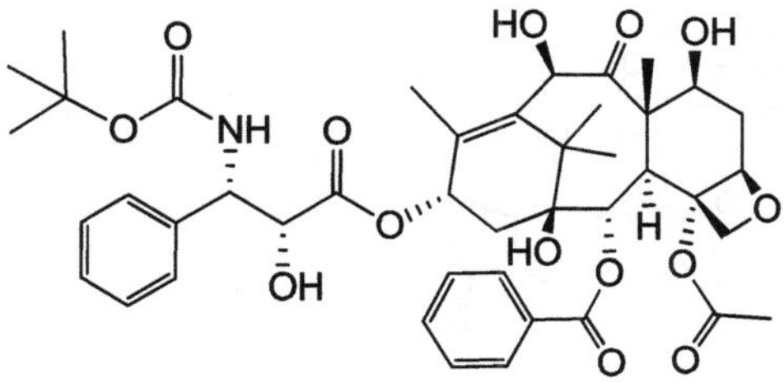

7 - docetaxel

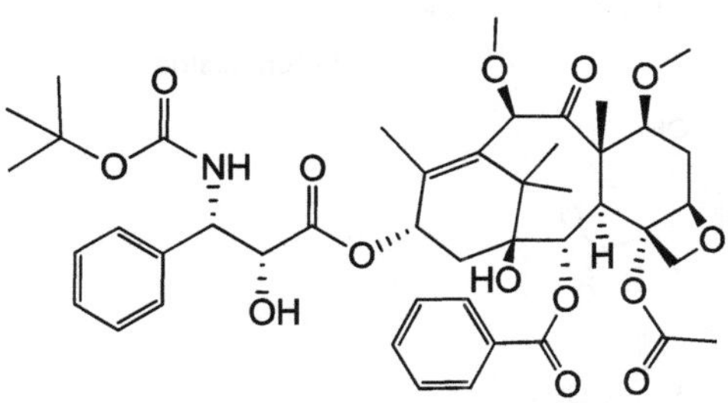

8 - cabazitaxel

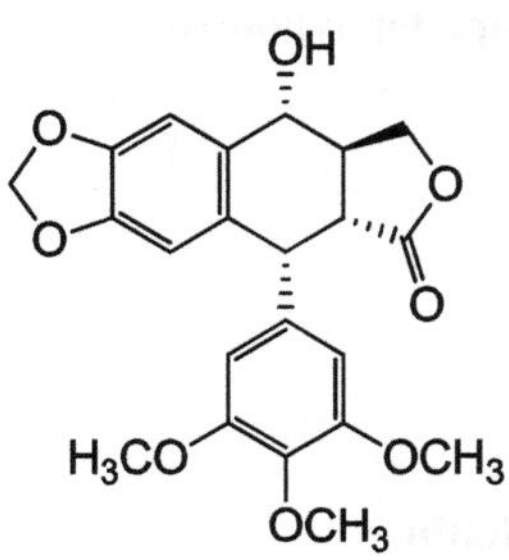

9 - podophyllotoxin

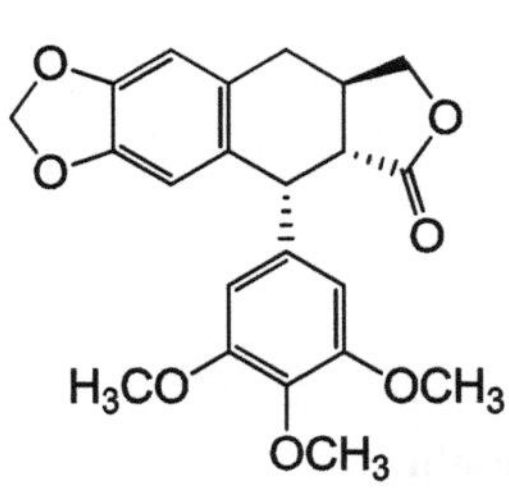

10 - deoxypodophyllotoxin

Fig. 5.1 (continued)

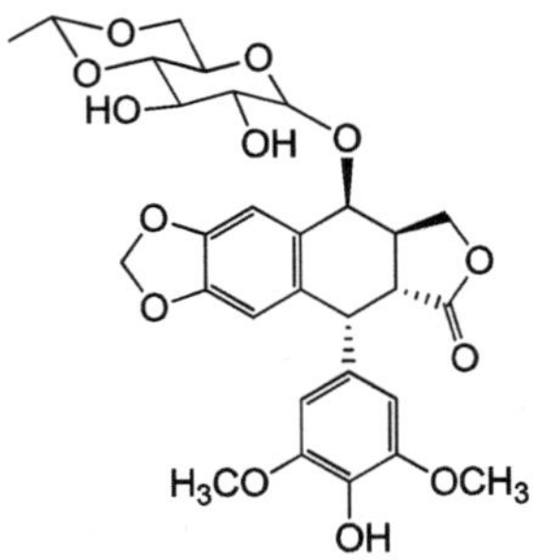

11 - etoposide

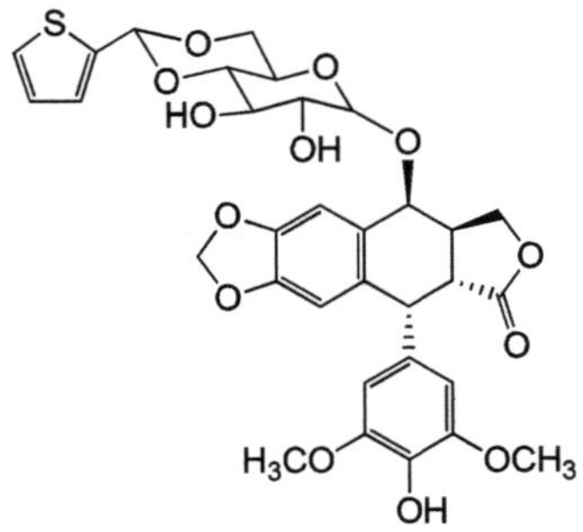

12 - teniposide

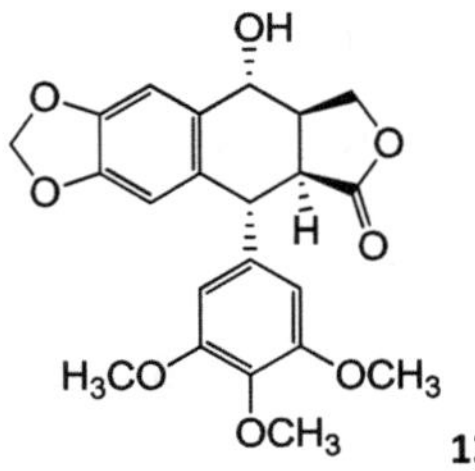

13 - picropodophyllotoxin

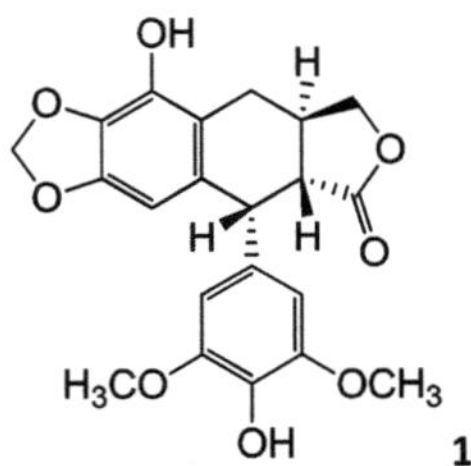

14 - α-peltatin

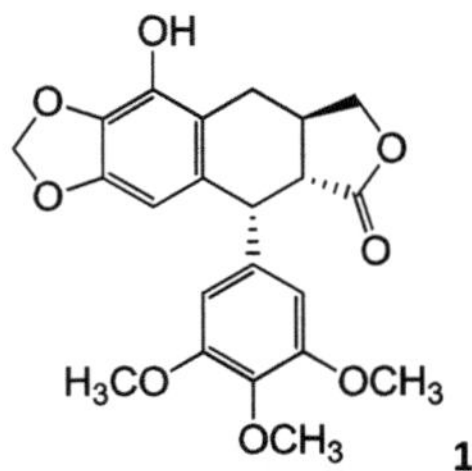

15 - β-peltatin

Fig. 5.1 (continued)

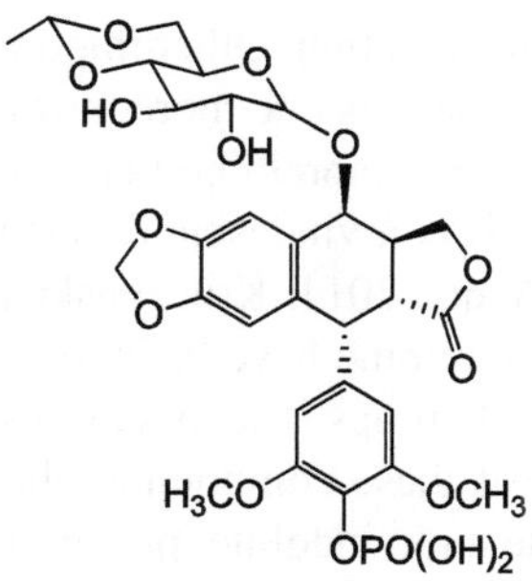

16 - etopophos

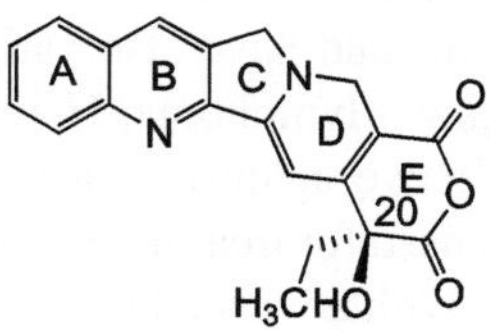

17 - camptothecin

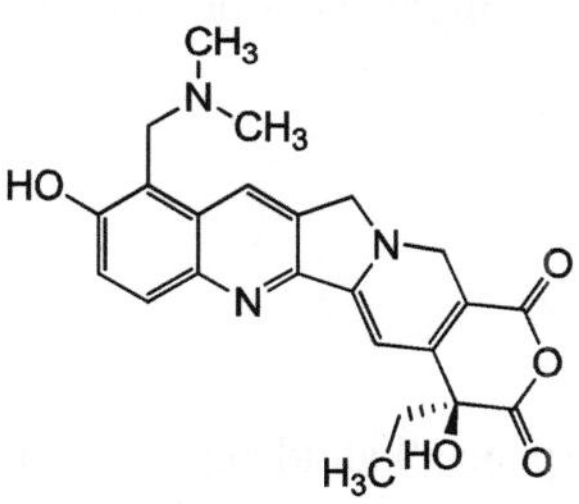

18 - topotecan

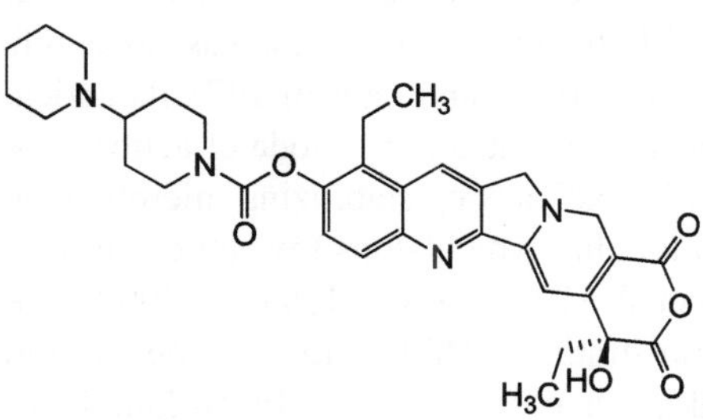

19 - irinotecan

Fig. 5.1 (continued)

vincristine has an aldehydic group at the same position. Other important vinca alkaloid is vindesine (**3**), a vinblastine analogue that has one more nitrogen forming an amide function and without one acetyl group. Vinorelbine (**4**) is a semisynthetic analogue from vinblastine and vinflunine (**5**) is a vindesine derivative with a bis-fluorinated group (Zhou et al. 2015; Dall'Acqua 2014; Kruczynski and Hill 2001). Vincristine, vinblastine, vindelsine, and vinorelbine have bisindole rings with differences in their carbon skeleton or functional groups. The loss of this bisindole ring results in decrease of anticancer property of these compounds; that is why other alkaloids from *C. roseus*, like catharanthine and vindoline, present low antimitotic action. Despite the similarity within the vinca alkaloids and their derivatives, these compounds exhibit different anticancer activities and have also variations in their toxicity. Vincristine is used to treat non-Hodgkin's lymphoma, Hodgkin's disease, and pediatric solid tumors. Vinblastine is effective against breast, testicular, and non-Hodgkin's lymphoma while vinorelbine is used for treatment of lung and breast cancer. Vinflunine is indicated to adult patients with advanced or metastatic urothelial cancer. These alkaloids can be used as a single agent or combined with another anticancer drug (Dall'Acqua 2014).

Taxol and Its Derivatives

Taxol (**6**) (generic name paclitaxel), a diterpene alkaloid derived from *Taxus* spp., is the most successful natural source of anticancer drug, with a high market and valued at above $1 billion per year (Malik et al. 2011). Taxol was isolated from the bark of *Taxus brevifolia* (Pacific yew) in 1962, and later in 1971 its anticancer properties were discovered (Wani et al. 1971). The unique mode of action was established by Schiff and coworkers in 1979; taxol acts by stabilizing microtubules and inhibiting their depolymerization into tubulin, which stops the cell cycle in the G2/M phase leading to cell death (Schiff et al. 1979; Horwitz 1994). In 1992, it was approved by the US Food and Drug Administration (FDA) for the treatment of refractory ovarian cancer and commercialized as Taxol® by Bristol-Myers-Squibb (BMS). After the success of taxol, the semisynthetic taxoid docetaxel (**7**) (registered by Sanofi-Aventis as Taxotere®), considered more potent, was approved by the FDA for the treatment of other types of cancers, such as breast cancer, squamous cancers of the head and neck, non-small-cell lung cancer, small-cell lung cancer, and AIDS-related Kaposi's sarcoma (Rowinsky et al. 1992; Rowinsky 1994; Singla et al. 2002).

Yew trees of *Taxus* spp. (Taxaceae) are very similar and mainly distributed in the Northern Hemisphere. The taxonomy is not well defined; therefore they are classically identified into eight species based on geographical location, i.e., *T. baccata* (European or English yew), *T. brevifolia* (Pacific or Western yew), *T. canadensis* (Canadian yew), *T. chinensis* (Chinese yew), *T. cuspidata* (Japanese yew), *T. floridana* (Florida yew), *T. globosa* (Mexican yew), and *T. wallichiana* (Himalayan yew) (Cope 1998; Gupta 2015).

Taxus species produces a range of taxanes and has attained a great importance due to increasing demand of taxol for anticancer treatment, but its low concentration in the bark (0.001–0.05% in *T. brevifolia*) and slow growth always represented a concern. Concomitantly with this, the extraction process is complicated and expensive and requires numerous trees to supply an adequate amount. To produce 1 kg of taxol, approximately 10,000 kg of *Taxus* bark or 3000 yew trees are required (Schippmann 2001; Malik et al. 2011). The extraction of *Taxus* from natural habitat has led this plant species to endangerment (Garawal 2016). Hence, several attempts have been made to find out the alternative methods for taxol production, such as total synthesis, semisynthesis, and plant cell culture in order to meet the ever-rising demand and availability of this compound at reasonable price.

The chemical synthesis of taxol was achieved in 1994 (Holton et al. 1994; Nicolau et al. 1994). However, it was not commercially viable due to complex pathway, low yield, and toxicity of secondary products. The semisynthetic preparation of taxol requires 10-deacetylbaccatin III (10-DAB), baccatin III or 9-dihydro-13-acetylbaccatin III, which are more abundant and can be found in renewable needles of *Taxus*, making it an attractive option. The synthesis of a C-13 side chain and linkage to 10-DAB produce taxol. Several methods of synthesis of this moiety were developed (Denis et al. 1988; Koskinen et al. 1994; Hamamoto et al. 2000; Borah et al. 2004). BMS and Indena have *Taxus* plantations to supply the companies with intermediates for taxol and analogue synthesis in satisfying yields (Malik et al. 2011).

Plant cell culture, especially at industrial level is considered as the most promising way for the sustainable production of taxol and related taxoids (Expósito et al. 2009). The advantages of this approach are renewable and environmentally friendly resource, and continuous and uniform-quality taxol, because it is not influenced by weather and season or contamination. The biotechnological production of taxol through *Taxus* cell culture has been started since early 1990s (Fett-Neto et al. 1992; Wicremesinhe and Arteca 1993). After that several approaches were made in an effort to increase the production of taxol and related taxanes, including optimization of culture conditions; selection of high-yield cell lines; induction by elicitors; use of adsorbants, additives, and precursors; employment of a two-phase culture system; immobilization; and use of combined strategies, i.e., use of two elicitors with different acting paths. In the literature, there are excellent reviews discussing in detail about all these strategies (Tabata 2004; Frense 2007; Expósito et al. 2009; Malik et al. 2013).

Other alternative sources of taxol have been explored, such as production by synthetic biology, from the endophytic fungi and taxol from other non-*Taxus* plants. Liu et al. (2016) reported the recent advances in taxol production.

In addition to the scarce availability of source material, the compound posed challenges on the physicochemical properties, limiting its use by several factors. It was difficult to formulate into a delivery system acceptable for human use. The drug is insoluble in water, and thus very difficult to formulate into a suitable delivery system. To overcome this problem, taxol was formulated with a carrier Cremophor EL (polyethoxylated castor oil or polysorbate 80), which is nonbiological and exerts biological effects, associated with severe anaphylactic hypersensitivity reactions (Gelderblom et al. 2001). To prevent these reactions, patients need to be pretreated with corticosteroids.

After many studies, a new delivery system for taxol was developed. Abraxane®, a nanoparticle albumin-bound (nab) formulation of paclitaxel, was approved in 2005. Nanoparticle paclitaxel is also called paclitaxel albumin-stabilized nanoparticle formulation (Pazdur 2013). Nowadays its sale controls 50–60% of the total paclitaxel market in the USA. Cabazitaxel (**8**) (registered as Jevtana® by Sanofi-Aventis), approved by the FDA in 2010, is another semisynthetic taxane anticancer agent, combined with prednisone. Since then, other technologies have been advanced to improve solubility, facilitate administration, and enhance the pharmacological profile and bioavailability (Bionap 2016).

Taxol®, Taxotere®, and Abraxane® are the most solicited taxanes in clinical oncological trials, which allowed the understanding of their relationships with tumor processes and the proper treatment of cancer patients. Many new taxanes, such as Taxoprexin® and Xytotax®, have been evaluated in phase I–III trials in different cancer treatments demonstrating good results (Fauzee et al. 2011). The success of taxol and its analogues encourages research to improve the natural source yield of taxanes or the discovery of novel potential anticancer compounds.

Podophyllotoxin and Its Derivatives

Podophyllotoxins are naturally occurring aryltetralin lignans, found in plants, particularly in the genus *Podophyllum.* The class includes several closely related chemical structures, particularly podophyllotoxin (**9**), deoxypodophyllotoxin (**10**), 4″-demethyl analogues, peltatins, and their corresponding glycosides. They also include some hemisynthetic compounds, etoposide (**11**) and teniposide (**12**), and a number of other derivatives (Botta et al. 2001).

The genus *Podophyllum* has been known for several centuries from a botanical point of view. Originally including American *Podophyllum*, called *Podophyllum peltatum* L., and Indian *Podophyllum* usually named *Podophyllum emodi* Wallich (syn. *P. hexandrum* Royce), they belong to family Berberidaceae and were used by the natives of both continents as cathartics, anthelminthics, and cholagogues (Imbert 1998).

Podophyllotoxin is the main constituent of podophyllin, resin obtained by extraction with alcohol from roots and dry rhizomes of the species *Podophyllum peltatum* L. and *Podophyllum emodi* Wallich. Kaplan (1942) demonstrated the curative topic effect of podophyllin in *Condyloma acuminate*, venereal condyloma, and benign tumor. This result created the possibility of using podophyllin against cancerous tissues. King and Sullivan (1946) demonstrated that podophyllotoxin was a potent antimitotic as colchicine and for blockage of cell division metaphase of mitosis.

After this discovery, the chemistry and antineoplastic activity of podophyllotoxin and many components of podophyllin were extensively studied (Imbert 1998;

Botta et al. 2001). In 1951 its structure was elucidated by Hartwell and Schrecker (1951), but it is noteworthy that podophyllotoxin had already been isolated in 1880. Studies have shown its high toxicity, which is related to chemical structure and activity (Seidlova-Massinova et al. 1957). The cytotoxic activity of podophyllotoxin is strictly connected with the unique configuration at C-2, C-3, and C-4, with its highly strained, *trans*-fused γ-lactone system. In the presence of mild base catalysis, podophyllotoxin epimerizes smoothly to the thermodynamically more stable C-2 epimer picropodophyllotoxin (**13**), which shows little or no cytotoxic activity (Brewer et al. 1979).

Although the routes of total synthesis of podophyllotoxin have been improved since 1966 (Gensler and Gatsonis 1966) it is not yet commercially viable due to its complex structure; therefore, the need for producing podophyllotoxin and derived anticancer drugs has stimulated the exploration of other higher plant sources (Gordaliza et al. 2004; Malik et al. 2014b).

Other lignans and their glycosides, such as deoxypodophyllotoxin, α-peltatin (**14**), and β-peltatin (**15**) were extracted from other *Podophyllum* species (Singh and Shah 1994; Qazi et al. 2011). Related lignans have been isolated from species of *Callitris* and *Juniperus* (Cupressaceae) (Renouard et al. 2015; Cantrell et al. 2013), *Linum* (Linaceae) (Doussot et al. 2016), *Dysosma* (Berberidaceae) (Karuppaiya and Tsay 2015), *Bursera* (Burseraceae) (Peña-Morán et al. 2016; Mojica et al. 2016), *Bridelia* (Euphorbiaceae) (Pettit et al. 2016), and *Leptohyptis* (Lamiaceae) (Brandão et al. 2017).

Podophyllotoxin and its derivative 6-methoxypodophyllotoxin have been obtained by in vitro cultures using differentiated organ cultures, mainly roots, undifferentiated callus, and suspension cell cultures of different species of *Podophyllum*, *Linum*, *Juniperus*, and *Callitris* (Chattopadhyay et al. 2002; Federolf et al. 2007).

The attempt to prepare podophyllotoxin derivatives that did not have a high gastrointestinal toxic effect and better solubility allowed the production of semisynthetic cytotoxic epipodophyllotoxins: etoposide and teniposide (Stähelin and von Wartburg 1991). They differ chemically only in the substitution of a methyl group (etoposide) for the thenylidene (teniposide) on the glucopyranoside sugar. Both are poorly soluble in water (Schacter 1996).

The advantages of these two semisynthetic derivatives of podophyllotoxin are their less pronounced side effects than those of the precursor. Both can differentiate cancer cells from normal cells because of the two forms of TOP2, α and β, which are very similar but genetically distinct. An etoposide analogue has been marketed under the name Etopophos (**16**), with the advantage of being water soluble. These derivatives are used clinically as potent chemotherapeutic agents for a variety of tumors including small-cell lung carcinoma, testicular cancer, and malignant lymphoma including testicular and small-cell lung cancers, lymphoma, leukemia, and Kaposi's sarcoma (Witterland et al. 1996). The pharmacodynamics of the epipodophyllotoxins etoposide and teniposide differ completely from those of podophyllotoxin for their interaction with topoisomerase (Bohlin and Rositn 1996).

The search for active anticancer derivatives of podophyllotoxin is still continuing; deoxypodophyllotoxin was found to be a potent antitumor and antiproliferative agent, in several tumor cells under in vitro conditions. However, deoxypodophyllotoxin has not been used clinically yet because of the lack of in vivo studies, which were limited by its insolubility in water. Preparation of solid inclusion complex of deoxypodophyllotoxin with a derivative of cyclodextrin revealed that this complex exhibited significant antitumor activity against various human cancer cell lines such as small-cell lung cancer, leukemia, cervical cancer, and gastric cancer (Zhu et al. 2010; Khaled et al. 2013).

Recently, Khaled et al. (2016) demonstrated that the antitumor effect of podophyllotoxin-HP-β-CD (20 mg/kg) in human breast cancer MDA-MB-231 xenograft was more effective than etoposide (20 mg/kg). These findings suggest that this drug is a promising chemotherapy candidate against human breast carcinoma.

Camptothecin and Its Derivatives

Camptothecin (CPT) (**17**) is a natural product first isolated from the bark of the Chinese tree *Camptotheca acuminate* (Nyssaceae), from which the semisynthetic analogues topotecan (**18**) and irinotecan (**19**) are derived. These compounds are potent anticancer agents widely used clinically throughout the world. CPT was discovered and developed by the US National Cancer Institute (NCI) at about the same time and by the same group that were also working on taxol (Wall and Wani 1996).

All camptothecins have a pentacyclic structure, consisting of a quinolone ring (rings A and B), a pyridone ring (ring D), and a terminal ring α-hydroxy-δ-lactone (ring E), which has a chiral center on carbon C20. This five-ring structure is essential for activity, since a tetracyclic analogue of camptothecin was tested and considered inactive (Slichenmyer et al. 1993; Garcia-Carbonero and Supko 2002).

Unmodified CPT molecule exhibits very low solubility in aqueous media and is rapidly inactivated through hydrolysis of lactone ring (reversible in acidic media) at physiological pH, leading to a water-soluble carboxylate, an inactive form (Fassberg and Stella 1992). Unfortunately, tests with the water-soluble salt camptothecin-sodium showed high toxicity, causing hemorrhagic cystitis and myelotoxicity (Gottlieb and Luce 1972; Moertel et al. 1972). After that, two water-soluble derivatives of camptothecin were successfully developed (Hsiang et al. 1985) and approved by the US Food and Drug Administration: topotecan for ovarian and lung cancers (Herzog 2002) and irinotecan for colorectal cancer (Garcia-Carbonero and Supko 2002). These semisynthetic derivatives have higher physicochemical stability and solubility at lower pH values, and did not cause hemorrhagic cystitis (Malonne and Atassi 1997). In contrast to the structurally related topotecan, irinotecan is a prodrug, which has to be converted by human hepatic carboxylesterase (CES) enzymes to 7-ethyl-10-hydroxycamptothecin (SN-38), its active form (Haaz et al. 1998; Humerickhouse et al. 2000; Mathijssen et al. 2001). Other drugs have been developed from CPT and are the subject of clinical trials, for example belotecan, 9-aminocamptothecin, exatecan, 9-nitrocamptothecin, and camptothecin glycoconjugates (Pizzolato 2003; Kim et al. 2015).

Camptothecins are anticancer agents of category, which target the nuclear DNA topoisomerase I (TOP1), an essential human enzyme. CPTs act by binding to the TOP1 cleavage complex, leading to an accumulation of DNA strand breaks upon replication, causing apoptosis during the S phase of the cell cycle (Hsiang et al. 1985; Hsiang et al. 1989).

C. acuminata is a woody tree commonly known as Xi Shu and Chinese Happy Tree. All parts of this plant species contain some amount of CPT which has the role to protect the plant against herbivores or pathogens (Adamovics et al. 1979; Hartmann and Lipp 2006). The CPT level varies depending on the plant tissue, age of the tissue, and environmental conditions (Li et al. 2002). The highest level is generally found in young leaves (about 4–5 mg/g dry weight) approximately 250% higher than in bark, which was previously used for CPT extraction (López-Meyer et al. 1994; Lorence and Nessler 2004).

CPT is also produced by other species of different families, e.g., *Nothapodytes foetida* (Aiyama et al. 1988), *Pyrenacantha klaineana* (Zhou et al. 2000), and *Merrilliodendron megacarpum* (Arisawa et al. 1981) (Icacinaceae); *Ophiorrhiza pumila* (Saito et al. 2001) (Rubiaceae); *Ervatamia heyneana* (Gunasekera et al. 1979) (Apocynaceae); and *Mostuea brunonis* (Dai et al. 1999) (Gelsemiaceae). However, the major source of CPT is still via happy tree that is now rare in its native habitat in China. There is an estimative existence of only less than 4000 trees in the wild. Because of the popularity of CPT derivatives, it has been estimated that at least 100 million young happy trees will be needed to meet the demands (Lorence and Nessler 2004). Therefore, the extraction of CPT from limited natural plant resources may result in environmental concerns and hence cannot meet the expanding demand of the market (Lorence and Nessler 2004; Sirikantaramas et al. 2007; Yamazaki et al. 2010; Kai et al. 2008, 2013, 2014). At present (2016) Sigma (Brazil) lists the price of CPT at US$ 1.371 per gram and for its derivative, topotecan, at US$ 1.473 per 25 mg.

Due to low content of CPT in plants (about 1 mg/g dry weight (DW) (López-Meyer et al. 1994), it is very important to increase CPT production and develop sustainable methods to obtain CPT for clinical applications (Ni et al. 2011; Cui et al. 2015). Some possible ways are through clonal propagation of elite cultures, and by the repeated harvest of only Xi Shu young leaves (Davis et al. 2011). Another promising approach is the use of plant biotechnology, which may provide the way for enhancement of CPT production by transferring key CPT biosynthetic genes (and/or transcript factor) into CPT-producing plant cell, and then large-scale culture of transgenic cell lines, hairy roots, or regenerated plants to obtain CPT (Cui et al. 2015; Kai et al. 2015).

New Potential Anticancer Drugs from Plants

The plant kingdom, specially the higher plants as angiosperms, has an important metabolism capable to produce a great variety of compounds with potential to be used by pharmaceutical industries to get new and innovative medicines: the ability to produce substances that in some cases are impossible to be synthesized in laboratories probably due to the presence of specific enzymes catalyzing the reaction in the metabolic pathway.

According to the papers and books published every year, we know that less than one-third of all higher plants in the world have been studied to identify the possible drugs. According to Veeresham (2012), only 35,000–70,0000 plant species have already been studied for the medicinal use. It has been estimated that there are around 270 thousand plants in the world. Hence, the potential to discover new drugs that can be used to produce medicines is extremely high. Regarding anticancer drugs, many studies have shown so many possibilities to new drugs that can be used to treat this problem. Some of these compounds are in clinical trial steps and can be approved by drug agencies in the next few years.

Combretastatin A4 was isolated from the ornamental plant *Combretum caffrum* and has shown the ability to inhibit the microtube polymerization of cancer cells. It is a phenolic secondary metabolite of stilbene group that has a C6C2C6 skeleton. This substance has a potential to be used as a cytotoxic against different types of cancer, including the multidrug resistance; however it (also analogues) has a low water solubility and low bioavailability that are limitations to the development of anticancer medicines. Further studies are required to understand the relationships between structure activities to solve the solubility and bioavailability issues (Marrelli et al. 2011; Dall'Acqua 2014).

Curcumin is another phenolic secondary metabolite isolated from rhizome of *Curcuma longa*, used in traditional Asian cuisine. The powder of this rhizome is known as turmeric preparation. According to the World Health Organization, curcumin might be related to the low rates of colorectal, lung, and prostate cancers in India. This compound is still in clinical trials to prove its effectiveness and safety (Dahmke et al. 2014).

In vitro and in vivo tests have shown the anticancer action of green tea from *Camellia sinensis* and one of its main constituents: epigallocatechin-3-gallate in several different tumor cell models. However, further studies are needed to approve their use as anticancer drugs (Rahmani et al. 2015).

Many plants and their secondary or even primary metabolites show a role in inhibition of the development of cancer cells but it is important to know that a substance should have studied, including preclinical and clinical trials, and also approved by a drug agency to be used as a drug.

Conclusions

Medicinal plants have proved to be a source of innumerous substances with therapeutic potential. Even with the modern ways to get synthetic drugs, the plants are still studied with the aim to find out new drugs. Among the anticancer drugs, there are many compounds isolated from plant species that have the ability to prevent or treat some kind of cancer cells, highlighting the alkaloids from vinca, taxol, epipodophyllotoxin, camptothecin, and their semisynthetic or synthetic derivatives. Many other compounds are now in clinical trial to attest its efficiency and safety. Facing the new challenges to design new drugs that can be used to treat multidrug-resistant cancers, plants represent one of the ways to discover new anticancer drugs.

Acknowledgments Authors would like to acknowledge the Foundation for the Support of Research, Scientific, and Technological Development of the state of Maranhão—FAPEMA for financial support.

References

Achan J, Talisuna AO, Erhart A, Yeka A, Tibenderana JK, Baliraine FN, Rosenthal PJ, D'Alessandro U (2011) Quinine, an old antimalarial drug in a modern world: role in the treatment of malaria. Malar J 1(10):1–14

Adamovics JA, Cina JA, Hutchinson CR (1979) Minor alkaloids of *Camptotheca acuminate*. Phytochemistry 18:1085–1086

Aiyama R, Nagai H, Nokata K, Shinohara C, Sawada S (1988) A camptothecin derivative from *Notapodytes foetida*. Phytochemistry 27:3663–3664

Arisawa M, Gunasekera SP, Cordell GA, Farnsworth NR (1981) Plant anticancer agents XXI. Constituents of *Merrilliodendron megacarpum*. Planta Med 43:404–407

BioNap (2016) Oasmia pharmaceuticals—advancing nanotechnology based chemotherapies. http://www.bionapcfa.com/2016/03/oasmia-pharmaceuticals-advancing.html. Accessed 24 Nov 2016

Bohlin L, Rositn B (1996) Podophyllotoxin derivatives: drug discovery and development. Drug Discov Today 1:343–351

Borah JC, Gogoi S, Boruwa J, Kalita B, Barua NC (2004) A highly efficient synthesis of the C-13 side-chain of taxol using Shibasaki's asymmetric Henry reaction. Tetrahedron Lett 45(18):3689–3691

Botta B, Delle Monache G, Misiti D, Vitali A, Zappia G (2001) Aryltetralin lignans: chemistry, pharmacology and biotransformations. Curr Med Chem 8:1363–1381

Brandão HN, Medrado HHS, David JP, David JM, Pastore JFB, Meira M (2017) Determination of podophyllotoxin and related aryltetralin lignans by HPLC/DAD/MS from Lamiaceae species. Microchem J 130:179–184

Brewer CF, Loike JD, Horwitz SD, Sternlicht H, Gensler WJ (1979) Conformational analysis of podophyllotoxin and its congeners. Structure—activity relationship in microtubule assembly. J Med Chem 22:215–221

Cantrell CL, Zheljazkov VD, Osbrink WLA, Castro A, Maddox V, Craker LE, Astatkie T (2013) Podophyllotoxin and essential oil profile of *Juniperus* andrelated species. Ind Crop Prod 43:668–676

Chattopadhyay S, Srivastava AK, Bhojwani SS, Bisaria VS (2002) Production of podophyllotoxin by plant cell cultures of *Podophyllum hexandrum* in bioreactor. J Biosci Bioeng 93:215–220

Cope EA (1998) Taxaceae: the genera and cultivated species. Bot Rev 64:291–322

Cragg GM, Newman DJ (2013) Natural products: a continuing source of novel drug leads. Biochim Biophys Acta 1830:3670–3695

Cui LJ, Ni XL, Ji Q, Teng XJ, Yang YR, Wu C, Zekria D et al (2015) Co-overexpression of geraniol-10-hydroxylase and strictosidine synthase improves anti-cancer drug camptothecin accumulation in *Ophiorrhiza pumila*. Sci Rep 5:8227

Dahmke IN, Boettcher SP, Groh M, Mahlkecht U (2014) Cooking enhances curcumin anti-cancerogenic activity through pyrolytic formation of deketene curcumin. Food Chem 151:514–519

Dai JR, Cardellina JH, Boyd MR (1999) 20-Ob-Glucopyranosyl camptothecin from *Mostuea brunonis*: a potential camptothecin pro-drug with improved solubility. J Nat Prod 62:1427–1429

Dall'Acqua S (2014) Natural products as antimitotic agents. Curr Top Med Chem 14:2272–2285

Davis C, Fanning K, Trueman S, Williams D (2011) Growing the Chinese happy tree (Xi Shu) in Australia-maximising the extraction of the anti-cancer compound camptothecin. RIRDC Publication, Australia

de Sa IM (2011) Chloroquine resistance and the search for antimalarial drugs from the 1960s to 1980s. Hist Cienc Saude Manguinhos 18(2):407–430

de Sant'Ana RS, de Castro Mattos JS, da Silva AS, de Mello LM, Nunes AA (2016) Associated factors with mammographic changes in women undergoing breast cancer screening. Einstein 14(3):324–329

Denis JN, Greene AE, Guenard D, Gueritte-Voegelein F, Mangatal L, Potier P (1988) Highly efficient, practical approach to natural taxol. J Am Chem Soc 110(17):5917–5919

Doussot J, Mathieu V, Colas C, Molinie R, Corbin C, Montguillon J, Moreno Y, Banuls L, Renouard S, Lamblin F, Dupre P (2016) Investigation of the lignan content in extracts from *Linum*, *Callitris* and *Juniperus* species in relation to their in vitro antiproliferative ctivities. Planta Med. doi:10.1055/s-0042-118711

Du Y, Kong G, You X et al (2012) Elevation of highly up-regulated in liver cancer (HULC) by hepatitis B virus X protein promotes hepatoma cell proliferation via down-regulating p18. J Biol Chem 287(31):26302–26311

Expósito O, Bonfill M, Moyano E, Onrubia M, Mirjalili MH, Cusido RM, Palazón J (2009) Biotechnological production of taxol and related taxoids: current state and prospects. Anti Cancer Agents Med Chem 9:109–121

Fassberg J, Stella VJ (1992) A kinetic and mechanistic study of the hydrolysis of camptothecin and some analogues. J Pharm Sci 81(7):676–684

Fauzee NJS, Dong Z, Wang YL (2011) Taxanes: promising anti-cancer drugs. Asian Pac J Cancer Prev 12:837–851

Federolf K, Alfermann AW, Fuss E (2007) Aryltetralin-lignan formation in two different cell suspension cultures of *Linum album*: deoxypodophyllotoxin 6-hydroxylase, a key enzyme for the formation of 6-methoxypodophyllotoxin. Phytochemistry 68:1397–1406

Fett-Neto AG, DiCosmo F, Reynolds WF, Sakata K (1992) Cell culture of Taxus as a source of the antineoplastic drug taxol and related taxanes. Bio/Technology 10:1572–1576

Fonseca JDS, Farías RAQ (2016) Obesity and cancer: pathophysiology and epidemiological evidence. Rev Med Ris 22(2):91–97

Frense D (2007) Taxanes: perspectives for biotechnological production. Appl Microbiol Biotechnol 73:1233–1240

Futreal PA, Coin L, Marshall M et al (2004) A census of human genes. Nat Rev Cancer 4(3):177–183

Garawal T (2016) Taxus: an endangered genus. Int J Agr Sci Res 6(2):175–178

Garcia-Carbonero R, Supko J (2002) Current perspectives on the clinical experience, pharmacology, and continued development of the camptothecins. Clin Cancer Res 8:641–661

Gelderblom H, Verweij J, Nooter K, Sparreboom A (2001) Cremophor EL: the drawbacks and advantages of vehicle selection for drug formulation. Eur J Cancer 37(13):1590–1598

Gensler WJ, Gatsonis CD (1966) Synthesis of podophyllotoxin. J Organomet Chem 31:4004–4008

Giachetti D, Monti L (2005) Medicinal plant in phytotherapy. Ann Ist Super Sanita 41(1):17–22

Golberg DR (2009) Aspirin: turn of the century miracle drug. https://www.chemheritage.org/distillations/magazine/aspirin-turn-of-the-century-miracle-drug. Accessed 30 Sep 2016

Gordaliza M, Garcia PA, Miguel del Corral JM, Castro MA, Gomez-Zurita MA (2004) Podophyllotoxin: distribution, sources, applications and new cytotoxic derivatives. Toxicon 44:441–459

Goss A (2014) Building the world's supply of quinine: Dutch colonialism and the origins of a global pharmaceutical industry. Endeavour 38(1):8–18

Gottlieb JA, Luce JK (1972) Treatment of malignant melanoma with camptothecin (NSC-100880). Cancer Chemother Rep 56(1):103–105

Gunasekera SP, Badawi MM, Cordell GA, Farnsworth NR, Chitnis M (1979) Plant anticancer agents X. Isolation of camptothecin and 9-methoxycamptothecin from *Ervatamia heyneana*. J Nat Prod 42:475–477

Gupta D (2015) An overview of taxus. J Drug Discov Therap BP 3(29):1–7

Haaz MC, Rivory L, Riché C, Vernillet L, Robert J (1998) Metabolism of irinotecan (CPT-11) by human hepatic microsomes: participation of cytochrome P-450 3A and drug interactions. Cancer Res 58:468–472

Hamamoto H, Mamedov VA, Kitamoto M, Hayashi N, Tsuboi S (2000) Chemoenzymatic synthesis of the C-13 side chain of paclitaxel (Taxol) and docetaxel (Taxotere). Tetrahedron Asymmetry 11(22):4485–4497

Hartmann JT, Lipp H-P (2006) Camptothecin and podophyllotoxin derivatives inhibitors of topoisomerase I and II—mechanisms of action, pharmacokinetics and toxicity profile. Drug Saf 29(3):209–230

Hartwell JL, Schrecker AW (1951) Components of podophyllin. V. The constitution of podophyllotoxin. J Am Chem Soc 73:2909–2916

Herzog TJ (2002) Update on the role of topotecan in the treatment of recurrent ovarian cancer. Oncologist 7(Suppl. 5):3–10

Holgado E, Perez M, Wren A, Cortes J, Gomez-Pinillo A (2015) Influencing cancer treatment. Lancet Oncol 16:1591–1593

Holton RA, Somoza C, Kim HB (1994) First total synthesis of Taxol. L Functionalization of the B ring. J Am Chem Soc 116:1597–1560

Hoogte ARVD, Pieters T (2014) Science in the service of colonial agro-industrialism: the case of cinchona cultivation in the Dutch and British East Indies, 1852–1900. Stud Hist Phil Biol Biomed Sci 47:12–22

Horwitz SB (1994) Taxol (paclitaxel): mechanisms of action. Ann Oncol 5(Suppl 6):S3–S6

Hsiang Y-H, Hertzberg R, Hecht S, Liu LF (1985) Camptothecin induces protein-linked DNA breaks via mammalian DNA topoisomerase I. J Biol Chem 260:14873–14878

Hsiang Y-H, Lihou MG, Liu LF (1989) Arrest of replication forks by drug-stabilized topoisomerase I-DNA cleavable complexes as a mechanism of cell killing by camptothecin. Cancer Res 49:5077–5082

Humerickhouse R, Lohrbach K, Li L, Bosron W, Dolan ME (2000) Characterization of CPT-11 hydrolysis by human liver carboxylesterase isoforms hCE-1 and hCE-2. Cancer Res 60:1189–1192

Huxtable RJ, Schwarz KV (2001) The isolation of morphine: first principles in science and ethics. Mol Interv 14:189–191

Imbert TF (1998) Discovery of podophyllotoxins. Biochimie 80:207–222

Kai GY, Dai LM, Mei XY, Zheng JG, Wang W, Lu Y, Qian ZY, Zhou GY (2008) *In vitro* plant regeneration from leaf explants of *Ophiorrhiza japonica*. Biol Plant 52(3):557–560

Kai GY, Li SS, Wang W, Lu Y, Wang J, Liao P, Cui LJ (2013) Molecular cloning and expression analysis of a new gene encoding 3-hydroxy-3-methylglutaryl-CoA synthase from *Camptotheca acuminata*. Russ J Plant Physiol 60(1):131–138

Kai GY, Teng XJ, Cui LJ, Li SS, Hao XL, Shi M, Yan B (2014) Effect of three plant hormone elicitors on the camptothecin accumulation and gene transcript profiling in *Camptotheca acuminata* seedlings. Inter J Sci 3(1):86–95

Kai G, Wu C, Gen L, Zhang L, Cui L, Ni X (2015) Biosynthesis and biotechnological production of anti-cancer drug Camptothecin. Phytochem Rev 14:525–539

Kaplan IW (1942) Condyloma acuminate. New Orleans Med Surg J 94:388–390

Karuppaiya P, Tsay HS (2015) Therapeutic values, chemical constituents and toxicity of Taiwanese *Dysosma pleiantha*—a review. Toxicol Lett 236:90–97

Khaled M, Jiang ZZ, Zhang LY (2013) Deoxypodophyllotoxin: a promising therapeutic agent from herbal medicine. J Ethnopharmacol 149:24–34

Khaled M, Belaaloui G, Jiang Z-Z, Zhu X, Zhang L-Y (2016) Antitumor effect of deoxypodophyllotoxin on human breast cancer xenograft transplanted in BALB/c nude mice model. J Infect Chemother 22:692–696
Kim YK, Koo NY, Yun PY (2015) Anticancer effects of CKD-602 (Camtobell®) via G2/M phase arrest in oral squamous cell carcinoma cell lines. Oncol Lett 9(1):136–142
King IS, Sullivan M (1946) The similarity of the effects of podophyllotoxin and colchicine and their use of in treatment *Condyloma acuminate*. Science 104:244–245
Kinsley-Scott TR, Norton SA (2003) Useful plants of dermatology. VII: *cinchona* and antimalarials. J Am Acad Derma 49(3):499–502
Koskinen AMP, Karvinen EK, Siirila JP (1994) Enantioselective synthesis of the Taxol and Taxotere side chains. J Chem Soc Chem Commun 1:21–22
Kruczynski A, Hill BT (2001) Vinflunine, the latest Vinca alkaloid in clinical development: a review of its preclinical anticancer properties. Crit Rev Oncol Hematol 40:159–173
Labreche F, Duguay P, Boucher A, Arcand R (2014) Estimating the proportion of occupational câncer with minimal resources: an example from Quebec. Occup Environ Med 71(6):A90
Li S, Yi Y, Wang Y, Zhang Z, Beasley RS (2002) Camptothecin accumulation and variations in *Camptotheca*. Planta Med 68:1010–1016
Liu WC, Gonga T, Zhu P (2016) Advances in exploring alternative Taxol sources. RSC Adv 6:48800–48809
López-Meyer M, Nessler CL, McKnight TD (1994) Sites of accumulation of the antitumor alkaloid camptothecin in *Camptotheca acuminata*. Planta Med 60:558–560
Lorence A, Nessler CL (2004) Camptothecin, over four decades of surprising findings. Phytochemistry 65:2735–2749
Mahdi JG, Mahdi AJ, Mahdi AJ, Bowen ID (2006) The historical analysis of aspirin discovery, its relation to the willow tree and antiproliferative and anticancer potential. Cell Prolif 39:147–155
Malik S, Cusido RM, Mirjalili MH, Moyano E, Palazon J, Bonfill M (2011) Production of the anticancer drug taxol in *Taxus baccata* suspension cultures: a review. Process Biochem 46:23–34
Malik S, Mirjalili MH, Fett-Neto A, Mazzafera P, Bonfill M (2013) Living between two worlds: two-phase culture systems for producing plant secondary metabolites. Crit Rev Biotechnol 33(1):1–22
Malik S, Bhushan S, Sharma M, Ahuja PS (2014a) Biotechnological approaches to the production of shikonins: a critical review with recent updates. Crit Rev Biotechnol 16:1–14
Malik S, Biba O, Gruz J, Arroo RRJ, Strnad M (2014b) Biotechnological approaches for producing aryltetralin lignans from *Linum* species. Phytochem Rev 13:893–913
Malonne H, Atassi G (1997) DNA topoisomerase targeting drugs: mechanisms of action and perspectives. Anti-Cancer Drugs 8:811–822
Marrelli M, Conforti F, Statti GA, Cachet X, Michel S, Tillequin F, Menichini F (2011) Biological potential and structure activity relationships of most recently developed vascular disrupting agents: an overview of new derivatives of natural combretastatin A-4. Curr Med Chem 18:3035–3081
Mathijssen RHJ, van Alphen RJ, Verweij J, Loos WL, Nooter K, Stoter G, Sparreboom A (2001) Clinical pharmacokinetics and metabolism of irinotecan (CPT-11). Clin Cancer Res 7:2182–2196
McWhirter JE, Hoffman-Goetz L (2016) Coverage of skin cancer risk factors and UV behaviors in popular U.S. magazines from 2000 to 2012. J Cancer Educ 31:382–388
Moertel CG, Schutt AJ, Reitemeier RJ, Hahn RG (1972) Phase II study of camptothecin (NSC-100880) in the treatment of advanced gastrointestinal cancer. Cancer Chemother Rep 56(1):95–101
Mojica MA, Leon A, Rojas-Sepulveda AM, Marquina S, Mendieta-Serrano MA, Salas-Vidal E, Villarreal ML, Alvarez L (2016) Aryldihydronaphthalene-type lignans from *Bursera fagaroides var. fagaroides* and their antimitotic mechanism of action. RSC Adv 6:4950–4959
Newman DJ, Cragg GM (2012) Natural products as sources of new drugs over the 30 years from 1981 to 2010. J Nat Prod 75:311–335

Ni XL, Wen SQ, Wang W, Wang XY, Xu H, Kai GY (2011) Enhancement of camptothecin production in *Camptotheca acuminata* hairy roots by overexpressing ORCA3 gene. J Appl Pharm Sci 1(8):85–88

Nicolau KC, Yang Z, Liu JJ, Ueno H, Nantermet PG, Guy RK, Clairborne CF et al (1994) Total synthesis of taxol. Nature 367:630–634

Noble RL (1990) The discovery of the vinca alkaloid schemo therapeutic agents against cancer. Biochem Cell Biol 68:1344–1351

O'Neill PM, Posner GH (2004) A medicinal chemistry perspective on artemisinin and related endoperoxides. J Med Chem 47:2945–2964

Patridge E, Gareiss P, Kinch MS, Hoyer D (2016) An analysis of FDA-approved drugs: natural products and their derivatives. Drug Discovery Today 21(2):204–207

Pazdur R (2013) Cancer drug information. http://www.cancer.gov/cancertopics/druginfo/fda-nanoparticle-paclitaxel. Accessed 24 Nov 2016

Peña-Morán OA, Villarreal ML, Álvarez-Berber L, Meneses-Acosta A, Rodríguez-López V (2016) Cytotoxicity, post-treatment recovery, and selectivity analysis of naturally occurring podophyllotoxins from *Bursera fagaroides var. fagaroides* on breast cancer cell lines. Molecules 21:1013–1028

Petter M (2015) Shining a spotlight on occupational cancers. Safet Health Pract 33(10):8

Pettit GR, Searcy JD, Tan R, Cragg GM, Melody N, Knight JC, Chapuis J-C (2016) Antineoplastic Agents. 585. Isolation of *Bridelia ferruginea* anticancer podophyllotoxins and synthesis of 4-Aza-podophyllotoxin structural modifications. J Nat Prod 79:507–518

Pizzolato LBS (2003) The camptothecins. Lancet 361:2235–2242

Polovich M, Olsen M, LeFebvre K (2014) Chemotherapy and biotherapy guidelines and recommendations for practices. Oncology Nursing Society, Pittsburg

Qazi PH, Rashid A, Shaw SA (2011) *Podophyllum hexandrum*—a versatile medicinal plant. Int J Pharm Pharm Sci 3:261–268

Rahmani A, Adebasis YA, ALY S (2015) Role of green tea and its constituent epigallocatechin-3-gallate in the health management. Int J Pharm Pharm Sci 7(3):6–12

Rao Y, Zhang D, Li R (2015) Tu Youyou and the discovery of artemisinin: 2015 nobel laureate in physiology or medicine. World Scientific, London

Renouard S, Corbina C, Colas C, Fidel T, Lopez T, Leclerc EA, Hendrawati O, Falguières A, Doussot J, Ferroud C, Maunit B, Lainé E, Hano C (2015) Aerial parts of *Callitris* species as a rich source of deoxypodophyllotoxin. Ind Crop Prod 63:53–57

Rodrigues DE, Moreira KFA, Oliveira TS (2016) Barriers to prevention of cervical cancer in the city of Porto Velho, Rondônia, Brazil. Invest Educ Enferm 34(1):58–66

Rowinsky EK (1994) Update on the antitumor activity of paclitaxel in clinical trials. Ann Pharmacother 28(5 Suppl):S18–S22

Rowinsky EK, Onetto N, Canetta RM, Arbuck SG (1992) Taxol: the first of the taxanes, an important new class of antitumor agents. Semin Oncol 19(6):646–662

Saito K, Sudo H, Yamazaki M, Koseki-Nakamura M, Kitajima M, Takayama H, Aimi N (2001) Feasible production of camptothecin by hairy root culture of *Phiorrhiza pumila*. Plant Cell Rep 20:267–271

Salim V, de Luca V (2013) Towards complete elucidation of monoterpene indole alkaloid biosynthesis pathway: *Catharanthus roseus* as a pioneer system. In: New light on alkaloid biosynthesis and prospects, 1st edn. Elsevier, London.

Schacter L (1996) Etoposide phosphate: what, why, where, and how? Semin Oncol 23:1–7

Schiff PB, Fant J, Horwitz SB (1979) Promotion of microtubule assembly in vitro by taxol. Nature 22:665–666

Schippmann U (2001) Medicinal plants significant trade study (CITES Project S-109). German Federal Agency for Nature Conservation, Bonn

Seidlova-Massinova V, Santavy M, Santavy J (1957) The biological effect of some podophyllotoxin compounds and their dependence on chemical structure. J Natl Cancer Inst 18:359–369

Sena JS, Girão RJS, Figueiredo SM et al (2016) Occupational skin cancer: systematic review. Rev Assoc Med Bras 62(3):280–286

Shankari V, Gurunathan S (2015) Drug discovery: an appraisal. Int J Pharm Pharm Sci 7(4):59–66
Singh J, Shah NC (1994) *Podophyllum*: a review. Curr Res Med Arom Plants 16:53–83
Singla AK, Garg A, Aggarwal D (2002) Paclitaxel and its formulations. Int J Pharm 235(1–2):179–192
Sirikantaramas S, Asano T, Sudo H, Yamazaki M, Saito K (2007) Camptothecin: therapeutic potential and biotechnology. Curr Pharm Biotechnol 8:196–202
Slichenmyer W, Rowinsky EK, Donehower RC, Kaufmann SH (1993) The current status of camptothecin analogues as antitumor agents. J Natl Cancer Inst 85:271–291
Stähelin HF, von Wartburg A (1991) The chemical and biological route from podophyllotoxin glucoside to etoposide: ninth cain memorial award lecture. Cancer Res 51:5–15
Sterwart BW, Wild CP (2014) World cancer report 2014. http://publications.iarc.fr/Non-Series-Publications/World-Cancer-Reports/World-Cancer-Report-2014. Accessed 25 Sep 2016
Tabata H (2004) Paclitaxel production by plant-cell-culture technology. Adv Biochem Eng Biotechnol 87:1–23
Veeresham C (2012) Natural products derived from plants as a source of drugs. J Adv Pharm Technol Res 3(4):200–201
Volgin VN, Stranadko EF, Kagoyants RV (2014) The results of the treatment of various morphological types of ENT skin cancer by photodynamic therapy. Saratov J Med Sci Res 10(3):555–558
Wall ME, Wani MC (1996) Camptothecin and taxol: from discovery to clinic. J Ethnopharmacol 51:239–254
Wang H, Khor TO, Shu L, Su Z, Fuentes F, Lee JH, Kon ANT (2012) Plants against cancer: a review on natural phytochemicals in preventing and treating cancers and their druggability. Anti Cancer Agents Med Chem 12(10):1281–1305
Wani M, Taylor H, Wall M, Coggon P, McPhail A (1971) Plant antitumor agents. VI. The isolation and structure of taxol, a novel antileukemic and antitumor agent from *Taxus brevifolia*. J Am Chem Soc 93:2325–2357
Weeks JC, Uno H, Taback N et al (2014) Interinstitutional Variation in management decisions for treatment of 4 common types of cancer: a multi-institutional cohort study. Ann Intern Med 161(1):20–30
Wicremesinhe ERM, Arteca RN (1993) Taxus callus cultures: initiation, growth optimization, characterization and taxol production. Plant Cell Tissue Organ Cult 35:181–193
Witterland AH, Koks CH, Beijnen JH (1996) Etoposide phosphate, the water soluble prodrug of etoposide. Pharm World Sci 18:163–170
World Health Organization (2015) Cancer 2015. http://www.who.int/mediacentre/factsheets/fs297/en/. Accessed 20 Sep 2016
Xue W, Chen S, Hao Y et al (2014) CRISPR-mediated direct mutation of cancer genes in the mouse liver. Nature 514:380–384
Yamazaki M, Asano T, Yamazaki Y, Sirikantaramas S, Sudo H, Saito K (2010) Biosynthetic system of camptothecin: an anticancer plant product. Pure Appl Chem 82(1):213–218
Yang EB, Wang Y, Zang HJ, Li ZM, Liu XH (2016) Accumulation effects of 3-N-benzyloxycarbonyl-aminobutyric salicylate on alkaloids in *Catharanthus roseus* cells. Lett Drug Des Discov 13(2):117–120
Zhou BN, Hoch JM, Johnson RK, Mattern MR, Eng WK, Ma J, Hecht SM, Newman DJ, Kingston DGI (2000) Use of COMPARE analysis to discover new natural product drugs: isolation of camptothecin and 9-methoxycamptothecin from a new source. J Nat Prod 63:1273–1276
Zhou P, Yang J, Zhu J et al (2015) Effects of β-cyclodextrin and methyl jasmonate on the production of vindoline, catharanthine, and ajmalicine in *Catharanthus roseus* cambial meristematic cell cultures. Appl Microbiol Biotechnol 99:7035–7045
Zhu X, Wu BJ, Luo HW, Tao L, Zhao Q, Guo QL, Wu CX (2010) Preparation of solid inclusion complex of deoxypodophyllotoxin-sbe-β-cd and its antitumor activity. J China Pharm Univ 41:447–450

Chapter 6
Cambial Meristematic Cells: A Sustainable Platform for the Production of Plant-Derived Anticancer Drugs

Verity I.P. Loake and Marisol Ochoa-Villarreal

Abstract For centuries, man has looked toward nature as a means to ameliorate a plethora of health conditions. Natural products (NPs), in particular those from plants, have proven to be a vital source of pharmaceuticals as in the case of the well-established anticancer drug, paclitaxel and the emerging ginsenoside triterpenoids. However, plant sources of high-value NPs can often be slow growing and are rarely domesticated and the target molecules are typically found in only low concentrations. Numerous strategies have been implemented for the production of these high-value molecules, including natural harvest from the source plant, total chemical synthesis or semi-synthesis from a more abundant precursor. Currently, plant cell culture arguably offers the most sustainable, robust, controllable and environmentally friendly platform to produce NPs for the pharmaceutical industry. In this context, the isolation of cambial meristematic cells (CMCs) and their potential utility to produce key anticancer drugs may serve to overcome some of the limitations previously associated with production of pharmaceuticals from traditional dedifferentiated plant cells (DDCs).

Keywords Anticancer drugs • Plant cell culture • Plant natural products • Secondary metabolites • Cambial meristematic cells • Plant stem cells

Introduction

For millennia, the chemical cornucopia found in the natural world has been harnessed to provide treatment for numerous ailments. Many plants, fungi, animals and marine invertebrates have been used in traditional medicine as herbal extracts, which has played an important role in the culture and history of different civilizations, often underpinning significant advances in medical science. To date, approximately 200,000

V.I.P. Loake • M. Ochoa-Villarreal (✉)
Institute of Molecular Plant Sciences, School of Biological Sciences,
University of Edinburgh, King's Buildings, Max Born Crescent, Edinburgh EH9 3BF, UK
e-mail: m.ochoa-villarreal@sms.ed.ac.uk

S. Malik (ed.), *Biotechnology and Production of Anti-Cancer Compounds*,
DOI 10.1007/978-3-319-53880-8_6

NPs have been identified (Fiehn 2002) and a number of these have been utilised as anti-inflammatory, antioxidant and antibacterial agents, among others. However, NPs have perhaps been most prominent in providing numerous blockbuster anticancer drugs.

Approximately, 70% of anticancer drugs approved from 1981 to 2010 were either natural products or derivatives thereof (Newman and Cragg 2012; Cragg and Newman 2013; Newman and Cragg 2007). The vast chemical diversity of plants, driven by the evolution of a profusion of secondary metabolites, offers an abundant source of potential pharmaceuticals, superior in complexity to that achieved by combinatorial chemistry. Recently, combinatorial chemistry has chiefly been employed by pharmaceutical companies to create chemical libraries to underpin the search for new lead compounds. However, the relative number of drug leads discovered and approved has been falling. This has primed a renaissance of NP screening with potential drug targets (Dias et al. 2012; Cragg and Newman 2005). However, exploiting the potential of plant-derived NPs for biomedical applications is often limited by insufficient quantities of the target molecule. Here we outline alternative routes for the production of NPs including extraction from the source plant, total chemical synthesis and semi-synthesis. Significantly, we highlight the great potential of plant cell culture in this space, especially the utility of cambial meristematic cells (CMCs) for the production of anticancer drugs.

Plants as a Source of Anticancer Drugs

Currently, 25% of the drugs on the market are plant-derived (Amin et al. 2009). Further, 10% of the World Health Organisation's list of essential medicines are obtained directly or indirectly from plants. These include the anticancer vinca alkaloids, vincristine and vinblastine, isolated from the Madagascar periwinkle, *Catharanthus roseus*. Especially prominent is paclitaxel, isolated from various *Taxus* species (yew trees), podophyllotoxin from *Podophyllum* species and camptothecin from *Camptotheca acuminata* (Fig. 6.1). All these molecules are derived from plant secondary metabolism, the evolution of which has been largely driven by environmental stresses, insect predation or attempted microbial infections (Miresmailli and Isman 2014; Wu and Chappell 2008). Typically, most plant NPs are produced in very low amounts (<0.1% of dry weight) and are often tissue, species and developmental state specific. Some of these NPs are produced through complex biosynthetic routes in a wide variety of different structures with different complexities (Oksman-Caldentey and Inzé 2004). It has been estimated that from 750,000 species of known higher plants a mere 2% have been screened for biological activities. Therefore, there is still an immense source of NPs to potentially exploit from plants, which may provide a continuing supply of anticancer molecules. In this context, the fair and equitable sharing of benefits arising out of the utilisation of such plant-based natural products has been enshrined in the Nagoya Protocol, agreed by 86 UN member states and active since 2014.

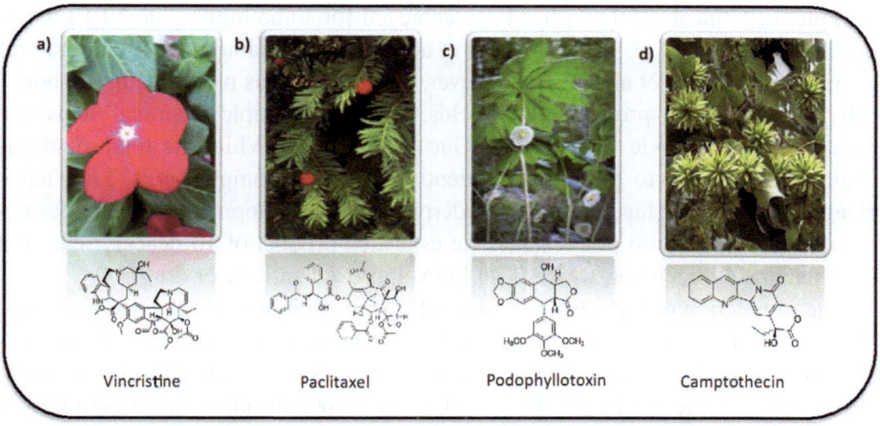

Fig. 6.1 Plant natural products used as anticancer. (**a**) Vincristine from *Catharanthus roseus*; (**b**) paclitaxel from *Taxus* spp.; (**c**) podophyllotoxin from *Podophyllum peltatum,* and (**d**) camptothecin from *Camptotheca acuminata*

Extraction of Natural Products from the Source Plant

The extraction of target NPs from the source plant is the first step and sometimes the longer term strategy to isolate the target molecule. The accumulation of these secondary metabolites can be highly variable according to the source, location, harvest season and environmental conditions. Typically, the source plant is not domesticated and often their wild population may be limiting and/or difficult to access. The flexibility of a plant to grow in a wider range of areas may be advantageous to increase the extent of cultivation. In this aspect, *C. roseus*, the source of the anticancer molecules vincristine and vinblastine, native to Madagascar, has been cultivated widely (Cragg and Newman 2005). Another limitation associated with the extraction of the NPs is the slow growth rate of some plant species, for example, *Taxus* species, the source of the anticancer drug, paclitaxel. It can take more than 30 years to grow a mature *Taxus* tree (Miresmailli and Isman 2014).

Total or Semi-synthesis of Target Natural Product

An alternative strategy for the production of NPs for drugs is total chemical synthesis. This is advantageous when the NP molecule is relatively simple with no complicated structure, such as either acetyl salicylic acid (aspirin) or ephedrine, utilized to prevent low blood pressure during spinal anaesthesia. However, many NPs have chiral centres with region-specific and stereo-specific properties necessary for their biological function (Wu and Chappell 2008), making chemical synthesis both difficult and often commercially unprofitable (Oksman-Caldentey and Inzé 2004).

Complete chemical synthesis has been achieved for some high-value NPs such as paclitaxel (Holton et al. 1994a; Holton et al. 1994b; Nicolaou et al. 1994) and vincristine (Yokoshima et al. 2002). However, this strategy has proved to be economically unviable as it produces low yields, involves multiple chemical steps and generates several toxic by-products (Guo et al. 2006). While the total synthesis strategy is unlikely to produce sufficient amounts of complex NPs, knowledge gleaned in this space has sometimes underpinned the development of synthetic analogues. This is the case of docetaxel, an esterified product of 10-deacetyl-baccatin III, a paclitaxel precursor, which is an FDA-approved anticancer drug.

The production of NPs using semi-synthesis may circumvent the disadvantages presented by total chemical synthesis. In this strategy, a more abundant precursor is harvested from its natural source followed by a series of chemical reactions to produce the final target molecule. This production route still relies on extraction from plant sources and often generates toxic chemical waste products (Wu and Chappell 2008). Despite this, semi-synthesis is often more economically viable than total chemical synthesis, especially for more complex molecules. For instance, a semi-synthesis process is used to produce irinotecan and topotecan, two analogues produced from the natural anticancer camptothecin. Other examples include the semi-synthetic etoposide, utilised for chemotherapy against Kaposi's sarcoma and teniposide, employed in the treatment of childhood acute lymphocytic leukaemia. Both of these anticancer molecules are produced from a biochemical precursor isolated from *Podophyllum emodi*, native to India and Pakistan (Cragg et al. 1995). This strategy is also used for the production of paclitaxel, in which a more abundant precursor, 10-deacetyl-baccatin III, is harvested (Howat et al. 2014).

Plant Cell Culture as a Platform for the Production of Anticancer Compounds

Plant cell culture (PCC) is a well-established platform used to produce NPs. This technology provides several advantages over natural harvest, total chemical synthesis or semi-synthesis, especially for complex molecules with multiple chiral centres (Roberts 2007). PCC is an environmentally friendly alternative that does not limit the production of an NP by seasonal availability (Rao and Ravishankar 2002). The production of NPs by PCC can also potentially alleviate the loss of plant biodiversity, especially for endangered wild populations (Roberts 2007; Wilson and Roberts 2012). Strikingly higher rates of biomass production can also be obtained in comparison to whole-plant cultivation given that the doubling time in PCC for some species is less than 24 h (Xu et al. 2011). Importantly, NP production by PCC can be undertaken under stringent Good Manufacturing Practice (GMP) conditions enabling the consistent synthesis of a high-quality product. Further, PCC facilitates efficient downstream purification processes, especially when the target NPs are secreted into the culture medium (Davies and Deroles 2014). PCC also offers high levels of containment enabling the possible application of gene editing technologies to increase yields. This production strategy also does not utilize farmland that could be employed for food cultivation and also decreases the use of water resources associated with contemporary agricultural production and plant nursery regimes.

PCC has been used in the production of several NPs and significantly some high-value anticancer drugs such as terpenoid indole alkaloids from *C. roseus* (Pasquali et al. 2006), taxanes from *Taxus* species (Tabata 2006; Malik et al. 2011), podophyllotoxin from *Podophyllum* species (Yousefzadi et al. 2010) and camptothecin from *C. acuminata* (Pi et al. 2010). The commercial development of PCC is now rapidly accelerating. Recent advances in the selection of high-performing cell lines, culture optimisation, use of improved and disposable bioreactors for scale-up and secretion and extraction of NPs (Ochoa-Villarreal et al. 2016) have all supported significant advances in PCC. There are currently seven PCC companies that produce commercial pharmaceutical products; some of the NPs synthesised include the anticancer drugs berberine, paclitaxel and podophyllotoxin (Wilson and Roberts 2012).

Generation of Dedifferentiated Cells for PCC

PCC can potentially be originated from any part of the plant, but traditionally embryos, roots and shoots have been utilised. The explants when placed in a medium solution with phytohormones generate a dedifferentiated cluster of cells. These dedifferentiated cells (DDCs) are considered totipotent due to their ability to regenerate a complete plant (Thorpe 2012). The DDCs originate a callus that can be placed in a liquid medium to create suspension cultures. These suspension cultures can be indefinitely maintained in a generally dedifferentiated state by using the phytohormones auxin and cytokinin. The generation of DDCs is not a simple reversal of the dedifferentiation process; it drives many changes in plant cells. This process results in the mitotic activation of the specialised cells into a multicellular heterogeneous mixture of proliferating cells (Grafi et al. 2007), resulting in genetic and epigenetic modifications (Roberts 2007; Grafi et al. 2007; Sugimoto et al. 2010). The resulting suspension cells characteristically show low and inconsistent NP yields (Roberts 2007).

The production of NPs is inversely related to cell growth. This growth inhibition during the stationary phase is associated with the production of enzymes related to secondary metabolism. Therefore, a two-phase system to produce NPs is often deployed (Malik et al. 2012). The first medium supports an increase in biomass while the second enables NP synthesis. Generally, a decrease in the carbon/nitrogen ratio, elimination of phytohormones such as the auxin, 2,4-D, a reduction in phosphate levels or a change in sugar levels may also support an increase in NP production (Bhojwani and Dantu 2013).

An important limiting factor for the commercial exploitation of plant suspension cultures is their tendency to form large aggregates. These clusters originate as the plant cells divide and with cells becoming linked together by the middle lamella of the cell wall, a pectin-rich polymer that binds to the new divided cells, forming large aggregates from 100 μm to 2 mm (Kolewe et al. 2008). The formation of clusters results in a high degree of variation in the cultured cells, resulting from exposure to heterogenous microenvironments where oxygen and nutrients are not evenly distributed (Naill and Roberts 2004). As a consequence, such suspension cells often exhibit decreased yields of the target NP.

Isolation and Characterisation of Cambial Meristematic Cells

The isolation, culture and characterisation of cambial meristematic cells (CMCs) are exciting developments in PCC, with these cells providing an attractive platform for the production of NPs (Lee et al. 2010). CMCs are innately undifferentiated cells that function as plant stem cells (Lee et al. 2010; Ochoa-Villarreal et al. 2015). The deployment of CMCs circumvents the dedifferentiation process previously associated with the generation of DDCs. These undifferentiated cells are isolated from the vascular cambium (Ye 2002). CMCs are defined by their abundant vacuoles, dense cytoplasm and thin cell walls (Frankenstein et al. 2005).

CMCs were first isolated from *Taxus cuspidata*, where paclitaxel biosynthesis is most conspicuous within the region containing these cells (Strobel et al. 1993). A recently developed twig was first collected from a wild yew, *T. cuspidata*. Subsequently, tissue that contained cambium, phloem, cortex and epidermis was peeled from the xylem. The resulting tissue was then cultured and actively proliferating cambium cells could be gently separated from the DDCs derived from phloem, cortex and epidermis. This mass of proliferating cells was morphologically distinct from adjacent DDCs. The production of CMCs from other plant species suggests that this technology has broad utility (Moon et al. 2015).

Microscopic evaluation of CMCs revealed small spherical abundant vacuoles that are characteristic of the vascular cambium. In contrast, DDCs have a single large vacuole typical of plant cells. The vascular cambium is typically composed of two forms: fusiform initials that produce traqueary elements and xylary fibres and ray initials that produce ray parenchyma cells (Ye 2002). CMCs were able to differentiate a high frequency into traqueary elements in response to inductive cues. The response of CMCs to ionising radiation was also evaluated. Stem cells are sensitive to programmed cell death triggered by ionising radiation to protect genome integrity within this pivotal cell type (Fulcher and Sablowski 2009). In CMCs exposure to γ-radiation triggered a high level of rapid programmed cell death, in contrast with little cell death in DDCs (Lee et al. 2010). In a similar fashion, CMCs exhibited rapid programmed cell death in response to the drug zeocin, unlike DDCs (Chankova et al. 2007). In addition, a combination of next-generation sequencing transcriptomics was utilised to confirm the identity of CMCs (Lee et al. 2010). The procambium marker, phloem intercalated with xylem (PXY) gene, encodes for a receptor-like kinase (RLK) whose function is to maintain cell polarity and orientation of cell division in the vascular cells (Fisher and Turner 2007). The expression of the PXY was clearly higher in CMCs, in comparison to DDCs. Another procambium marker, wooden leg (WOL), which encodes a two-component histidine kinase expressed in the cambium, was also up-regulated in CMCs. These experiments are consisted with the stem cell-like nature of the isolated CMCs.

Advantages of CMCs for Production of Anticancer Molecules

The performance of CMCs with respect to growth and production of the target high-value molecule was strikingly superior to that of the DDCs. Following growth on solid media with subcultures every 2 weeks for 22 months, the dry cell weight (dcw) of *T. cuspidata* CMCs was 1250 g. By contrast, *T. cuspidata* embryo- or needle-derived DDCs produced a total of 0.32 and 0.41 g dcw, respectively. Aggregation of plant cells within suspension cultures is a common feature which decreases the supply of oxygen and nutrients within the cell cluster, thereby decreasing NP biosynthesis (Joshi et al. 1996). In this context, the reduced aggregation size of CMCs, between 2 and 3 cells per cluster, in contrast to the large cell aggregates typically found in DDCs, represents a significant advantage for the culture of these cells. Cell aggregation also promotes sensitivity to shear stress especially when cells are cultured within a bioreactor (Joshi et al. 1996). Shear stress often constitutes a significant limitation for the scale-up of plant suspension cells. When cultured in a 10 L stirred tank bioreactor *T. cuspidata* CMCs were considerably less sensitive to shear stress than DDCs derived from this plant species. Presumably, this increased shear stress resistance exhibited by CMCs is a consequence of reduced aggregation, abundant small vacuoles and thinner cell walls.

Enhanced Elicitation of Natural Product Biosynthesis in CMCs

The addition of microbial extracts or endogenous plant immune activators, so-called elicitors, is a well-established strategy to trigger the biosynthesis of a target NP (Pauwels et al. 2009), as many of these molecules are integral to the plant defence response against microbes and insects. The application of the elicitor, methyl jasmonate (MeJA), to *T. cuspidata* CMCs increased the production of paclitaxel by 14,000% (Lee et al. 2010), in comparison with 220 and 433% of needle- or embryo-derived DDCs, respectively. Thus, CMCs are significantly more sensitive to elicitation than traditional DDCs.

The advantages of CMCs with respect to DDCs for plant cell suspension culture facilitate significant benefits during the scale-up from laboratory to commercial levels of production. After 10 days elicitation with MeJA, using a 3 L airlift bioreactor *T. cuspidata* CMCs, produced 98 mg/kg of fresh cell weight (fcw) compared with only 11 and 13 mg/kg of fcw of needle- or embryo-derived DDCs. Similar results were found when using a 20 L airlift bioreactor. With respect to paclitaxel production, while *T. cuspidata* DDCs produced no measurable levels of this molecule under these conditions, CMCs produced 268 mg/kg of paclitaxel (Lee et al. 2010). Interestingly, most of this synthesised paclitaxel was released into the culture media. Secretion of the target NP into the culture medium is an important aspect in PCC because it facilitates the facile purification of the final product (Wilson and Roberts 2014). Collectively, CMCs therefore provide an attractive alternative to produce NPs, especially those with a complex structure such as paclitaxel.

Utility of CMCs for Paclitaxel Production

Paclitaxel was discovered during a major bio-screening program initiated by the National Cancer Institute (CNI) between the years 1958 and 1980, in which more than 110,000 plant-derived compounds were examined for anticancer activity (Kingston 2000; He et al. 2000). The history has been addressed in many reviews (Wilson and Roberts 2012; Patel 1998; Mountford 2010), which describe the monumental effort undertaken by many scientific groups and organisations from the discovery of a potentially antitumour agent to the development and its commercialisation for cancer treatment. Paclitaxel has been approved by the FDA for the treatment of many diseases including breast, ovarian and lung cancer, AIDS-related Kaposi's sarcomas and coronary artery disease in over 50 countries (Mountford 2010). The diterpenoid paclitaxel belongs to the class of mitotic disrupter anticancer drugs. Its mechanism of action is based on the stabilisation of the microtubule network, blocking the cell cycle-inducing apoptosis in affected cells (Manfredi et al. 1982). The core structure of paclitaxel contains four tetracyclic rings with numerous functional groups including hydroxyl and benzoyl acetyl groups and an oxetane ring. The antitumor activity of paclitaxel is mainly due to the C13 side chain (Kingston 2000). The interaction of the C3' amide-acyl group in the C13 chain and the hydroxyl group at C2' with β-tubulin of the microtubule produces cytotoxicity and microtubule stabilisation (He et al. 2000), blocking the cell cycle at the G2/M phase (Schiff and Horwitz 1980).

Paclitaxel was first obtained from the inner bark of *Taxus brevifolia* (Pacific yew) (Wani et al. 1971), and has subsequently been identified in the 12 known *Taxus* species which are distributed throughout the northern hemisphere. It accumulates in very low concentrations (about 0.001–0.05% of dry weight) and its natural extraction is both expensive and not environmentally friendly (Roberts 2007). For instance, it is necessary to take 10,000 kg of *Taxus* bark or 3000 yew trees to produce 1 kg of paclitaxel (Patel 1998). Further, one cancer patient needs approximately 3 g of paclitaxel per treatment (Malik et al. 2011). Despite this, in China paclitaxel continues to be traditionally harvested from *Taxus chinensis* (Malik et al. 2011).

Paclitaxel can be chemically synthesised but the method is far from being economically feasible (Holton et al. 1994a, 1994b; Nicolaou et al. 1994;). Instead, a semi-synthesis process was developed to deal with the increasing demand for this molecule. The process requires a precursor, 10-deacetyl-Baccatin III, which is subjected to a series of 11 chemical transformations, using 13 different solvents and some steps that require cryogenic temperatures, to produce paclitaxel (Mountford 2010).

In past decades, significant effort has been focused on the biotechnological production of paclitaxel based on PCC. A number of research groups have extensively investigated the optimisation of this process. This involved comparing paclitaxel production among a variety of different *Taxus* species, optimisation of culture conditions including media composition, selection of high-yielding cell lines, development of a two-stage culture system, cell immobilisation strategies, in situ product removal, use of different adsorbants and addition of precursors and elicitors (Wilson and Roberts 2012; Tabata 2006; Malik et al. 2011; Cusido et al. 2014; Expósito et al. 2009; Vongpaseuth and Roberts 2007; Sabater-Jara et al. 2010). Many elicitors have been used to increase paclitaxel production in PCC (Malik et al. 2011; Vongpaseuth and Roberts 2007); however, the immune-activator methyl jasmonate (MeJA) is thought to be most effective (Mirjalili and Linden 1996). Reported values for *T. cuspidata* treated with MeJA show a range from 20 to 84 mg/kg fcw of paclitaxel (Mirjalili and Linden 1996). In 2002, Bristol-Myers Squibb, one of the main paclitaxel suppliers, switched production methods entirely from the semi-synthetic route to PCC, for environmental and safety reasons (Mountford 2010). Presently, two companies have established PCC for profitable paclitaxel production, Phyton Biotech in the USA and Samyang Genex in Korea (Nosov 2012).

Significantly, some of the main problems associated with the employment of cell suspension cultures of traditional DDCs, including low paclitaxel yield, slow rate of growth, extensive cell aggregation, relative insensitivity to elicitation and susceptibility to shear stress, were all largely overcome with CMCs from *T. cuspidata* (Table 6.1). Thus, the use of CMCs provides an attractive technology platform for the production of paclitaxel and potentially other important plant-derived anticancer drugs.

Several paclitaxel biosynthetic genes have been expressed functionally in *Saccharomyces cerevisiae* and employed to establish a five-step biosynthesis of the paclitaxel intermediate taxadiene, the first committed paclitaxel intermediate, producing 0.85 mg/L (DeJong et al. 2006; Engels et al. 2008). Furthermore, efforts are also being undertaken to establish paclitaxel production in *Escherichia coli* (Huang et al. 2001). More recently, the production of taxadiene has been increased towards 1 g/L in this microbe. This was achieved by partitioning taxadiene biosynthesis into two modules: a native upstream methylerythritol-phosphate pathway forming isopentenyl pyrophosphate and a heterologous downstream terpenoid-forming pathway (Ajikumar et al. 2010). However, there is still a long way to go to move from the production of the intermediate, taxadiene, to the end product, paclitaxel, and many obstacles remain. Significantly, the full list of enzymes responsible for paclitaxel biosynthesis remains to be determined. Also, many of the enzymes involved are P450s, which are notoriously difficult to express in heterologous microbial hosts, especially *E. coli*. Further, even if paclitaxel could be produced, it is likely the titres of this molecule would not be high enough to make this process economically viable. Industrial scale bioproduction requires a level of approximately >50 g/L, which is far from the less than 1 g/L for even taxadiene. So while promising, synthetic biology as a route to the production of plant-derived anticancer drugs currently remains in the developmental stage.

Table 6.1 Comparison of parameters evaluated for paclitaxel production in CMCs and DDCs

Properties evaluated	CMCs		DDCs Needle derived		DDCs Embryo derived	
Dry cell weight[a]	1250 g		0.32 g		0.41 g	
Aggregate size[b]	95%	<0.5 mm	60%	>3 mm	50%	>3 mm
			30%	1–3 mm	38%	1–3 mm
			8%	0.5–1 mm	7%	0.5–1 mm
			2%	<0.5 mm	5%	<0.5 mm
Paclitaxel yield[c]	102 mg/kg		23 mg/kg		39 mg/kg	
Paclitaxel yield[d] 3 L airlift bioreactor	98 mg/kg		11 mg/kg		13 mg/kg	
Paclitaxel yield[e] 3 L perfusion bioreactor	264 mg/kg		6 mg/kg		10.8 mg/kg	
Paclitaxel yield[f] 20 L airlift bioreactor	268 mg/kg		n.d.		n.d.	

Modified from Lee et al. (2010)
n.d.: Not detected
[a]Growth of CMCs and DDCs on solid media after 22 months of isolation
[b]Aggregation size in suspension culture
[c]Paclitaxel production 10 days after elicitation, 3-month-old cell suspension, following batch culture in a flask format
[d]Paclitaxel yield 10 days after elicitation of the indicated 6-month-old repeatedly subcultured cell suspensions, after batch culture in a 3 L airlift bioreactor
[e]Paclitaxel production following 45 days of perfusion culture
[f]Paclitaxel production following elicitation of 28-month-old in a 20 L airlift bioreactor

CMCs as a Platform for Ginsenoside Production

Panax meaning "all-healing" in Greek, shares the same origin as "panacea" and was applied to this genus because the Swedish botanist Linnaeus, the developer of the binomial nomenclature system, was well informed of the utility of this plant to traditional Chinese medicine. Further, the name ginseng comes from the Chinese words "Jen Sheng", meaning "man-herb", because of the humanoid shape of the root or rhizome of the plant, which is most commonly consumed.

Ginsenosides, triterpene saponins, are the major pharmacologically active molecules of *P. ginseng*. Ginsenosides Rg1, Rc, Rd, Re, Rb1, Rb2 and Rb0 are quantitatively the most important. Pharmaceutical activities associated with ginseng include anticancer, antioxidant and neuroprotection (Leung and Wong 2010; Keum et al. 2000).

P. ginseng is one of the most valuable oriental plants used in traditional Chinese medicine. The collection of rare wild mountain *P. ginseng* can be a lucrative activity with single plants often selling for $10,000 or more. Not surprisingly, this endeavour has driven the wild *P. ginseng* populations to the point of extinction. Thus, most *P. ginseng* is obtained by field cultivation. However, under these conditions the levels of ginsenosides are typically low, constituting one of the major hurdles that needs to be addressed in order to advance applications of ginsenosides. The use of PCC for the production of ginsenosides therefore constitutes an attractive alternative for the cost-effective and sustainable production of ginsenosides.

In this context, CMCs from wild mountain ginseng have therefore been isolated, cultured and characterised (Lee et al. 2010). As expected, these *P. ginseng* CMCs outperformed related DDCs with respect to growth under various formats and synthesis of the active NP. Following culture in a 3 L airlift bioreactor, *P. ginseng* CMCs produced 791 and 4425 mg/kg fcw of ginsenosides F2 and gypenoside XVII, respectively. Significantly, this is 23.8- and 24.1-fold more ginsenosides F2 and gypenoside XVII, respectively, than previously reported sources (Lee et al. 2010). In addition, *P. ginseng* CMCs also produce high levels of the major ginsenosides. For example, the levels of Rb1 and Rc produced were 19.7 and 7.14 mg/g fcw (Ochoa-Villarreal et al. 2015). Therefore, *P. ginseng* CMCs also provide an effective and sustainable platform for the commercial production of ginsenosides.

Conclusion

Plants represent a vast source of NPs with a high degree of diverse chemical structures. Recently, there has been renewed interest in drug screening using plant NPs, as libraries of molecules synthetised by combinatorial chemistry have not delivered the expected burst of new drugs. The production of plant NPs for the pharmaceutical industry harnesses chemical synthesis or semi-synthesis technology, when the target molecules are simple and easy to produce such as ephedrine and acid acetylsalicylic. PCC technology represents an excellent alternative platform for the production of complex NPs with multiple chiral centres such as paclitaxel. However, traditional DDCs also have significant limitations for the synthesis of NPs. Instability of product yield, high levels of cell aggregation and sensitivity to shear stress represent significant obstacles. In this context, CMCs, innately undifferentiated cells, circumvent many of the current problems associated with the use of DDCs. Thus, CMCs may provide a route for the robust, sustainable and cost-effective production of plant anticancer drugs. In addition, the use of CMCs in combination with new advances in PCC technology (Ochoa-Villarreal et al. 2016) offers exciting future opportunities in this space.

Acknowledgments M.O.V. is supported by a scholarship from CONACYT, Mexico.

References

Ajikumar PK, Xiao W-H, Tyo KEJ, Wang Y, Simeon F, Leonard E et al (2010) Isoprenoid pathway optimization for taxol precursor overproduction in *Escherichia coli*. Science 330(6000): 70–74

Amin A, Gali-Muhtasib H, Ocker M, Schneider-Stock R (2009) Overview of major classes of plant-derived anticancer drugs. Int J Biomed Sci 5(1):1–11

Bhojwani SS, Dantu PK (2013) Production of industrial phytochemicals. In: Plant tissue culture: an introductory text, 1st ed. Springer India. pp 275–286

Chankova SG, Dimova E, Dimitrova M, Bryant PE (2007) Induction of DNA double-strand breaks by zeocin in *Chlamydomonas reinhardtii* and the role of increased DNA double-strand breaks rejoining in the formation of an adaptive response. Radiat Environ Biophys 46(4):409–416

Cragg GM, Newman DJ (2005) Plants as a source of anti-cancer agents. J Ethnopharmacol 100(1–2):72–79

Cragg GM, Newman DJ (2013) Natural products: a continuing source of novel drug leads. Biochim Biophys Acta 1830(6):3670–3695

Cragg PGM, Boyd MR, Schepartz SA (1995) Pharmaceutical prospecting and the potential for pharmaceutical crops. Natural product drug discovery and development at the United States National Cancer Institute. Ann Mo Bot Gard 82:47–53

Cusido RM, Onrubia M, Sabater-Jara AB, Moyano E, Bonfill M, Goossens A et al (2014) A rational approach to improving the biotechnological production of taxanes in plant cell cultures of *Taxus* spp. Biotechnol Adv 32(6):1157–1167

Davies KM, Deroles SC (2014) Prospects for the use of plant cell cultures in food biotechnology. Curr Opin Biotechnol 26:133–140

DeJong JM, Liu Y, Bollon AP, Long RM, Jennewein S, Williams D et al (2006) Genetic engineering of taxol biosynthetic genes in *Saccharomyces cerevisiae*. Biotechnol Bioeng 93(5):212–224

Dias DA, Urban S, Roessner U (2012) A historical overview of natural products in drug discovery. Metabolites 2(2):303–336

Engels B, Dahm P, Jennewein S (2008) Metabolic engineering of taxadiene biosynthesis in yeast as a first step towards taxol (paclitaxel) production. Metab Eng 10(3–4):201–206

Expósito O, Bonfill M, Moyano E, Onrubia M, Mirjalili MH, Cusidó RM et al (2009) Biotechnological production of taxol and related taxoids: current state and prospects. Anti Cancer Agents Med Chem 9(1):109–121

Fiehn O (2002) Metabolomics—the link between genotypes and phenotypes. Plant Mol Biol 48(1–2):155–171

Fisher K, Turner S (2007) PXY, a receptor-like kinase essential for maintaining polarity during plant vascular-tissue development. Curr Biol 17(12):1061–1066

Frankenstein C, Eckstein D, Schmitt U (2005) The onset of cambium activity—a matter of agreement? Dendrochronologia 23(1):57–62

Fulcher N, Sablowski R (2009) Hypersensitivity to DNA damage in plant stem cell niches. Proc Natl Acad Sci U S A 106(49):20984–20988

Grafi G, Ben-Meir H, Avivi Y, Moshe M, Dahan Y, Zemach A (2007) Histone methylation controls telomerase-independent telomere lengthening in cells undergoing dedifferentiation. Dev Biol 306(2):838–846

Guo BH, Kai GY, Jin HB, Tang KX (2006) Taxol synthesis. Afr J Biotechnol 5:15–20

He L, Jagtap PG, Kingston DGI, Shen H-J, Orr GA, Horwitz SB (2000) A common Pharmacophore for taxol and the Epothilones based on the biological activity of a taxane molecule lacking a C-13 side chain. Biochemistry 39(14):3972–3978

Holton R, Somoza A, Kim C, Liang HB, Biediger F, Boatman RJ, Shindo PD, Smith M, Kim CC, First S (1994a) Total synthesis of taxol. 2. Completion of the C and D rings. J Am Chem Soc 116(4):1599–1600

Holton R, Somoza A, Kim C, Liang HB, Biediger F, Boatman RJ, Shindo PD, Smith M, Kim CC, First S (1994b) Total synthesis of taxol. 1. Functionalization of the B ring. J Am Chem Soc 116(4):1597–1598

Howat S, Park B, Oh IS, Jin Y-W, Lee E-K, Loake GJ (2014) Paclitaxel: biosynthesis, production and future prospects. New Biotechnol 31(3):242–245

Huang Q, Roessner CA, Croteau R, Scott AI (2001) Engineering *Escherichia coli* for the synthesis of taxadiene, a key intermediate in the biosynthesis of taxol. Bioorg Med Chem 9(9):2237–2242

Joshi JB, Elias CB, Patole MS (1996) Role of hydrodynamic shear in the cultivation of animal, plant and microbial cells. Chem Eng J Biochem Eng J 62(2):121–141

Keum Y-S, Park K-K, Lee J-M, Chun K-S, Park JH, Lee SK et al (2000) Antioxidant and anti-tumor promoting activities of the methanol extract of heat-processed ginseng. Cancer Lett 150(1):41–48

Kingston DGI (2000) Recent advances in the chemistry of taxol. J Nat Prod 63:726–734
Kolewe ME, Gaurav V, Roberts SC (2008) Pharmaceutically active natural product synthesis and supply via plant cell culture technology. Mol Pharm 5(2):243–256
Lee E-K, Jin Y-W, Park JH, Yoo YM, Hong SM, Amir R et al (2010) Cultured cambial meristematic cells as a source of plant natural products. Nat Biotechnol 28(11):1213–1217
Leung KW, Wong AS-T (2010) Pharmacology of ginsenosides: a literature review. Chin Med 5:20
Malik S, Cusidó RM, Mirjalili MH, Moyano E, Palazón J, Bonfill M (2011) Production of the anticancer drug taxol in *Taxus baccata* suspension cultures: a review. Process Biochem 46(1):23–34
Malik S, Hossein Mirjalili M, Fett-Neto AG, Mazzafera P, Bonfill M (2012) Living between two worlds: two-phase culture systems for producing plant secondary metabolites. Crit Rev Biotechnol 33:1–22
Manfredi JJ, Parness J, Horwitz SB (1982) Taxol binds to cellular microtubules. J Cell Biol 94:688–696
Miresmailli S, Isman MB (2014) Botanical insecticides inspired by plant-herbivore chemical interactions. Trends Plant Sci 19(1):29–35
Mirjalili N, Linden JC (1996) Methyl Jasmonate induced production of taxol in suspension cultures of *Taxus cuspidata*: ethylene interaction and induction models. Biotechnol Prog Am Chem Soc 12(1):110–118
Moon SH, Venkatesh J, Yu JW, Park SW (2015) Differential induction of meristematic stem cells of *Catharanthus roseus* and their characterization. Comp Rend Biol Acad Sci 338(11):745–756
Mountford PG (2010) The taxol® story–development of a green synthesis via plant cell fermentation. In: Green chemistry in the pharmaceutical industry, pp 145–160
Naill MC, Roberts SC (2004) Preparation of single cells from aggregated *Taxus* suspension cultures for population analysis. Biotechnol Bioeng 86(7):817–826
Newman DJ, Cragg GM (2007) Natural products as sources of new drugs over the last 25 years. J Nat Prod 70(3):461–477
Newman DJ, Cragg GM (2012) Natural products as sources of new drugs over the 30 years from 1981 to 2010. J Nat Prod 75(3):311–335
Nicolaou KC, Yang Z, Liu JJ, Ueno H, Nantermet PG, Guy RK et al (1994) Total synthesis of taxol. Nature 367(6464):630–634
Nosov AM (2012) Application of cell technologies for production of plant-derived bioactive substances of plant origin. Appl Biochem Microbiol 48(7):609–624
Ochoa-Villarreal M, Howat S, Jang MO, Kim IS, Jin Y-W, Lee E-K et al (2015) Cambial meristematic cells: a platform for the production of plant natural products. New Biotechnol 32(6):581–587
Ochoa-Villarreal M, Howat S, Hong S, Jang MO, Jin Y-W, Lee E-K et al (2016) Plant cell culture strategies for the production of natural products. BMB Rep 49(3):149–158
Oksman-Caldentey K-M, Inzé D (2004) Plant cell factories in the post-genomic era: new ways to produce designer secondary metabolites. Trends Plant Sci 9(9):433–440
Pasquali G, Porto DD, Fett-Neto AG (2006) Metabolic engineering of cell cultures versus whole plant complexity in production of bioactive monoterpene indole alkaloids: recent progress related to old dilemma. J Biosci Bioeng 101(4):287–296
Patel RN (1998) Tour de paclitaxel: biocatalysis for semisynthesis. Annu Rev Microbiol 52:361–395
Pauwels L, Inzé D, Goossens A (2009) Jasmonate-inducible gene: what does it mean? Trends Plant Sci 14(2):87–91
Pi Y, Jiang K, Hou R, Gong Y, Lin J, Sun X et al (2010) Examination of camptothecin and 10-hydroxycamptothecin in *Camptotheca acuminata* plant and cell culture, and the affected yields under several cell culture treatments. Biocell 34(3):139–143
Rao SR, Ravishankar GA (2002) Plant cell cultures: chemical factories of secondary metabolites. Biotechnol Adv 20(2):101–153
Roberts SC (2007) Production and engineering of terpenoids in plant cell culture. Nat Chem Biol 3(7):387–395
Sabater-Jara AB, Tudela LR, López-Pérez AJ (2010) In vitro culture of *Taxus* sp.: strategies to increase cell growth and taxoid production. Phytochem Rev 9(2):343–356

Schiff PB, Horwitz SB (1980) Taxol stabilizes microtubules in mouse fibroblast cells. Proc Natl Acad Sci U S A 77(3):1561–1565
Strobel GA, Stierle A, Hess WM (1993) Taxol formation in yew—*Taxus*. Plant Sci 92(1):1–12
Sugimoto K, Jiao Y, Meyerowitz EM (2010) Arabidopsis regeneration from multiple tissues occurs via a root development pathway. Dev Cell 18(3):463–471
Tabata H (2006) Production of paclitaxel and the related taxanes by cell suspension cultures of *Taxus* species. Curr Drug Targets 7(4):453–461
Thorpe T (2012) History of plant cell culture. In: Loyola-Vargas VM, Ochoa-Alejo N (eds) Methods in molecular biology, vol 877. Humana Press, Totowa, NJ, pp 169–180
Vongpaseuth K, Roberts SC (2007) Advancements in the understanding of paclitaxel metabolism in tissue culture. Curr Pharm Biotechnol 8(4):219–236
Wani MC, Taylor HL, Wall ME, Coggon P, McPhail AT (1971) Plant antitumor agents. VI. Isolation and structure of taxol, a novel antileukemic and antitumor agent from *Taxus brevifolia*. J Am Chem Soc 93(9):2325–2327
Wilson SA, Roberts SC (2012) Recent advances towards development and commercialization of plant cell culture processes for the synthesis of biomolecules. Plant Biotechnol J 10(3):249–268
Wilson SA, Roberts SC (2014) Metabolic engineering approaches for production of biochemicals in food and medicinal plants. Curr Opin Biotechnol 26:174–182
Wu S, Chappell J (2008) Metabolic engineering of natural products in plants; tools of the trade and challenges for the future. Curr Opin Biotechnol 19(2):145–152
Xu J, Ge X, Dolan MC (2011) Towards high-yield production of pharmaceutical proteins with plant cell suspension cultures. Biotechnol Adv 29(3):278–299
Ye Z-H (2002) Vascular tissue differentiation and pattern formation in plants. Annu Rev Plant Biol 53:183–202
Yokoshima S, Ueda T, Kobayashi S, Sato A, Kuboyama T, Tokuyama H et al (2002) Stereocontrolled total synthesis of (+)-vinblastine. J Am Chem Soc 124(10):2137–2139
Yousefzadi M, Sharifi M, Behmanesh M, Moyano E, Bonfill M, Cusido RM et al (2010) Podophyllotoxin: current approaches to its biotechnological production and future challenges. Eng Life Sci 10(4):281–292

Chapter 7
Family Fabaceae: A Boon for Cancer Therapy

Ashita Sharma, Rajwant Kaur, Jatinder Kaur Katnoria, Rajinder Kaur, and Avinash Kaur Nagpal

Abstract Cancer is a deadliest disease which has been a challenging prorogue for human health and civilization. Legumes or plants of family Fabaceae have been part of our diet since times immemorial. Plants of this family are rich in phytoconstituents which make them effective therapeutic agents for various diseases. The synergistic effects of different phytoconstituents are far more than the effect of individual compounds. Studies have revealed that those parts of the world which consume legumes as a staple diet have less mortality due to cancer as compared to other parts of the world. Also, it has been reported that phytochemical components of plants of this family inhibit carcinogenesis at various stages. The use of phytoconstituents derived from these plants along with traditional chemotherapy can have synergistic effects which help in fighting cancer. This review focuses on various compounds present in plants of family Fabaceae and their anticarcinogenic potential.

Keywords Legumes • Cancer • Lectins • Apoptosis • Carcinogenesis

Introduction

Cancer is among the dreadful diseases having the incurable attributes, thus posing threat to human health and existence. There have been significant advances in the field of cancer research, yet cancer is the major disease responsible for human deaths throughout the world (Siegel et al. 2011). Genetic, dietary, and environmental factors have been associated with the increase in incidences of carcinogenesis. A study has revealed that about 30–35% of cancer cases are the result of carcinogenesis induced through diet (Doll and Peto 1981). From times immemorial, medicinal plants are known to have the phytoconstituents that are being used for therapeutic purposes. Ingestion of phytoconstituents in the form of diet is known to fight and prevent cancer (Kaur and Kapoor 2001). Legumes or plants of family Fabaceae

A. Sharma • R. Kaur • J.K. Katnoria • R. Kaur • A.K. Nagpal (✉)
Department of Botanical and Environmental Sciences, Guru Nanak Dev University, Amritsar, Punjab 143001, India
e-mail: avnagpal@rediffmail.com

S. Malik (ed.), *Biotechnology and Production of Anti-Cancer Compounds*,
DOI 10.1007/978-3-319-53880-8_7

have played an important role in traditional diet in many parts of the world and are known to have important phytoconstituents that can help in cancer chemoprevention. Fabaceae is the third largest family of flowering plants comprising over 750 genera and more than 19,000 species which range from herbs to large canopy trees. The members of family Fabaceae are found to be growing in a variety of climatic zones like temperate, humid, arid, highlands, and savannahs (Morris 1999). The plants of this family are of industrial and pharmaceutical importance owing to their organic constituents which are used as raw material in many industries.

Fabaceae has a diverse fossil record and oldest fossils have been reported to be from late Palaeocene era and the study of fossils has suggested that these plants evolved in arid or semiarid areas near Tethys sea and have been very closely associated with human civilization in Asia, Europe, and America (Schrire et al. 2005). Being the rich source of proteins, these have become a staple diet in these areas since 6000 BC. Presently, members of family Fabaceae are distributed worldwide except the poles. Different kinds of beans play important role in cuisines of various regions including Asia, Middle East, South America, and Mexico (Messina 1999). The phytochemicals present in the plants of this family are important in terms of both pharmaceutical and industrial use. The plants of this family are rich in chemicals like rotenone, tephrosin, and deguelin which have the antitumor potential and are also used as pesticides in many regions of the world. Also, some plant species are source of alkaloids, antibiotics, and glycosides; thus they have always been important for the pharmaceutical industry (Morris 1999). As a result of wide variation in germplasm and extensive economic importance, efforts are being made for the conservation and preservation of legume germplasm so as to ensure possible availability of improved cultivars with added value (Morris 1999).

Dietary intake of legumes is higher in vegetarian diet than in nonvegetarian one. Nutritionally legumes are rich source of proteins, dietary fibers, and micronutrients. Also, they have varieties of flavonoids, phytosterols, phytates, saponins, and phenolic acids which are known for anticancer potential (Messina 1999). The members of this widely distributed and large family can thus be an answer to the fatal disease, cancer.

Traditional Uses

A lot of traditional knowledge is associated with the use of various plants for cure of many diseases. The indigenous healers and traditional medicine practitioners have been using the plants of family Fabaceae for the treatment of various diseases. Traditionally, consumption of some plants from this family was associated with the general well-being and longevity of human population. Bark of plants like *Acacia catechu, Acacia nilotica, Albizia lebbeck* and *Bauhinia variegata* have been traditionally used for the treatment of diarrhea, ulcers and inflammations (Devi et al.

2012). Leaves of *Acacia nilotica, Albizia lebbeck, Bauhinia acuminata, Cajanus cajan, Cassia fistula, Desmodium triflorum, Erythrina variegata* and *Mimosa diplotricha* have been used for various skin diseases, menstrual complications, lungs and gastric problems (Rahman and Parvin 2014). Ancient literatures as Vedas, Bible and Quran also have quoted the use of these plants for medicines and other ethnobotanical uses. Many phytoconstituents present in plants protect the cells from oxidative stress as they have antioxidant and antimutagenic properties (Bravo 1998; Ferguson 2001; Tedesco et al. 2001; Aruoma 2003; Surh and Ferguson 2003).

The use of leguminous plants for cancer chemoprevention has recently gained a lot of importance owing to the presence of many phytochemical groups which have antitumor properties. Low mortality rate due to breast cancer in Asian countries is attributed to the consumption of leguminous plants in routine diet (Messina 1999). Rodent studies have revealed that soy products when added to diet can reduce the risk of breast and prostate cancer. Saponins, isoflavones, phytates and phenolic acids are known to cause apoptosis in tumor cells (Morris 1999; Yuenyongsawad et al. 2013).

Various Phytoconstituents and Pathway of Action

As the research progressed in the field of molecular biology, the scientific community gained deeper understanding of the process of carcinogenesis. Sporn et al. (1976) suggested that the process of carcinogenesis can be divided into three phases, i.e., initiation, promotion and progression (Fig. 7.1). As the carcinogen enters the body, it either gets activated through various metabolic pathways or gets detoxified within the human body. If activated through metabolic mechanisms, these carcinogenic agents increase the oxidative stress and thus cause DNA damage; this leads to the initiation of carcinogenesis. During promotion phase the initiated cells start proliferating actively resulting in accumulation of preneoplastic cells. During progression stage preneoplastic cells start invading and spreading to various parts of the body. This is an irreversible phase (Surh 2003).

There are multiple pathways involved in cancer occurrence and progression; thus treatment or prevention of cancer by aiming at single pathway doesn't prove to be an effective strategy (Vogelstein and Kinzler 2004; Dandawate et al. 2014). There are many hindrances in treatment strategies to be followed for cancer, viz. drug resistance and side effects of chemotherapy. These hindrances affect the efficiency of cancer treatment, making it challenging for the scientific community to develop strategies and drugs for cancer treatment (Payne et al. 2006; Meguid et al. 2009; Rybak et al. 2009; Gomez-Veiga et al. 2012). Another approach for cancer treatment is referred as cancer chemoprevention which involves inhibition of carcinogenesis at initiation phase or at promotion/progression stage (Sporn et al. 1976).

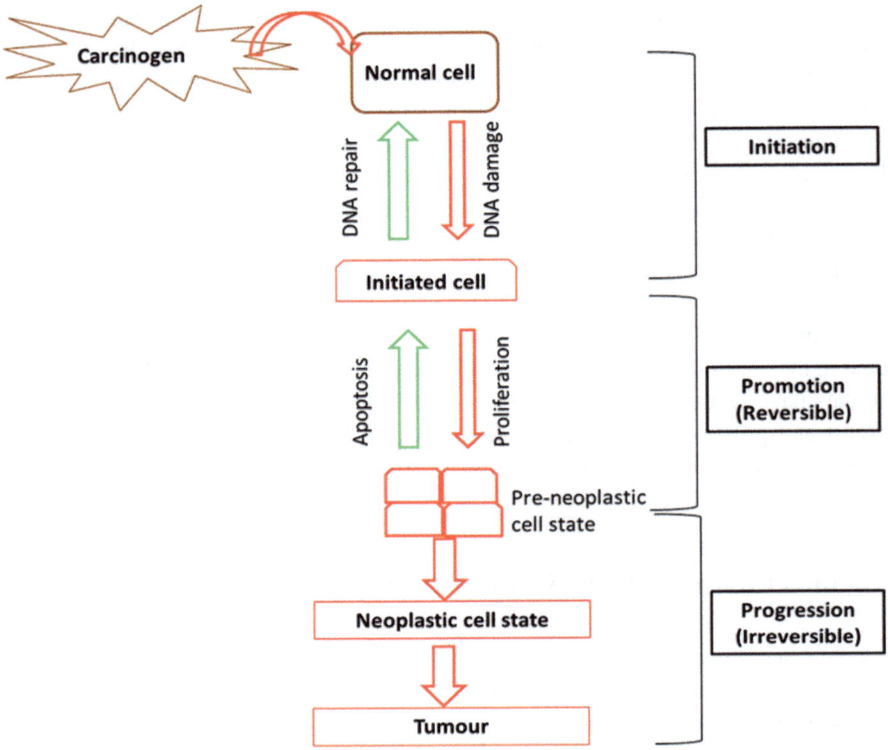

Fig. 7.1 Diagrammatic representation of process of carcinogenesis

Chemopreventive agents are thus classified into two categories: blocking and suppressive agents. Blocking agents inhibit the activation of carcinogen through metabolic pathways and restrict the interaction of carcinogenic agents with biomolecules. Suppressive agents suppress promotion or progression of cancer cells (Wattenberg 1985). Chemopreventive agents can have antioxidative and antiproliferative effects or they modulate enzyme activity, signal transduction pathways, cell cycle, etc., thus affecting the process of carcinogenesis (Walaszek et al. 2004; De Flora and Ferguson 2005; Steward and Brown 2013).

A wide range of chemoprotective components are present in legumes. These are nutrients like, resistant starch, non-starch polysaccharides, oligosaccharides, folates, selenium, zinc, or nonnutritive bioactive constituents such as protease inhibitors, saponins, phytosterols, lectins and phytates (Mathers 2002). Legumes contain a large amount of these components which when consumed in appropriate quantities can provide chemopreventive benefits. The resistant starch present in legumes is known to have a strong preventive effect on the colorectal cancer (Cassidy et al.

1994). It is documented that non-starch polysaccharides and oligosaccharides present in pulses provide a substrate in the bowel for the fermentation of microorganisms and thus enhancing the production of butyrate (Key and Mathers 1995). The direct relationship between substrate and butyrate production is not established but it has been observed that the increase in transit time of substrate in the bowel increases the butyrate production (Mathers and Dawson 1991). Extensive in vitro studies have been carried out that have shown antineoplastic effect of butyrate, thus resulting in suppression of tumor cells (Hague et al. 1996). Butyrate is regarded as histone deacetylase inhibitor and direct administration of butyrate reduced tumor size and proliferation in the carcinogen-treated rats (D'Argenio et al. 1996; Wu et al. 2001). Other components present in legumes like selenium, zinc and folates help in protection of DNA by management of oxidative stress (Mathers 2002).

Lectins present in members of family Fabaceae also induce apoptosis in tumor cells (Yau et al. 2015). Lectins in legumes have been extensively studied in past decades to understand the molecular level interactions with proteins and carbohydrates (Damodaran et al. 2008). Recent studies showed that lectins present in members of this family can bind to glycoconjugates present specifically on tumor cells and thus can act as efficient cancer chemopreventive agents (Ueno et al. 2000). It has been reported that legume lectin, Concanavalin A (ConA), affects mitochondria and results in release of cytochrome C leading to death of murine macrophage PU5–1.8 cells (Suen et al. 2000). Also, studies have shown that ConA induced cell death in human melanoma A375 cells through caspase-dependent pathway. Thus, apoptosis caused through ConA is induced through mitochondrial damage, through release of cytochrome or caspase activation (Liu et al. 2009).

Another lectin from *Sophora flavescens* is known to induce apoptosis in cancer cells through activation of caspase-dependent pathway (Liu et al. 2008). Also, lectin derived from seeds of *Phaseolus coccineus* L. is known to induce programmed cell death through caspase activation in murine fibrosarcoma L929 cells. It has been mentioned that the antineoplastic activity of this lectin could be because of sugar-binding-specific activity as the apoptosis decreased immediately with the inhibition of sialic-acid-specific activity (Chen et al. 2009). These studies suggest that legume lectins possess anticancer properties which are linked to the molecular mechanisms associated to them. These studies would provide new dimension to the cancer research using legume lectins as anticancer agents (Liu et al. 2010).

Another group of phytochemicals in leguminous plants which has gained considerable importance in recent years is isoflavones. They are recently known for their therapeutic properties against many diseases including cancer (Fritz et al. 2016). There are primarily two types of isoflavones in legumes, daidzein and genistein. Numerous in vitro studies have been carried out which have showed the effect of these isoflavones on the cancer cells like breast, prostate, skin, and intestinal (Peterson and Barnes 1991, 1993, 1996; Naik et al. 1994; Pagliacci et al. 1994; Peterson et al. 1996; So et al. 1996; Clark et al. 1996; Zava and Duwe 1997; Kyle et al. 1997; Kuo et al. 1997). It has been suggested in some studies that genistein can be used for the treatment of tumors along with the conventional chemotherapy (Messina 1999;). Genistein prevents invasion of collagen cells from bovine microvascular cells and leads to the growth of capillary-like structures when supported with fibroblast growth factor. This kind of anti-angiogenic activity is significant in cancer chemoprevention as it halts the development of new blood vessels and thus restricts the size of tumor to 1–2 mm, thus making it clinically insignificant (McAllister et al. 1995).

Saponins are another group of phytochemicals chiefly present in legumes. The legumes are the major source of intake of saponins in our diet which are known to have antitumor properties and they inhibit the formation of hormone-induced tumors in colon and lungs (Messina 1999; Kerwin 2004; Kapoor 2015). Saponins are effective in the cancer chemoprevention because of their antioxidant and antimutagenic properties. It has been reported that they may have direct effect on the proliferation of cancer cells or inhibit the enzymes related to tumor growth (Kerwin 2004).

Legume seeds are though the part of our diet but other parts of plants like roots, leaves, and barks are also traditionally used for various medicinal purposes. In recent times, plethora of scientific studies have been carried out to estimate the anticancer potential of these plants. Various in vitro and in vivo studies have shown the remarkable potential of plants of Fabaceae family in cancer chemoprevention. Table 7.1 summarizes various studies carried out to assess the anticancer potential of plants of Fabaceae.

Conclusions

The uncontrolled division of cells leading to the dreadful disease named cancer threatens human civilization. Surgical removal, chemotherapy and radiation therapy are the different kinds of treatments employed for cancer management. The drugs used for the treatment have many side effects, thus affecting the normal functioning of human body. Another approach of cancer chemoprevention involves natural or biological agents which can reverse, suppress or prevent tumor formation at different stages of carcinogenesis. Plant wealth is being explored throughout the world to identify the phytoconstituents which can be used for treatment of cancer, alone or along with the regular chemotherapy. The family Fabaceae is known for its medicinal properties helping to combat various diseases. The phytochemicals present in plants of this family are also found to be effective against carcinogenesis at various stages. Intake of legumes as staple diet helps in reducing the mortality due to cancer. Also, these phytochemicals can be extracted and used for the treatment of this dreadful disease.

Table 7.1 Summary of literature on anticancerous potential of some plant species of Fabaceae family

S. No.	Name of plant	Plant part used (solvent/s used)	Active compound/s	Target cell line/s	Results	References
1.	*Abrus precatorius*	Seeds (petroleum ether)	Flavonoids; alkaloids; tannins; phenols; saponins; and triterpenoids	EAC cells (Ehrlich ascites carcinoma in mice)	Anticancer properties of the extract were considered to be due to flavonoids, alkaloids, and terpenoids	Anbu et al. (2011)
2.	*Abrus precatorius*	Leaves (water)	Alkaloids; flavonoids (flavones); tannins; coumarins; sterol; triterpenoids; and saponins	P815 (murine mastocytoma cell line)	IC50 value of the extract was found to be 200 μg/mL, lower doses of extracts, viz., 3.12; 6.25; 12.15; 25; and 50 μg/mL, showed 43–47% inhibition	Lebri et al. (2015)
3.	*Aeschynomene fascicu*	Root and bark (methanol, hexane, dichloromethane, and ethyl acetate fractions)	Spinochalcone C; spinochalcone A; isocordoin; and secundiflorol G (compounds were isolated from hexane fraction)	KB (nasopharynx cells), Hep-2 (laryngeal carcinoma), HeLa (cervix adenocarcinoma), SiHa (squamous cells), DU-145 (prostate carcinoma) and PC-3 (prostate adenocarcinoma), normal cell line (MDCK)	Spinochalcone C did not show cytotoxicity on tested cell lines. Spinochalcone A exhibited cytotoxic activity against DU-145 cell line. Secundiflorol G showed strong cytotoxic activity against KB and Hep-2 cell lines, while isocordoin showed moderate activity on KB, Hep-2, and DU-145 cell lines	Caamal-Fuentes et al. (2015)
4.	*Astragalus gombiformis*	Leaves (petroleum ether, dichloromethane, and methanol)	–	A549 (human lung epithelial carcinoma cell)	Methanol extract showed maximum activity	Teyeb et al. (2012)

(continued)

Table 7.1 (continued)

S. No.	Name of plant	Plant part used (solvent/s used)	Active compound/s	Target cell line/s	Results	References
5.	*Bauhinia strychnifolia*	Stem (ethanol)	Quercetin; 3,5,7,3′,5′-pentahydroxy-flavanonol-3-O-α-L-rhamnopyranoside; 3,5,7-trihydroxy-chromone-3-O-α-L-rhamnopyranoside; and mixture of β-sitosterol and stigmasterol	HT-29 (human colon cell), HeLa (cervix cancer cell), MCF-7 (breast cancer), and KB (oral carcinoma)	Among five compounds, 3,5,7,3′,5′-pentahydroxyflavanonol-3-O-α-L-rhamnopyranoside possessed maximum activity against all cell lines	Yuenyongsawad et al. (2013)
6.	*Bauhinia variegata*	Leaves (methanol)	Alkaloids; steroids; triterpenoids; and flavonoids	EAC (Ehrlich ascites carcinoma)	Extract showed 63% cytotoxicity due to presence of alkaloids, steroids, triterpenoids, and flavonoids	Sinha and Verma (2012)
7.	*Bidens pilosa, Centella asiatica, Cnicus benedictus, Dicoma capensis, Hypoxis hemerocallidea*, and *Sutherlandia frutescens*	Leaves (water)	–	DU–145 (prostate cancer cells), MDA-MB-231 and MCF-7 (breast cancer cells), and MCF-12A (nonmalignant breast cell line)	Only *D. capensis* exhibited pronounced cytotoxic effects against two cell lines MCF-7 and MCF-12A	Steenkamp and Gouws (2006)
8.	*Butea monosperma*	Flowers (water)	–	Human hepatoma cells (Huh7), hepatic cancer (HepG2), and AML12 cells (immortalized mouse hepatocytes, CRL 2254)	Extract arrested cell growth and caused pro-apoptotic death of cancer cells	Choedon et al. (2010)

S. No.	Name of plant	Plant part used (solvent/s used)	Active compound/s	Target cell line/s	Results	References
9.	*Caesalpinia bonduc*	Stem bark (ethanol, petroleum ether, chloroform, ethyl acetate, and water)	Phenolics and flavonoids	DLA (Dalton's lymphoma ascites cells in Swiss mice) cells	Ethanol extract showed maximum (100%) activity	Sandhia and Bindu (2015)
1.	*Canthium Parviflorum*	Leaves (ethanol)	Flavonoids	HeLa (cervix cancer cell) and DLA (Dalton's lymphoma ascites cells in Swiss mice) cells	Extract showed more cytotoxicity on HeLa cells	Prabhu et al. (2011)
11.	*Cassia alata*	Leaves (methanol)	Kaempferol	A549 (lung cancer cells)	Plant extract exhibited cytotoxicity	Levy and Lewis (2011)
12.	*Cassia tora*	Leaves (methanol, ethyl acetate, and water)	Flavonoids; saponins; tannins; glycosides; anthraquinone; amino acids; coumarins; and steroids	MCF-7 (breast cancer), SiHa (carcinoma of uterus), IMR-32 (human neuroblastoma), HT-29 (human colon adenocarcinoma), SK.N.SH (human neuroblastoma cell line), and OVCAR-5 (ovarian cancer)	Ethyl acetate extract showed maximum cytotoxic effect which corresponded to presence of highest flavonoid compounds	John et al. (2012)
13.	*Cytisus purgans*	Aerial parts (methanol)	Sparteine and lupanine	MDA-MB-231 (human breast cancer) and A549 (human lung cancer)	No cytotoxic activity of extract on both cell lines was observed	Benaiche et al. (2015)
14.	*Delonix elata*	Leaves (water)	Sugars; steroids; flavonoids; saponins; tannins; and glycosides	MCF-7 (breast cancer) and HepG2 (hepatic cancer)	Extract exhibited more cytotoxic activity on MCF-7	Kumar et al. (2014)

(continued)

Table 7.1 (continued)

S. No.	Name of plant	Plant part used (solvent/s used)	Active compound/s	Target cell line/s	Results	References
15.	*Derris indica*	Fruits (hexane)	Derrivanone; derrischalcone; tunicatachalcone; obovatachalcone; glabrachromene; ovalichalcone; pongamol; desmethoxykanugin; karanjin; lanceolatin B; kanjone; pongaglabrone; 8-methoxyfurano (6,7:4″5″)-flavone; pongaflavone; (2R,3R)-3-hydroxy-5-methoxy-2″,2″-dimethylpyrano [7,8:5″,6″]-flavanone; and candidone	M156 (cholangiocarcinoma cell line) and HepG2 (human hepatoma cells)	Among 16 compounds, (2R,3R)-3-hydroxy-5-methoxy-2″,2″-dimethylpyrano [7,8:5″,6″]-flavanone and candidone exhibited high cytotoxicity	Decharchoochart et al. (2014)
16.	*Glycyrrhiza glabra*	Roots (chloroform, methanol and water)	18 β-glycyrrhetic acid	Vero (African green monkey kidney) and MCF7 (Breast cancer)	Chloroform extract showed high cytotoxic activity due to presence of high amount of 18 β-glycyrrhetic acid	Rathi et al. (2009)
17.	*Lotus corniculatus*	Aerial parts (*n*-butanol and chloroform)	Benzoic acid; transilin; isosalicin; soya saponin I; dehydrosoya saponin I; medicarpin-3-O-β-D-glucopyranoside; pharbitoside A; and p-coumaric acid	J774A1 (murine monocyte/macrophage), HEK-293 (human epithelial kidney cells) and WEHI-164 (murine fibrosarcoma)	Chloroform extract showed more antiproliferative activity than *n*-butanol extract	Abderrahmane et al. (2014)

S. No.	Name of plant	Plant part used (solvent/s used)	Active compound/s	Target cell line/s	Results	References
18.	*Lotus polyphyllos*	Aerial parts (ethanol, *n*-hexane, methylene chloride, *n*-butanol, ethyl acetate)	Oleanolic acid; 3,4-caffeic acid; dimethylether; methyl ferulate; ferulic acid; fumaric acid; luteolin 3′,4′-dimethyl ether; β-sitosterol-3-O-β-D-glucopyranoside; butyl-O-α-L-rhamnopyranoside; ethyl-O-α-L-rhamnopyranoside; and ethyl-O-β-D-glucopyranoside	HepG2 (hepatic cancer), PC3 (prostate cancer), HeLa (cervical cancer), MCF-7 (breast cancer), and HCT-116	Methylene chloride fraction showed the highest cytotoxicity	Osman et al. (2015)
19.	*Mucuna cochinchinensis*	Leaves (methanol and ethyl acetate)	Phenols	HeLa (cervix cancer cell), Hep2 (human Black, cervix carcinoma), MCF7 (breast cancer), NIH 3T3 (fibroblast cell line)	Ethyl acetate extract showed significant anticancer activity as compared to methanol extract	Natarajan et al. (2013)
20.	*Mucuna pruriens*	Seeds (methanol)	–	EAC (Ehrlich ascites carcinoma in mice)	Extract exhibited significant cytotoxicity against EAC cells	Rajeshwar et al. (2005)
21.	*Ononis sicula* and *O. hirta*	Aerial parts (ethanol, methanol, chloroform, *n*-hexane, distilled water, and butanol	Alkaloids; flavonoids; and terpenoids	Hep-2 (larynx carcinoma), MCF7 (breast epithelial adenocarcinoma), and Vero (African green monkey kidney)	Aqueous and butanol extracts of *O. hirta* exhibited more cytotoxic effect	Talib and Mahasneh (2010)

(continued)

Table 7.1 (continued)

S. No.	Name of plant	Plant part used (solvent/s used)	Active compound/s	Target cell line/s	Results	References
22.	*Parkinsonia aculeata*	Leaves (ethyl acetate)	C-glycosides; flavonoids; terpenoids; phenols; and sugars	B16F10 (mice melanoma)	Extract showed cytotoxic activity due to presence of flavonoids	Singh et al. (2013)
23.	*Pisum sativum* cv. *Stratus* (green peas), *P. sativum* cv. *Golden* (yellow peas), *Cicer arietinum* cv. Small Amits (chickpeas), *Lens culinaris* cv. Morton (lentils), *Phaseolus vulgaris* (red kidney beans), *P. vulgaris* (small red beans.), *P. vulgaris* (pinto beans), *P. vulgaris* cv. Turtle Eclipse (black beans), *Glycine max* cv. *Proto* (yellow soybeans), *G. max* Merr. cv C-1 (black soybeans), *Phaseolusaureus* (mung beans), *Vigna angularis* (adzuki beans) and *Vignaunguiculata* (black-eyed pea)	Seeds (water, acetone, and acetic acid)	Phenols; procyanidin; saponin; and phytic acid	CAL27, AGS, HepG2, SW480, and Caco-2 (digestive system cancer cell lines), SK-OV-3 (ovary cancer cell), and MCF-7 (breast cancer cell)	Black bean, pinto bean, red kidney bean and small red bean, black soybean, lentil, adzuki bean, and mung bean exhibited stronger cancer cell proliferation inhibitory effects as compared to green and yellow peas, chickpea, yellow soybean, and black-eyed pea	Xu and Chang (2012)

S. No.	Name of plant	Plant part used (solvent/s used)	Active compound/s	Target cell line/s	Results	References
24.	*Pithecellobium dulce*	Bark (acetone and methanol)	–	HTB-22 (breast mammary cell carcinoma), HeLa (cervix cancer cells), skin (epithelial malignant melonoma), and tongue (epithelial squamous carcinoma)	Acetone extract showed maximum effect on breast cancer cells. Both extract showed no effect on skin cancer cells.	Cates et al. (2014)
25.	*Pseudarthria viscida*	Whole plant (ethanol)	Phenolic acids; flavonoids; terpenoids; and tannins	HT29 (human colorectal cancer cell line), C2C12 (mouse muscle cell line), and 3T3L1 (mouse, embryo fibroblast) cell lines	Plant exhibited maximum cytotoxicity on HT29 cancer cells	Vijayabaskaran et al. (2010)
26.	*Sesbania grandiflora*	Leaves and flowers (ethanol)	–	EAC cells (Ehrlich ascites carcinoma in mice)	Leaf extract of plant was formed more potent than flower extract	Sreelatha et al. (2011)
27.	*Sesbania grandiflora*	Flowers (methanol)	Steroids; alkaloids; terpenoids; glycosides; flavonoids; and carbohydrates	HEL, HeLa (cervix cancer cell), CRFK (Crandell Rees feline kidney) and Vero cells (kidney epithelial cells of African monkey)	Extract showed 100% cytotoxicity against all cell lines except Vero cells (20% cytotoxicity)	Arthanari et al. (2012)
28.	*Sophora pachycarpa*	Roots (methanol, chloroform, ethyl acetate, *n*-butanol, and water)	–	A549, HeLa, HL-60, MCF-7, and PC3	Chloroform and ethyl acetate extracts exhibited strong cytotoxic effects	Mousavia et al. (2014)

(continued)

Table 7.1 (continued)

S. No.	Name of plant	Plant part used (solvent/s used)	Active compound/s	Target cell line/s	Results	References
29.	*Thermopsis rhombifolia*	Leaves (ethanol)	–	HT-29 (colon cancer cells) and SH-SY5Y (brain cancer cells)	Extracts were more effective on brain cells than colon cells	Kerneis et al. (2015)
30.	*Trigonella foenum-graecum*	Whole plant (chloroform)	–	MCF-7 (breast cancer cells)	Extract promotes apoptosis in MCF-2 cells and causes cell death	Khoja et al. (2011)
		Seeds (methanol)	Squalene and Naringenin	HepG2 (hepatocellular carcinoma)	Extract induced apoptosis in HepG2 cells	Khalil et al. (2015)
		Seeds (water)	Gingerol; cedrene; zingerone; vanillin; and eugenol	T-cell lymphoma (TCP), B-cell lymphomas, thyroid papillary carcinoma (FRO), and human breast cancer (MCF7)	Extract had significant cytotoxicity against all cell lines	Alsemari et al. (2014)

References

Abderrahmane M, Bouheroum M, Malafronte N, D'Ambola M, Meriem A, Severino L (2014) Phytochemical investigation and cytotoxic activity of *Lotus corniculatus*. Pharmacologyonline 3:222–225

Alsemari A, Alkhodairy F, Aldakan A, Al-Mohanna M, Bahoush E, Shinwari Z, Alaiya A (2014) The selective cytotoxic anti-cancer properties and proteomic analysis of *Trigonella Foenum-Graecum*. Complementary Altern Med 14:114–112

Anbu J, Ravichandiran V, Sumithra M, Chowdary BS, Kumar KSSLVVSN, Kannadhasan R, Kumar RS (2011) Anticancer activity of petroleum ether extract of *Abrus precatorius* on ehrlich ascitis carcinoma in mice. Int J Pharm Bio Sci 2:24–31

Arthanari SK, Vanitha J, Ganesh M, Venkateshwaran K, Clercq D (2012) Evaluation of antiviral and cytotoxic activities of methanolic extract of *S. grandiflora* (Fabaceae) flowers. Asian Pac J Trop Biomed 2(2):S855–S858

Aruoma OI (2003) Methodological considerations for characterizing potential antioxidant actions of bioactive components in plant foods. Mutat Res 523–524:9–20

Benaiche G, Belattar N, Trifunnovic S, Vukovic N, Todorovic D, Todorovic M, Baskic D, Vukic M (2015) Isolation of alkaloids and anti-tumor activity of the crude methanolic extract of Algerian *Cytisus purgans*. Orient J Chem 31:1943–1948

Bravo L (1998) Polyphenols: chemistry, dietary sources, metabolism and nutritional significance. Nutr Rev 11:317–333

Caamal-Fuentes EE, Peraza-Sanchez SR, Torres-Tapia LW, Moo-Puc RE (2015) Isolation and identification of cytotoxic compounds from *Aeschynomene fascicularis*, a Mayan medicinal plant. Molecules 20:13563–13574

Cassidy A, Bingham SA, Cummings JH (1994) Starch intake and colorectal cancer risk: an international comparison. Br J Cancer 69:937–942

Cates RG, Thompson A, Brabazon H, McDonald S, Lawrence M, Williams S, Penciallilo P, Soria JAF, Espinoza LV, Martinez JV, Arbizu DA, Villagran E, Ancheta F (2014) Activities of Guatemalan medicinal plants against cancer cell lines and selected microbes: evidence for their conservations. J Med Plant Res 8:1040–1050

Chen J, Liu B, Ji N, Zhou J, Bian HJ, Li CY, Chen F, Bao JK (2009) A novel sialic acid-specific lectin from *Phaseolus coccineus* seeds with potent antineoplastic and antifungal activities. Phytomedicine 16:352–360

Choedon T, Shukla SK, Kumar V (2010) Chemopreventive and anti-cancer properties of the aqueous extract of flowers of *Butea monosperma*. J Ethnopharmacol 129:208–213

Clark JW, Santos-Moore A, Stevenson LE, Frackelton AR (1996) Effects of tyrosine kinase inhibitors on the proliferation of human breast cancer lines and proteins important in the RAS signaling pathway. Int J Cancer 65:186–191

D'Argenio G, Cosenza V, Delle-Cave M, Iovino P, Delle-Valle N, Lombardi G, Mazzacca G (1996) Butyrate enemas in experimental colitis and protection against large bowel cancer in a rat model. Gastroenterology 110:1727–1734

Damodaran D, Jeyakani J, Chauhan A, Kumar N, Chandra NR, Surolia A (2008) Cancer lectin DB: a database of lectins relevant to cancer. Glycoconj J 25:191–198

Dandawate P, Ahmad A, Deshpande J, Swamy KV, Khan EM, Khetmalas M, Padhye S, Sarkar F (2014) Anticancer phytochemical analogs 37: synthesis, characterization, molecular docking and cytotoxicity of novel plumbagin hydrazones against breast cancer cells. Bioorg Med Chem Lett 24:2900–2904

De Flora S, Ferguson LR (2005) Overview of mechanisms of cancer chemopreventive agents. Mutat Res 591:8–15

Decharchoochart P, Suthiwong J, Samatiwat P, Kukongviriyapan V, Yenjai C (2014) Cytotoxicity of compounds from the fruits of *Derris indica* against cholangiocarcinoma and HepG2 cell lines. J Nat Med 68:730–736

Devi S, Gupta AK, Singh M (2012) Ethano-medicinal use of plants belonging to families fabaceae and solanaceae, Hamirpur district (H.P.) Int J Sci Res Pub 2:1–4

Doll R, Peto R (1981) The causes of cancer: quantitative estimates of avoidable risks of cancer in the United States today. J Natl Cancer Inst 66:1191–1308

Ferguson LR (2001) Role of plant polyphenols in genomic stability. Mutat Res 475:89–111

Fritz H, Seely D, Flower G, Skidmore B, Fernandes R, Vadeboncoeur S, Kennedy D (2016) Soy, red clover and isoflavones and breast cancer: a systematic review. PLoS One 8(11):e81968. doi:10.1371/journal.pone.0081968

Gomez-Veiga F, Marino A, Alvarez L, Rodriguez I, Fernandez C, Pertega S, Candal A (2012) Brachytherapy for the treatment of recurrent prostate cancer after radiotherapy or radical prostatectomy. BJU Int 109:17–21

Hague A, Butt AJ, Paraskeva C (1996) The role of butyrate in human colonic epithelial cells: an energy source or inducer of differentiation and apoptosis? Proc Nutr Soc 55:937–943

John J, Mehta A, Mehta P (2012) Evaluation of antioxidant and anticancer potential of *Cassia tora* leaves. Asian J Tradit Med 7:118–125

Kapoor S (2015) Bioactives and therapeutic potential of legumes: a review. Int J Pharm Bio Sci 5:65–74

Kaur C, Kapoor HC (2001) Antioxidants in fruits and vegetables-the millennium's health. Int J Food Sci Technol 36:703–725

Kerneis S, Swift LH, Lewis CW, Bruyere C, Oumata N, Colas P, Ruchaud S, Bain J, Golsteyn RM (2015) Natural product extracts of the Canadian prairie plant, *Thermopsis rhombifolia*, have anti-cancer activity in phenotypic cell-based assays. Nat Prod Res 29:1026–1034

Kerwin SM (2004) Soy saponins and the anticancer effects of soybeans and soy-based foods. Curr Med Chem: Anti-Cancer Agents 4:263–272

Key FB, Mathers JC (1995) Digestive adaptations of rats given white bread and cooked haricot beans (*Phaseolus vulgaris*): large-bowel fermentation and digestion of complex carbohydrates. Br J Nutr 74:393–406

Khalil MM, Ibrahim MM, El-Gaaly GA, Sultan AS (2015) *Trigonella foenum* (Fenugreek) induced apoptosis in hepatocellular carcinoma cell line, HepG2, mediated by upregulation of p53 and proliferating cell nuclear antigen. Biomed Res Int 2015:1–11

Khoja KK, Shafi G, Hasan TN, Syed NA, Al-Khalifa AS, Al-Assaf AH, Alshatwi AA (2011) Fenugreek, a naturally occurring edible spice, kills MCF-7 human breast cancer cells via an apoptotic pathway. Asian Pac J Cancer Prev 12:3299–3304

Kumar PK, Prasad SK, Sudeepthi LN, Reddy RCSD, Rahaman AS, Ranjith MP (2014) *In vitro* cytotoxic activity of aqueous extract of *Delonix elata* (L.) gamble (fabaceae) leaves on MCF-7 and HepG-2 cell lines. Int J Pharmacol Toxicol 2:70–75

Kuo SM, Morehouse HFJ, Lin CP (1997) Effect of antiproliferative flavonoids on ascorbic acid accumulation in human colon adenocarcinoma cells. Cancer Lett 116:131–137

Kyle E, Neckers L, Takimoto C, Curt G, Bergan R (1997) Genistein-induced apoptosis of prostate cancer cells is preceded by a specific decrease in focal adhesion kinase activity. Mol Pharmacol 51:193–200

Lebri M, Tilaoui M, Bahi C, Achibat H, Akhramez S, Fofie YBN, Gnahoue G, Lagou SM, Zirihi GN, Coulibaly A, Zyad A, Hafid A, Khouili M (2015) Phytochemical analysis and in vitro anticancer effect of aqueous extract of *Abrus precatorius* Linn. Der Pharma Chemica 7:112–117

Levy A, Lewis A (2011) *Cassia alata* Leaf extract induces cytotoxicity in a 549 lung cancer cells via a mechanism that is caspase 8 dependent. West Indian Med J 60:608–614

Liu Z, Liu B, Zhang ZT, Zhou TT, Bian HJ, Min MW, Liu YH, Chen J, Bao JK (2008) A mannose-binding lectin from *Sophora flavescens* induces apoptosis in HeLa cells. Phytomedicine 15:867–875

Liu B, Li CY, Bian HJ, Min MW, Chen LF, Bao JK (2009) Antiproliferative activity and apoptosis-inducing mechanism of Concanavalin A on human melanoma A375 cells. Arch Biochem Biophys 482:1–6

Liu B, Bian H, Bao J (2010) Plant lectins: potential antineoplastic drugs from bench to clinic. Cancer Lett 287:1–12

Mathers JC (2002) Pulses and Carcinogenesis : potential for the prevention of colon, breast and other cancers. Br J Nutr 88:273–279

Mathers JC, Dawson LD (1991) Large bowel fermentation in rats eating processed potatoes. Br J Nutr 66:313–329

McAllister KA, Baldwin MA, Thukkani AK (1995) Six novel mutations in the endoglin gene in hereditary hemorrhagic telangiectasia type 1 suggest a dominant-negative effect of receptor function. Hum Mol Genet 4:1983–1985

Meguid RA, Hooker CM, Taylor JT, Kleinberg LR, Cattaneo SM, Sussman MS, Yang SC, Heitmiller RF, Forastiere AA, Brock MV (2009) Recurrence after neoadjuvant chemoradiation and surgery for esophageal cancer: does the pattern of recurrence differ for patients with complete response and those with partial or no response? J Thorac Cardiovasc Surg 138:1309–1317

Messina MJ (1999) Legumes and soybeans: overview of their nutritional profiles. Am J Clin Nutr 70:439–450

Morris JB (1999) Legume genetic resources with novel "value added " industrial and pharmaceutical use. In: Janick J (ed) Perspectives on new crops and new uses. ASHS Press, Alexandria

Mousavia SH, Motaeza M, Zamiri-Akhlaghib A, Emamic SA, Tayarani-Najarand Z (2014) *In-Vitro* evaluation of cytotoxic and apoptogenic properties of *Sophora Pachycarpa*. Iran J Pharm Res 13:665–673

Naik HR, Lehr JE, Pienta KJ (1994) An in vitro and in vivo study of anti-tumor effects of genistein on hormone refractory prostate cancer. Anticancer Res 14:2617–2620

Natarajan K, Narayanan N, Ravichandran N, Lalith KG (2013) Antioxidant and anticancer activity of *Mucuna Cochinchinensis* (Lour.) cheval seeds. Int J Pharm Pharm Sci 5:646–648

Osman SM, El-Khalik SMA, Saadeldeen AM, Koheil MA, Wink M (2015) Activity guided phytochemical study of Egyptian *Lotus polyphyllos* E.D. Clarke (Fabaceae). Int J App Res Nat Prod 8:18–26

Pagliacci MC, Smacchia M, Migliorati G, Grignana F, Riccardi C, Nicoletti I (1994) Growth-inhibitory effects of the natural phytoestrogen genistein in MCF-7 human breast cancer cells. Eur J Cancer 30:1675–1682

Payne AS, James WD, Weiss RB (2006) Dermatologic toxicity of chemotherapeutic agents. Semin Oncol 33:86–97

Peterson G, Barnes S (1991) Genistein inhibition of the growth of human breast cancer cells: independence from estrogen receptors and the multi-drug resistance gene. Biochem Biophys Res Commun 179:661–667

Peterson G, Barnes S (1993) Genistein and biochanin A inhibit the growth of human prostate cancer cells but not epidermal growth factor receptor autophosphorylation. Prostate 22:335–345

Peterson G, Barnes S (1996) Genistein inhibits both estrogen and growth factor-stimulated proliferation of human breast cancer cells. Cell Growth Differ 7:1345–1351

Peterson G, Coward L, Kirk M, Falany C, Barnes S (1996) The role of metabolism in mammary epithelial growth inhibition by the isoflavones genistein and biochanin A. Carcinogenesis 17:1861–1869

Prabhu PT, Panneerselvam P, Selvakumari S, Sivaraman D (2011) In vitro and in vivo anticancer activity of ethanolic extract of *Canthium Parviflorum* lam on DLA and Hela cell lines. Int J Drug Dev Res 3:280–285

Rahman AHMM, Parvin MIA (2014) Study of medicinal uses on Fabaceae family at Rajshahi, Bangladesh. Res Plant Sci 2:6–8

Rajeshwar Y, Gupta M, Mazumder UK (2005) Antitumor activity and in vivo antioxidant status of *Mucuna pruriens* (fabaceae) seeds against Ehrlich ascites carcinoma in Swiss albino mice. Iran J Pharmcol Ther 4:46–53

Rathi SG, Suthar M, Patel P, Bhaskar VH, Rajgor NB (2009) In-vitro cytotoxic screening of *Glycyrrhiza glabra* L. (Fabaceae): a natural anticancer drug. J Young Pharm 1:239–243

Rybak LP, Mukherjea D, Jajoo S, Ramkumar V (2009) Cisplatin ototoxicity and protection: clinical and experimental studies. Tohoku J Exp Med 219:177–186

Sandhia KG, Bindu AR (2015) Phytochemical, anti-inflammatory and *in vitro* anticancer activities of *Caesalpinia bonduc* stem bark. Int J Pharm Sci Res 6:50–56

Schrire BD, Lewis GP, Lavin M (2005) Biogeography of the Leguminosae. In: Lewis G, Schrire G, Mackinder B, Lock M (ed) Legumes of the world. Kew, England.

Siegel R, Ward E, Brawley O, Jemal A (2011) Cancer statistics, 2011: the impact of eliminating socioeconomic and racial disparities on premature cancer deaths. CA Cancer J Clin 61:212–236

Singh P, Shrivastava R, Sharma M, Singh M (2013) In vivo antitumor, antioxidant activities and toxicity profile of ethyl acetate crude leaf extract of *Parkinsonia aculeata* L. (fabaceae) on B16F10 melonoma. Int Res J Pharm 4:90–93

Sinha K, Verma AK (2012) Evaluation of antimicrobial and anticancer activities of methanol extract of in vivo and in vitro grown *Bauhinia variegata* L. Int Res J Biol Sci 1:26–30

So FV, Guthrie N, Chambers AF, Moussa M, Carroll KK (1996) Inhibition of human breast cell proliferation by flavonoids and citrus juice. Nutr Cancer 26:167–181

Sporn MB, Dunlop NM, Newton DL, Smith JM (1976) Prevention of chemical carcinogenesis by vitamin a and its synthetic analogs. Fed Proc 35:1332–1338

Sreelatha S, Padma PR, Umasankari E (2011) Evaluation of anticancer activity of ethanol extract of *Sesbania grandiflora* (Agati Sesban) against Ehrlich ascites carcinoma in Swiss albino mice. J Ethnopharmacol 134:984–987

Steenkamp V, Gouws MC (2006) Cytotoxicity of six South African medicinal plant extracts used in the treatment of cancer. S Afr J Bot 72:630–633

Steward WP, Brown K (2013) Cancer chemoprevention: a rapidly evolving field. Br J Cancer 109:1–7

Suen YK, Fung KP, Choy YM, Lee CY, Chan CW, Kong SK (2000) Concanavalin A induced apoptosis in murine macrophage PU5–1.8 cells through clustering of mitochondria and release of cytochrome c. Apoptosis 5:369–377

Surh YJ (2003) Cancer chemoprevention with dietary phytochemicals. Nat Rev Cancer 3:768–780

Surh YJ, Ferguson LR (2003) Dietary and medicinal antimutagens and anticarcinogens: molecular mechanisms and chemopreventive potential-highlights of a symposium. Mutat Res 523–524:1–8

Talib WH, Mahasneh AM (2010) Antiproliferative activity of plant extracts used against cancer in traditional medicine. Sci Pharm 78:33–45

Tedesco I, Russo GL, Nazzaro F, Russo M, Palumbo R (2001) Antioxidant effect of red wine anthocyanins in normal and catalase inactive human erythrocytes. J Nutr Biochem 12:505–511

Teyeb H, Zanina N, Neffati M, Douk W, Najjar MF (2012) Cytotoxic and antibacterial activities of leaf extracts of *Astragalus gombiformis* Pomel (Fabaceae) growing wild in Tunisia. Turk J Biol 36:53–58

Ueno T, Ohtawa K, Sakurai Y, Kodera Y, Hiroto M (2000) Polyethylene glycol-modified Concanavalin A as an effective agent to stimulate anti-tumour cytotoxicity. Cancer Detect Prev 24:100–106

Vijayabaskaran M, Venkateswaramurthy V, Arif Pasha MD, Babu G, Sivakumar P, Perumal P, Jayakar B (2010) In vitro cytotoxic effect of ethanolic extract of *Pseudarthria viscida* Linn. Int J Pharm Pharm Sci 2:93–94

Vogelstein B, Kinzler KW (2004) Cancer genes and the pathways they control. Nat Med 10:789–799

Walaszek Z, Hanausek M, Slaga TJ (2004) Mechanisms of chemoprevention. Chest 125:128–133

Wattenberg LW (1985) Chemoprevention of cancer. Cancer Res 45:1–8

Wu JT, Archer SY, Hinnebusch B, Meng S, Hodin RA (2001) Transient vs. prolonged histone hyperacetylation: effects on colon cancer cell growth, differentiation, and apoptosis. Am J Physiol: Gastrointest Liver Physiol 280:482–490

Xu B, Chang SKC (2012) Comparative study on antiproliferation properties and cellular antioxidant activities of commonly consumed food legumes against nine human cancer cell lines. Food Chem 134:1287–1296

Yau T, Dan X, Cheuk C, Ng W, Ng TB (2015) Lectins with potential for anti-cancer therapy. Molecules 20:3791–3810

Yuenyongsawad S, Bunluepuech K, Wattanapiromsakul C, Tewtrakul S (2013) Anti-cancer activity of compounds from *Bauhinia strychnifolia* stem. J Ethnopharmacol 150:765–769

Zava DT, Duwe G (1997) Estrogenic and antiproliferative properties of genistein and other flavonoids in human breast cancer cells in vitro. Nutr Cancer 27:31–40

Chapter 8
Small Cells for Big Ideas: The Cytotoxic Podophyllotoxin and the Long Journey in Discovering Its Biosynthetic Pathway

Pavlina Sasheva and Iliana Ionkova

Abstract For more than 70 years the plant-derived cytotoxic podophyllotoxin is still a focal point for many research groups. The relationship between the structure of podophyllotoxin and podophyllotoxin derivatives, and their biological activity, allowed the design of novel drugs to combat malignancies. The quests for safer drugs and biomass production through aseptic cultivation have been gaining momentum in the last decade or two, supported by the advances in the computer technologies, information sharing and bioinformatics/molecular biology. The recently discovered enzymatic conversions that lead directly to etoposide aglycone not only filled the gaps in the lignan biosynthetic pathway, but also offered an exciting direction for future research. This book chapter opens with a short summary of the structure of podophyllotoxin and podophyllotoxin-based drugs, and their biological targets. Compelling discoveries and unexpected turns of the lignan biosynthetic pathway are discussed next, followed by benchmark studies in the aseptic cultivation of the plants producing it.

Keywords Podophyllotoxin • Biosynthetic pathway • Etoposide • Aseptic cultivation • Structure-activity relationships • *Podophyllum* • *Linum*

Introduction

Podophyllotoxin (**1**) is an aryltetralin lignan mostly used in the anticancer therapy. Its antiviral properties are well documented; however the lignan is for topical use only due to its side effects in the gastrointestinal tract. The cytotoxic activity, though, has been the hotspot for the researchers, who in the last 70 years have put efforts in finding sustainable sources of podophyllotoxin, elucidating its biosynthetic pathway and devising efficient podophyllotoxin-based chemotherapeutics with limited side effects (Fig. 8.1).

P. Sasheva (✉) • I. Ionkova (✉)
Department of Pharmacognosy, Medical University of Sofia,
2 Dunav street, 1000 Sofia, Bulgaria
e-mail: polly.sasheva@gmail.com; ionkova@pharmfac.acad.bg

S. Malik (ed.), *Biotechnology and Production of Anti-Cancer Compounds*,
DOI 10.1007/978-3-319-53880-8_8

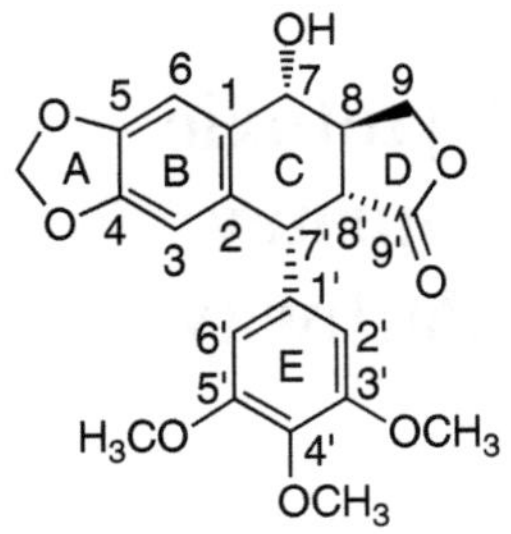

1

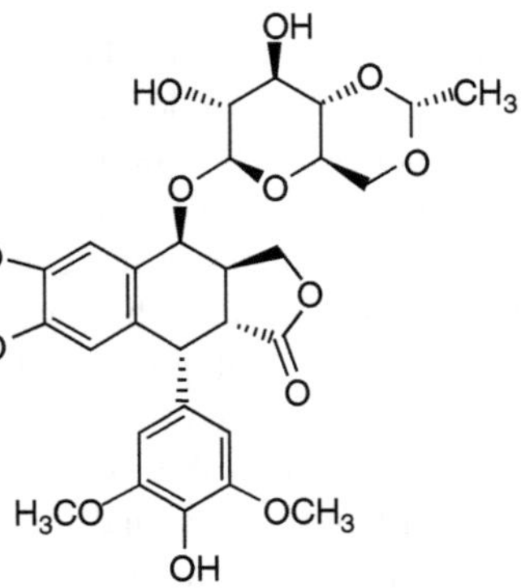

2

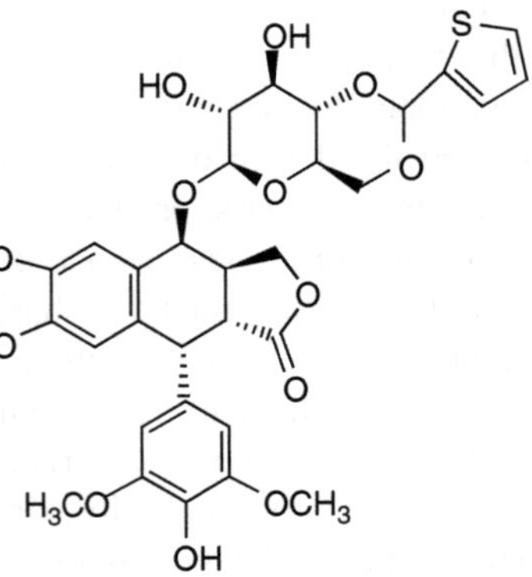

3

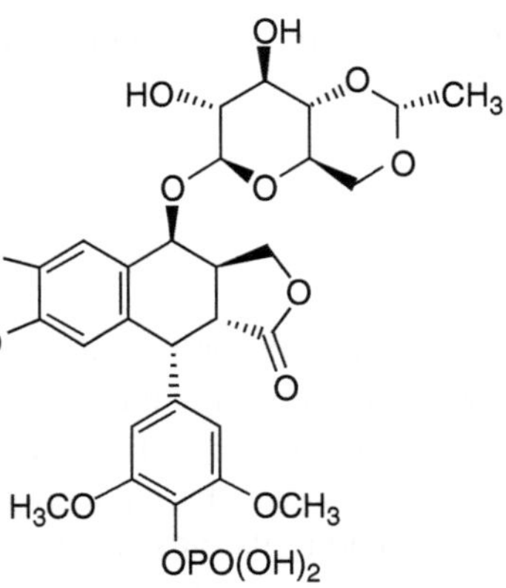

4

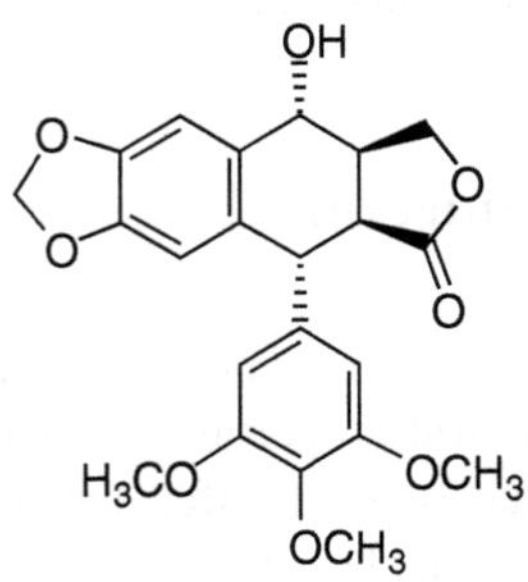

5

Fig. 8.1 Structures of podophyllotoxin (1), etoposide (2), teniposide (3), etopophos (4), picropdophyllotoxin (5), podophyllinic acid (6) and epipodophyllotoxin (7)

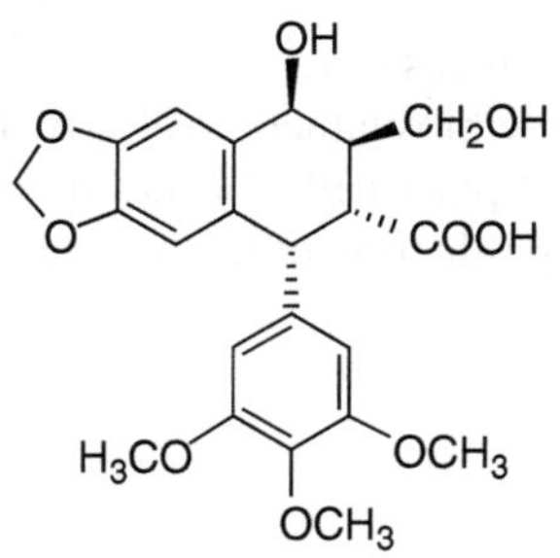

6

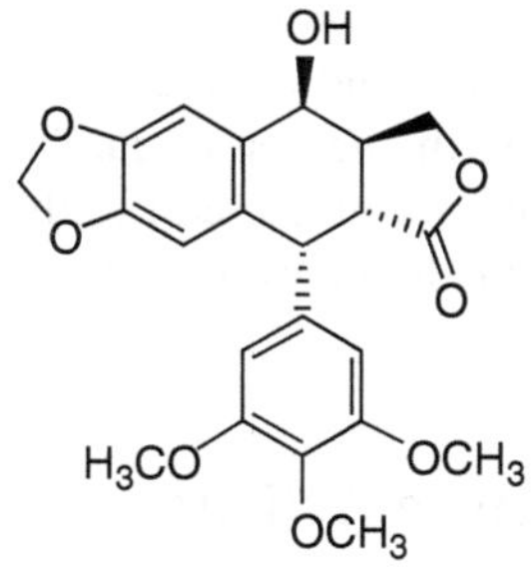

7

Fig. 8.1 (continued)

The first commercially available anticancer products using podophyllotoxin skeleton as a starting building block were developed by Sandoz Pharma AG. The time- and effort-consuming derivatisation of 600 podophyllotoxin structures spanning over 20 years of research led to the development of the commercially short-lived SP-I and SP-G in the 1960s, and the superior etoposide (**2**) and teniposide (**3**) in the 1970s (Stahelin and von Wartburg 1991). With Sandoz Pharma AG switching its research and development priorities in the late 1970s, etoposide and teniposide were licensed to the US-based Bristol-Myers Squibb. Using the etoposide structure, the company developed a readily water-soluble pro-drug etoposide phosphate (etopophos) (**4**), which was commercialised in 1996 (Stahelin and von Wartburg 1991; Botta et al. 2001).

The great success of the already established clinical practice drugs etoposide (**2**), teniposide (**3**) and etopophos (**4**) did not halt the development of efficacious and safe chemotherapeutics based on the podophyllotoxin skeleton. Quite the opposite, this aryltetralin lignan is still a lead structure used in the computer-aided drug design for in silico analysis of promising derivatives. Podophyllotoxin, though, did not prove promising for chemical synthesis, establishing plant sourcing as the only alternative at the moment. Metabolite screening programs identified many species as producers of this natural product, but very few showed potential for large-scale cultivation. In the mean time, the body of knowledge and techniques in molecular biology advanced to the extent that enabled the research of non-model plants at genomic and transcriptomic levels. The knowledge of the biosynthetic pathway leading to the podophyllotoxin had slowly and progressively accumulated. A significant contribution was the recent research by Lau and Sattely (Lau and Sattely 2015). The authors discovered the enzymatic conversions leading to a structure that is an immediate precursor of etoposide.

In the next pages, we focus on the aryltetralin lignan podophyllotoxin, asking three questions: what is podophyllotoxin, how is podophyllotoxin formed in plants and how to produce more. Much of the information that has been extensively covered elsewhere is omitted, and where relevant, we give suggested sources for further reading. In "structure-activity relationships" we give an overview of the chemical structure and biological targets of podophyllotoxin and its derivatives. Next, we discuss how and when different enzymes and intermediates were discovered, shortly covering the newest research on the regulation of the biosynthetic pathway. In the last part, we give credit to the aseptic cultivation of plant cells for advancing our understanding how a small cell can orchestrate the production of life-saving drugs.

Structure-Activity Relationships

The aryltetralin lignan podophyllotoxin (**1**) initially isolated by Podwyssotzki in 1880 from podophyllin, an alcohol extract of *Podophyllum* rhizomes (Podwyssotzki 1880), was off the scientific radar until 1940s. Shortly after the pharmacological properties (Kaplan 1942) and the mechanism of action (King and Sullivan 1946) of the podophyllotoxin-containing podophyllin was published, Hartwell and Schrecker proposed the correct structure of podophyllotoxin (Hartwell and Schrecker 1951).

Consisted of two coupled C_6C_3 propylbenzene units, the podophyllotoxin skeleton is optically active and hosts four chiral centres, located in rings C and D.[1] The biological activity of the aryltetralin lignan is tightly connected to the correct stereochemistry of the chiral carbon atoms. For example, the epimerisation of the fused *trans*-lactone (ring D) to its *cis*-configuration yielded the virtually inactive picropodophyllotoxin (**5**); the same was observed for podophyllinic acid (**6**) after opening the *trans*-lactone ring (Stahelin and von Wartburg 1991). The importance of the stereochemistry of the chiral centres for the biological activity was further demonstrated by Loike and colleagues, who showed that the epimerisation of the hydroxy group at position C7 afforded ten times less cytotoxic epipodophyllotoxin (**7**) (Loike et al. 1978).[2]

In the process of etoposide (**2**) and teniposide (**3**) derivatisation, the small equatorial substituent at position C7 of the podophyllotoxin skeleton (**1**) is epimerised and replaced by a bulky glucoside moiety in axial orientation (Ter Haar et al. 1996), affording drugs with rather different activity. The first biological tests of etoposide and teniposide showed that the drugs caused arrest in the late S/G2 phase of the cell cycle, as opposed to arrest in metaphase observed for the parent molecule podophyllotoxin (Stahelin and von Wartburg 1991). An interesting fact is that teniposide was commercialized (Stahelin and von Wartburg 1991) before the discovery of its molecular target, topoisomerase 2 in 1982 (Long and Minocha 1983). Etoposide was launched in the USA the following year, 1983 (Stahelin and von Wartburg 1991). The apparent distinction in the mechanism of action between podophyllotoxin and its derivatives has been observed as early as 1960s, but information about the exact target was missing. A breakthrough study in 1974 by Loike and colleagues

---

[1] In this work we use the IUPAC recommendations for numbering lignans and neolignans (Moss 2000). The reader should be aware that different systems may be encountered in other papers.

[2] More comprehensive read about the structure-activity relationships can be found in Loike et al. 1978; Stahelin and von Wartburg 1991; Damayanthi and Lown 1998; and Botta et al. 2001.

showed that etoposide and teniposide induced DNA fragmentation in HeLa cells. The authors proposed that the glucoside moiety at position C7 impedes the interaction with the microtubules, and that the presence of 4′-hydroxy group in ring E is necessary for the activity of the drugs (Loike et al. 1974; Loike et al. 1978).

Fast forward, with the onset of the twenty-first century, many exciting discoveries pushed further the anticancer drug research. In particular, the atomic models of the targets of podophyllotoxin and its semi-synthetic derivatives were solved, allowing in silico modelling of molecular interactions and advanced computer screening of superior drugs. In 2004, the 4.2 Å structure of podophyllotoxin-tubulin complex (PDB code: 1SA1) was published, confirming that the podophyllotoxin competes with colchicine for the colchicine-binding domain of the β-tubulin chain (Ravelli et al. 2004). Podophyllotoxin binds within the β-subunit of an α/β tubulin heterodimer, preventing importance for the microtubule dynamic instability switch between curved and straight conformation. Failure to lock the right conformation results in outward curling of the α/β heterodimer, weakening the lateral interactions and destabilising the neighbouring tubulins within the microtubule (Dorleans et al. 2009; Ravelli et al. 2004).[3] Similar to other microtubule-binding drugs, only small amount of podophyllotoxin is needed to interfere with the microtubule polymerisation, thus causing mitotic arrest (Stanton et al. 2011). Unfortunately, podophyllotoxin binds equally well to the microtubules from cancerous and normal cells, setting a complex task to derivatise a structure that targets tubulin isotypes expressed in cancer cells. A highly conserved protein, the human tubulin β-chain has several isotypes with small structural differences in the binding pocket (Tseng et al. 2010). The expression profile can also be different in neoplasia. Isotype βIII, for example, has been associated with poor prognosis in cancer patients (Portyanko et al. 2015), but according to research conducted in Tuszynski's group (Tseng et al. 2010; Huzil et al. 2010; Ravanbakhsh et al. 2013), the differences in tubulin geometry and biochemistry may enable the design of antimitotic drugs that have altered affinities and target-specific tubulin isotypes (Ravanbakhsh et al. 2013).

While podophyllotoxin is an antimitotic agent, its derivative etoposide is a topoisomerase 2 (Top2) poison that increases the levels of covalent complexes of Top2 with DNA (Nitiss 2009). Top2 has two isoforms: α (Top2α) and β (Top2β), the first found to be specific for the fast-proliferating cancerous cells, while the second is constitutively expressed (Capranico et al. 1992). The increasing amount of data that showed correlation between poisoning of Top2β and the occurrence of severe side effects (Nitiss 2009) led to the commonly expressed notion that there is a need of drugs that target specifically Top2α (Toyoda et al. 2008; Nitiss 2009; Wendorff et al. 2012). In 2011, a high-resolution atomic structure of Top2β complexed with etoposide (**2**) and DNA was published by Wu and colleagues (PDB code: 3QX3, Wu et al. 2011), bringing the research closer to identifying specific Top2α inhibitors. The authors showed that while the etoposide aglycone interacts strongly with both DNA and protein; the glucoside moiety protrudes to the major groove, forming very little in number van der Waals and hydrogen bonds. The large space the C7 substituent

[3] We highly recommend reading the papers from Ravelli et al. 2004 and Dorleans et al. 2009.

occupied led Wu and colleagues to the suggestion that modifications that strengthen the bonding or introduce specific Top2α interactions at this position could be important for new drug design. At the same time, the presence of the C4′–OH group in the pendant ring of the aglycone is essential—a replacement with methyl group at C4′, or hydrogen at C3′ and C5′ is unfavourable for the interactions within the complex (Wu et al. 2011). Research led by different groups identified Met762 (Wu et al. 2011; Wendorff et al. 2012; Drwal et al. 2014), Ser763 (Drwal et al. 2014) and Ser800 (Wendorff et al. 2012) as promising sites, where novel and specific inhibitors could bind. Drwal and colleagues created a homology model of Top2α to identify amino acids in a 10 Å proximity to the ligand, concluding that Met762 and Ser763 are good targets, while Ser800 could be too far for favourable interaction (Drwal et al. 2014). A crystal structure of Top2α complexed with DNA, but no ligand, became available in 2012 (PDB code: 4FM9, Wendorff et al. 2012). The authors of the research suggested that better Top2α selectivity may be achieved through modifications in the lactone group of the etoposide aglycone, which will strengthen the bond with Met762, while modifications in the glycosidic moiety of the drug will ensure stronger bonds with Ser800 (Wendorff et al. 2012).

In this section we outline the structure-activity relationships of podophyllotoxin drugs and their targets that would allow the design of safer and better therapeutics. Manufacturing superior drugs though depends on the availability of the crude substrate as well. For the past 40 years, the supply of podophyllotoxin has been a hot topic, where efforts were channelled to reconstitute the biosynthetic pathway to podophyllotoxin, and to identify plant species suitable for large-scale biomass production coupled with high yield of the target metabolite. The advances in the biosynthetic pathway research are discussed next, followed by advances in the aseptic cultivation of promising plant species.

The Biosynthetic Pathway

Two molecules coniferyl alcohol (**8**) afford the first lignan of the biosynthetic pathway, (+)-pinoresinol (**9**). (+)-Pinoresinol is sequentially reduced to (+)-lariciresinol (**10**) and (−)-secoisolariciresinol (**11**), and further converted to (−)-matairesinol (**12**) (Scheme 8.1). The last of the upstream lignans, (−)-matairesinol, is considered a central molecule in the lignan biosynthetic pathway, and a branch point for many lignan structures, including that of podophyllotoxin (**1**) (Suzuki and Umezawa 2007). The biosynthetic steps to coniferyl alcohol, which is formed in the general phenylpropanoid pathway (Humphreys and Chapple 2002), are not discussed in this chapter. The metabolic grid involving conversions between different hydroxycinnamyl alcohols, including coniferyl alcohol (Humphreys and Chapple 2002) will be of greater interest for discussions involving lignin biosynthesis, where a combination between different precursors, like coniferyl, sinapyl and *p*-coumaryl alcohols, leads to the formation of the cell wall polymer. Lignin polymerisation involves diverse inter-unit coupling rendering heterogenous and optically inactive structure

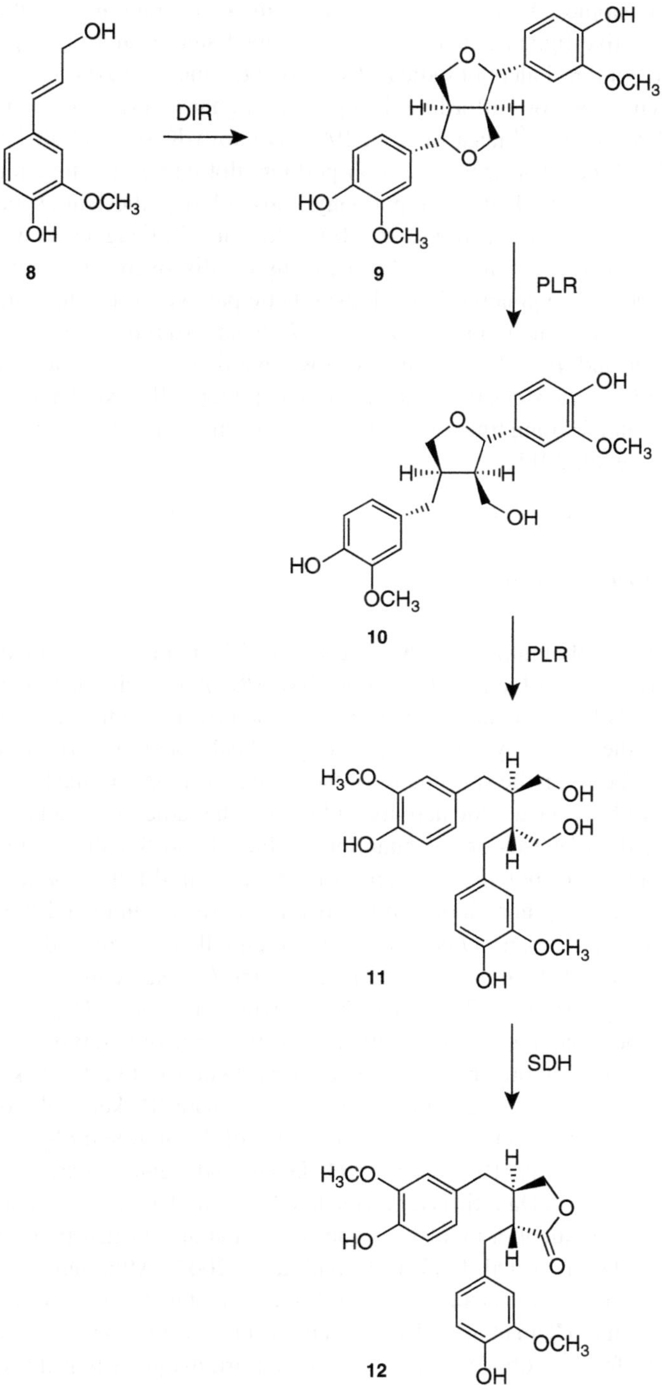

Scheme 8.1 Biosynthesis of the upstream lignans (+)-pinoresinol (*9*), (+)-lariciresinol (*10*), (−)-secoisolariciresinol (*11*) and (−)-matairesinol (*12*), starting from coniferyl alcohol (*8*). The enzymes involved in the conversions are DIR (dirigent protein), PLR (pinoresinol/lariciresinol reductase) and SDH (secoisolariciresinol dehydrogenase)

(Weng and Chapple 2010). On the other hand, the entry point to the diverse world of optically active lignan structures is a controlled stereo- and regiospecific coupling of two coniferyl alcohol radicals (Suzuki and Umezawa 2007).

The enzyme reactions affording the upstream lignans were revealed for the first time in *F. intermedia* (Umezawa et al. 1991; Dinkova-Kostova et al. 1996; Davin et al. 1997).[4] Since *Forsythia* is a non-podophyllotoxin-containing plant, it has been of great interest whether the podophyllotoxin-biosynthesising species utilise the same sequence of reactions. In 2000, Xia and colleagues confirmed that matairesinol was formed in *Linum flavum* using the discovered *Forsythia* pathway (Xia et al. 2000). The general lignan biosynthetic pathway was later confirmed for many plant species, including *Arabidopsis thaliana*, *Arctium lappa*, *L. album*, *L. usitatissimum* and *P. peltatum* and is now regarded as canonical (Suzuki and Umezawa 2007). The steps from matairesinol to podophyllotoxin have been mostly hypothetical and for long time the enzymes catalysing the reactions and their genes, enigma (Kim et al. 2015).

Forming Pinoresinol

Two coniferyl alcohol molecules (**8**) are used by dirigent protein (DIR) to afford the enantiomeric excess of either (+)-pinoresinol (**9**) or (−)-pinoresinol (Umezawa 2003). Radicals from the substrate are first formed by an oxidising agent, putative laccase, and then used by DIR for phenoxy radical coupling forming a covalent C8–C8′ bond between the side chains of the coniferyl alcohol radicals. The enzyme itself does not have a catalytic activity, but orients the radicals for selective dimerisation to yield either (+)- or (−)-enantiomer (Pickel and Schaller 2013). In some species, formation of both enantiomers was observed in different organs: *A. lappa* petioles formed (+)-pinoresinol, while ripening seeds accumulated the (−)- antipode (Suzuki et al. 2002a). Other species preferentially accumulated one antipode: (−)-pinoresinol in *A. thaliana* (Kim et al. 2012) and *L. usitatissimum* (Dalisay et al. 2015), and (+)-pinoresinol (**9**) in *Forsythia* (Davin et al. 1997), *Thuja plicata* (Kim et al. 2002) and *L. flavum* (Xia et al. 2000). The existence of two types of DIR, each forming a specific pinoresinol enantiomer, was demonstrated by Pickel and colleagues using *A. thaliana* recombinant DIR recombinant (Pickel et al. 2010).

DIR was discovered in the cell wall residues of *F. intermedia* by the group of L. Davin and N. Lewis (Pare et al. 1994; Davin and Lewis 1995). Further work indicated that the 78 kDa DIR is a glycosylated protein that lacks oxidising activity and uses coniferyl alcohol radicals as substrates that are formed in a one-electron oxidation event (Davin et al. 1997; Halls and Lewis 2002). Although initially identified as a homodimer (Halls and Lewis 2002), the crystal structure showed that the enzyme is a trimer (Kim et al. 2015). Each pocket has been shown to accommodate a pair of coniferyl alcohol radicals with spatial arrangement that does not allow

[4] Excellent reviews are provided by Davin and Lewis 2003 and Suzuki and Umezawa 2007.

interaction between substrates from different pockets. The authors speculated that specific residues from the binding sites may be responsible for the stereospecificity, but since the protein was not crystallised with its substrate (Kim et al. 2015), future analyses will be able to determine the exact position of the substrates within the binding site, as well as the interacting and the stereospecificity guiding amino acids of the active site.

From Pinoresinol to Lariciresinol and Secoisolariciresinol

A sequential reduction of (+)-pinoresinol (**9**) via (+)-lariciresinol (**10**) to (−)-secoisolariciresinol (**11**) is catalysed by a bifunctional cytosolic NADPH-dependent pinoresinol/lariciresinol reductase (PLR). PLR form *F. intermedia* converted exclusively (+)-pinoresinol into (+)-lariciresinol and (−)-secoisolariciresinol when cell-free extracts were incubated with (±)-pinoresinol (Katayama et al. 1993; Chu et al. 1993). Two isoforms found in *F. intermedia* were functionally characterised and the respective genes cloned; both proteins are able to catalyse the formation of (−)-secoisolariciresinol (Dinkova-Kostova et al. 1996). In *Thuja plicata* two PLRs were able to catalyse opposite reactions: PLR_Tp2 produced (−)-secoisolariciresinol from (+)-pinoresinol via (+)-lariciresinol, while PLR_Tp1 produced (+)-secoisolariciresinol from (−)-pinoresinol via (−)-lariciresinol (Fujita et al. 1999). The reductive conversion proposed by Dinkova-Kostova and colleagues (Dinkova-Kostova et al. 1996) was supported and further reinforced after the crystal structure of PLR_Tp1 was published by Min and colleagues (PDB code 1QYD, Min et al. 2003). The atomic structure showed seven β-strands and six α-helices with folding that is typical for the NADPH-dependent enzymes (Min et al. 2003). By using structural data for similar proteins, the authors deduced the putative cofactor-binding site and substrate-binding site, which they propose to be in close proximity to each other. According to the model, (+)-pinoresinol is stabilised in the hydrophobic pocket via aromatic stacking with the nicotinamide ring for [4R]-hydride transfer, where the amino acid Lys138 from PLR serves as a base for the catalysis. The involvement of the amino acid in the catalytic reaction was shown after mutating Lys138 to Ala138, which rendered defective enzyme, unable to reduce (+)-pinoresinol. The results also supported the idea that the symmetric substitutions observed in the PLR_Tp1 and PLR_Tp2 binding pocket may be responsible for the enantiospecificity (Min et al. 2003).

To Matairesinol

Closure of ring D to yield levorotary matairesinol (**12**) is a two-step process catalysed by secoisolariciresinol dehydrogenase (SDH) (Davin and Lewis 2003). The enzymatic conversion of (−)-secoisolariciresinol to (−)-matairesinol was described in 1991 by Umezawa and colleagues (Umezawa et al. 1991), and in 2001 the protein and the corresponding gene encoding the protein were published (Xia et al. 2001). The atomic structure of a recombinant SDH_Pp7 form *P. peltatum* became available in 2004, showing that the enzyme is NAD^+-dependent bifunctional homotetramer, consisting of seven β-strands and eight α-helices with a typical dehydrogenase structure (Youn et al. 2004). The crystal structure of complexed tetramer with NAD^+ and substrate showed that the catalysis occurs in a hydrophobic pocket, where NAD^+ and the substrate are well positioned for a B-face-specific hydride transfer (Youn et al. 2004). Three conserved amino acids in the binding site, Ser153, Lys171 and Tyr167, are involved in the reduction of (−)-secoisolariciresinol (**11**). Tyr167, in particular, serves as a base for the deprotonation of (−)-secoisolariciresinol, and in the subsequent nucleophilic attack an intermediate (−)-lactol is afforded. A second hydride transfer from (−)-lactol to a newly bound positively charged NAD yields (−)-matairesinol (Youn et al. 2004).

Optical purity in the lignan biosynthetic pathway is achieved at this step. The upstream lignans like pinoresinol, lariciresinol and secoisolariciresinol were afforded in reactions that yielded racemic mixtures that vary greatly between organs and plant species (Umezawa 2003). For the production of podophyllotoxin, only (−)-matairesinol, which started from (+)-pinoresinol, is utilised. In species like *L. usitatissimum*, where the main lignan accumulated is (+)-secoisolariciresinol, the steps involve (−)-pinoresinol and (−)-lariciresinol; however the plant cannot biosynthesise (−)-podophyllotoxin.

Matairesinol to Podophyllotoxin

The enzymatic conversions to matairesinol are well studied and comprise the early biosynthetic pathway of lignans observed in many plant species: *Forsythia*, *Linum*, *Podophyllum* and *Anthriscus* (Suzuki et al. 2002b; Satake et al. 2013). Until 2013, the conversions were mostly hypothetical with two intermediate metabolites identified in the early 1990s that lead to (−)-podophyllotoxin (**1**): (−)-yatein (**16**) and (−)-deoxypodophyllotoxin (**20**) (Broomhead et al. 1991). The reactions from (−)-matairesinol (**12**) were elucidated just recently in *Podophyllum* using modern techniques like transcriptome mining and massive parallel sequencing. Research of the lignan biosynthetic pathway in the (−)-yatein-accumulating plant, *Anthriscus*, showed that the plant species does not share the same sequence of reactions with *Podophyllum*, and converts (−)-matairesinol (**12**) to (−)-yatein (**16**) via different intermediates (Scheme 8.2).

Feeding experiments performed by Sakakibara and colleagues revealed the sequence of the reactions in *Anthriscus*: a hydroxylation of (−)-matairesinol (**12**) to (−)-thujaplicatin (**17**), subsequent methylation to 5-*O*-methylthujaplicatin (**18**) and 4,5-*O*,*O*-dimethylthujaplicatin (**19**), followed by a methylenedioxy bridge formation yielding (−)-yatein (**16**) (Sakakibara et al. 2003). The first methylation event yielding 5-*O*-methylthujaplicatin (**18**) was confirmed in 2013 by Ragamustari and

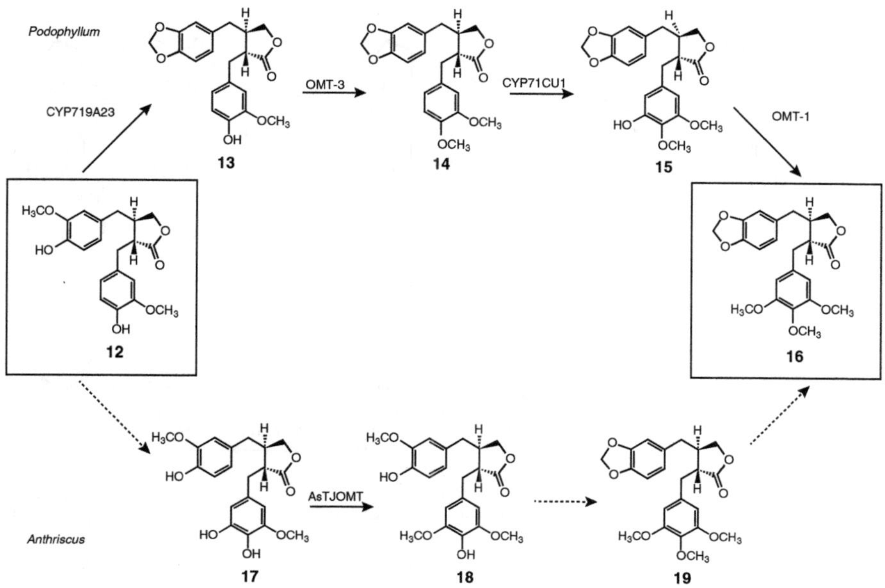

Scheme 8.2 Biosynthesis of (−)-yatein (***16***) from (−)-matairesinol (***12***) in *Podophyllum* (*up*) and *Anthriscus* (*below*). The reactions in *Podophyllum* involve formation of (−)-pluviatolide (***13***), (−)-5′-desmethoxy-yatein (***14***) and (−)-5′-desmethyl-yatein (***15***). The reactions in *Anthriscus* involve formation of (−)-thujaplicatin (***17***), 5-*O*-methylthujaplicatin (***18***) and 4,5-*O*,*O*-dimethylthujaplicatin (***19***). The putative enzymatic conversions are given in *dashed lines*. The enzymes involved in the conversions are CYP719A23, OMT-3 (*O*-methyltransferase-3), CYT71CU1, OMT-1 (*O*-methyltransferase-1) and AsTJOMT (*A. sylvestris* thujaplicatin *O*-methyltransferase)

colleagues. The authors cloned and expressed *A. sylvestris* thujaplicatin *O*-methyltransferase (AsTJOMT) that performs regioselective methylation. They suggested that the second methylation could be catalysed by an *O*-methyltransferase as well; however the rest of the conversions are still hypothetical and no enzymes or genes encoding the enzymes are confirmed (Scheme 8.2, dashed lines) (Ragamustari et al. 2013).

In *Podophyllum*, Marques and colleagues demonstrated that the methylenedioxy bridge formation takes place before the hydroxylation and methylation events (Marques et al. 2013). Performing massive parallel sequencing of *P. hexandrum* and *P. peltatum* transcriptomes, the authors discovered two enzymes catalysing the ring A closure in *Podophyllum*. The enzymes, CYP719A23 from *P. hexandrum* and CYP719A24 from *P. peltatum*, were able to convert (−)-matairesinol (**12**) to (−)-pluviatolide (**13**) (Marques et al. 2013).

In 2015, the biosynthetic pathway between (−)-pluviatolide (**13**) and (−)-deoxy-podophyllotoxin (**17**) was confirmed by Lau and Sattely (Lau and Sattely 2015). The study used transcriptome mining and combinatorial expression that enabled the authors to reconstitute the enzymatic conversions leading to the formation of

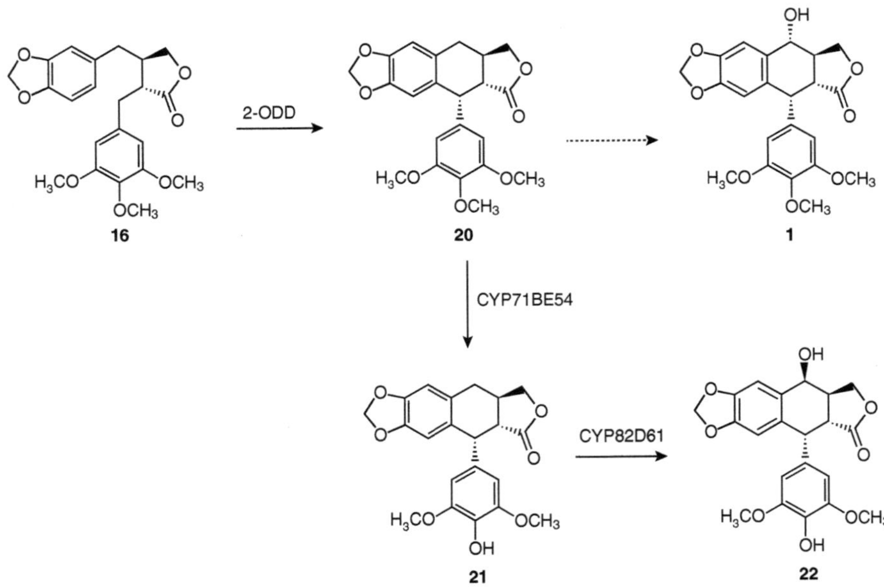

Scheme 8.3 Downstream reactions from (−)-yatein (*16*) via (−)-deoxypodophyllotoxin (*20*). The first branch (*up*) shows a single conversion to (−)-podophyllotoxin (**1**). The second branch (*below*) shows the biosynthesis of (−)-4′-desmethyl-deoxypodophyllotoxin (**21**) and (−)-4′-desmethyl-epipodophyllotoxin (**22**). The enzymes involved in the conversions are 2-ODD (2-oxoglutarate/Fe(II)-dependent dioxygenase), CYP71BE54 and CYP82D61

(−)-deoxypodophyllotoxin (**20**). Although the authors were not able to identify the enzyme hydroxylating (−)-deoxypodophyllotoxin (**20**) at position C7, they identified two enzymes that afford the immediate precursor of etoposide—a discovery, which from commercial point of view could be significantly more important. The total arsenal included two *O*-methyltransferases, three Cyt P450s, and one 2-oxoglutarate/Fe(II)-dependent dioxygenase (Lau and Sattely 2015). Infiltrating (−)-matairesinol (**12**) to transformed *N. benthamiana* leaves, Lau and Sattely were able to discover that *O*-methyltransferase-3 (OMT-3) utilised (−)-pluviatolide (**13**) to afford (−)-5′-desmethoxy-yatein (**14**) by transferring a methyl group at position C4′ of the pendant ring. The next reaction was hydroxylation to (−)-5′-desmethyl-yatein (**15**), followed by a methylation at position C5′ to (−)-yatein (**16**). The respective conversions were catalysed by CYP71CU1 and *O*-methyltransferase-1 (OMT-1) (Scheme 8.2). To afford the aryltetralin skeleton, a 2-oxoglutarate/Fe(II)-dependent dioxygenase (2-ODD) catalysed C–C bond formation, closing ring C (Lau and Sattely 2015) (Scheme 8.3).

The last enzyme converting (−)-deoxypodophyllotoxin (**20**) to (−)-podophyllotoxin (**1**) is yet to be elucidated. A reaction with putative (−)-deoxypodophyllotoxin-7H affording (−)-podophyllotoxin (**1**) has been described in 1999 by Henges and colleagues (Henges 1999). Although the authors detected a catalytic activity, they could not fully characterise the enzyme (Henges 1999). In 2007, Federolf and colleagues studied *L. album* cell lines with different metabolite profiles, proposing that deoxypodophyllotoxin-7H is not a cytochrome P450 monooxygenase as it cannot be effectively inhibited by specific inhibitors (Federolf et al. 2007). The transcriptome mining by Lau and Sattely was targeting the deoxypodophyllotoxin-7H, but instead of identifying the last step in the (−)-podophyllotoxin biosynthetic pathway, they discovered two enzymes that convert (−)-deoxypodophyllotoxin (**20**) to (−)-4′-desmethyl-deoxypodophyllotoxin (**21**) and (−)-4′-desmethyl-epipodophyllotoxin (**22**). The two enzymes are cytochrome P450s: CYP71BE54 and CYP82D61, performing sequential C4′ demethylation and C7 hydroxylation (Lau and Sattely 2015).

Despite the fact that the last step in the podophyllotoxin biosynthetic pathway is not yet deciphered, the work by Lau and Sattely is a milestone in the lignan research. More importantly, they proposed a direct way of biosynthesising the etoposide aglycone. The successful combinatorial expression in tobacco plants yielded 10.3 ng of (−)-4′-desmethyl-epipodophyllotoxin (**22**) on a dry weight basis when infiltrated with 100 μM (+)-pinoresinol (**9**) (Lau and Sattely 2015).

Regulation of the Biosynthetic Pathway

Studies focusing on the regulation of the podophyllotoxin biosynthetic pathway have accumulated in the past 10 years. Since the biosynthetic pathway was completed just recently, we would expect to see an increase in the research that investigates the pathway regulation via transcription factors, microRNAs and *cis*-acting genetic elements.

The promoter of *L. usitatissimum PLR* (*LuPLR1*) has been of particular interest in several studies, showing that the gene promoter contains *cis*-acting elements ABRE and MYB2 that are involved in the abscisic acid response (Hano et al. 2006; Renouard et al. 2012; Corbin et al. 2013). Putative *cis*-elements were found in the promoter regions of *SDH* in *P. hexandrum*, showing that MYB and WRKY may also be considered for further research (Kumar et al. 2016).

Different transcription factors have been identified in *P. hexandrum* and *P. peltatum*. In 2013, Bhattacharyya and colleagues assembled de novo a cDNA library from *P. hexandrum* cell culture, identifying members from transcription factor families potentially involved in the secondary metabolism regulation: AP2-EREBP, NAC, bHLH, MYB, bZIP, nTERF, WRKY, C2C2-CO-like and C2C2-Dof (Bhattacharyya et al. 2013). In 2016, the authors extended their research and challenged the cell cultures of *P. hexandrum* with methyl jasmonate to identify putative up-regulated genes involved in the regulation of the podophyllotoxin biosynthetic pathway (Bhattacharyya et al. 2016). A comparison between the 2013 and 2016 study showed that members from six transcription factors families were up-regulated upon application of the elicitor: *AP2-EREBP*, *NAC*, *bHLH*, *MYB*, *nTERF* and *WRKY* (Bhattacharyya et al. 2013; Bhattacharyya et al. 2016). Another study, authored by Kumar and colleagues, reported unique transcription factors in *P. hexandrum* (bZIP and MYB) and *P. peltatum* (bHLH and MYB/SANT) that correlated with the high podophyllotoxin content (Kumar et al. 2016).

Recently, one in silico-based study of *P. hexandrum* transcriptome was conducted by Biswas and colleagues, offering a glimpse of possible elements involved in the pathway regulation (Biswas et al. 2016). Although not definitive in answering how miRNAs are involved in the regulation of lignan biosynthetic pathway, the paper provides preliminary information and direction for future research.

Platforms to Produce Podophyllotoxin

The name "podophyllotoxin" derives from the plants it has been most commonly extracted from: *P. hexandrum* (Himalayan mayapple) and *P. peltatum* (mayapple, American mandrake) (Botta et al. 2001). The ethnobotanical use of these two species is described in detail elsewhere (Moraes et al. 2002; Ionkova 2010), but it is worth mentioning that the roots and rhizomes of *P. hexandrum* are the richest source of podophyllotoxin, reaching 4.3% on a dry weight basis (Gordaliza et al. 2004). Many plant species have been tested since to uncover new promising producers of the lignan; however no wild plant able to produce more podophyllotoxin on a dry weight basis than that of *P. hexandrum* has been described.

Stripping the plant matrix to extract pure podophyllotoxin from *Podophyllum* rhizomes to manufacture anticancer drugs is the only employed method so far, as chemical synthesis proved to be a low-yielding approach achieving racemic mixtures or involving too many steps (Botta et al. 2001; Bruschi et al. 2010). After decades of intensive collection of wild plants, *P. hexandrum* has been declared endangered (Bhadula et al. 1996; Airi et al. 1997). The interest in finding alternative sources of podophyllotoxin identified members of genus *Linum* to be promising. They were able to biosynthesise aryltetralin lignans, at the same time demonstrating excellent aseptic cultivation parameters.

So far, the most explored platform for podophyllotoxin sourcing is the native *P. hexandrum* plants. The research in the last 40 years aimed at identifying sustainable and renewable approach to deliver crude material for etoposide and teniposide production that would decrease the pressure over natural systems. *P. hexandrum* is endangered (Bhadula et al. 1996), while *Linum* species are small and could not possibly account for the increased demand for podophyllotoxin, if grown in the field. For many years the research focused on shifting the collection of native plants (Platform 1) to aseptic cultivation (Platform 2). Platform 2 largely relied on optimised protocols for biotechnological scale-up of high-producing cell lines. The roadblocks to provide commercial cell lines were many. Some of them still exist, including poor understanding of the biosynthetic pathway regulation, difficulties to switch from low to high yields of the target product and lack of genetic markers to assess the cell culture stability.

Cultivating plant cells can be laborious and time-consuming approach as it involves finding the best growth medium, selecting a high-producing cell line and establishing conditions for consistent growth and target metabolite production.

Table 8.1 Prevalent approaches used in the past four decades to increase the content of podophyllotoxin through aseptic cultivation of podophyllotoxin-accumulating plants

Decade	Major areas	References
1980s	*P. peltatum* and *P. hexandrum* callus	Kadkade (1981), Kadkade (1982), and van Uden et al. (1989)
	Effect of red light	
1990s	*L. album* in vitro cultures	Smollny et al. (1992) and Smollny et al. (1998)
	Effect of light/dark	
2000s	Medium optimisation	Chattopadhyay et al. (2001, 2002, 2003), van Furden et al. (2005), Shams-Ardakani et al. (2005), Vasilev and Ionkova (2006), Mohagheghzadeh et al. (2006), Baldi et al. (2008a, b), Farkya and Bisaria (2008), Ionkova (2009), and Ionkova and Fuss (2009)
	Use of elicitors	
	Agrobacterium rhizogenes-mediated transformation	
	Bioreactor studies	
	Biotransformation studies	
2010s	Use of elicitors	Ionkova et al. (2010), Yousefzadi et al. (2010b, c), Chashmi et al. (2011), Ionkov et al. (2012), Yousefzadi et al. (2012), Bahabadi et al. (2012), Sasheva et al. (2013), Tahsili et al. (2014), Sasheva and Ionkova (2015), and Sasheva et al. (2015)
	Agrobacterium rhizogenes-mediated transformation	
	Bioreactor studies	
	Gene expression studies	
	Metabolic engineering studies	

Many plant species have been tested in vitro for the production of podophyllotoxin, but most of them showed potential to accumulate lignans different than the target metabolite. Cell cultures can also show unreliable patterns as even cell lines derived from same species, *L. album*, had large differences in the ability to produce podophyllotoxin—a behaviour that was attributed to the genotype of the accessions used to start the cultures (Federolf et al. 2007). The disadvantage, however could be turned into an opportunity, as Federolf and colleagues did, to characterise enzymes involved in the downstream lignan biosynthetic pathway (Federolf et al. 2007).

The early research on cultivating podophyllotoxin-producing plants was initiated in the 1980s and 1990s (Table 8.1). The period marked the onset of the fermentation of *Podophyllum* and *Linum* cells, and offered a slow progress in delivering effortless systems for podophyllotoxin production. The attempts were concentrated on establishing the optimal medium composition and physical factors to achieve satisfactory culture growth and podophyllotoxin production. Kadkade showed that *P. peltatum* callus produce podophyllotoxin (Kadkade 1981), and that its content can be increased if the cell lines received red light (Kadkade 1982). Ten years later, Smollny and colleagues initiated callus from *L. album* (Smollny et al. 1992) and demonstrated that more podophyllotoxin was biosynthesised when the cells were grown in the dark (Smollny et al. 1998). Despite the attempts to scale up the cultivation of *P. hexandrum* in vitro, research groups reported slow growth, darkening and clumping of the cells, causing hypoxia (Fujii 1991; Giri and Narasu 2000; Chattopadhyay et al. 2001). With the addition of PVP and pectinase, the cell browning was solved, and despite the consistently slow growth, the cells have been successfully scaled up for a bioreactor cultivation in 2003 (Chattopadhyay et al. 2001).

After the initiation of *L. album* cultures, it became obvious that flax is another promising plant that can be used for podophyllotoxin production. Cells grew rapidly, the cultivation was easy and the podophyllotoxin production was comparable to that of *P. hexandrum* cells (Fuss 2003; Malik et al. 2014). Although podophyllotoxin was found in the in vitro cultures of *Callitris drummondii* (van Uden et al. 1990), *P. peltatum* (Fujii 1991), *Juniperus chinensis* (Muranaka et al. 1998), *L. persicum* (Mohagheghzadeh et al. 2003), *L. bulgaricus* (Vasilev and Ionkova 2006), *L. linearifolium* (Ionkova et al. 2010) and *L. thracicum* (Sasheva et al. 2013), it is *P. hexandrum* and *L. album* that have been accepted as the benchmark for in vitro studies.

While the early research focused on increasing the podophyllotoxin yield by varying the components of the growth medium and/or physical factors, later studies took advantage of the upstream lignan biosynthetic pathway discoveries.[5] As a result of an intense research programme conducted by Laurence Davin and Norman Lewis group, by 2001 the biosynthetic pathway from coniferyl alcohol to matairesinol was reconstituted; the enzymes DIR, PLR and SDH were discovered; and the genes encoding the respective enzymes were cloned (Davin and Lewis 2003). Elicitation, one of the heavily used techniques in the 2000s, initially employed to increase the podophyllotoxin production, was later coupled with transcriptomic studies to unravel the expression of *DIR*, *PLR* and *SDH* under conditions that mimic a biological attack. Among the elicitors, methyl jasmonate and *Fusarium graminearum* extracts proved to be the most successful agents to boost the podophyllotoxin production. An increase in the podophyllotoxin yields has been reported for *L. album* (van Furden et al. 2005) and *L. thracicum* (Sasheva et al. 2015); however the authors found that the high-producing cell lines had in general a weak response when challenged with methyl jasmonate. Despite the fact that methyl jasmonate was able to provoke a modest response, van Furden and colleagues observed accumulation of 7.69 mg podophyllotoxin on a dry weight basis (van Furden et al. 2005): much higher than the podophyllotoxin content of 0.14 mg on a dry weight basis, accumulated after *F. graminearum* elicitation (Bahabadi et al. 2011). A drastic eightfold increase in podophyllotoxin content after *L. album* elicitation with *F. graminearum* (Bahabadi et al. 2011; Bahabadi et al. 2012; Tahsili et al. 2014) was associated with up-regulation of *PLR* (Tahsili et al. 2014). Wounding and treatment with methyl jasmonate were also found to trigger over-expression of *DIR*, *PLR* and *SDH* in *P. hexandrum* (Wankhede et al. 2013).

Other techniques that dominated the 2000s included *Agrobacterium rhizogenes*-mediated transformation and bioreactor cultivation. Hairy roots, which are genetically stable, grow fast and do not require plant hormones (Chashmi et al. 2011), proved to be efficient system to produce podophyllotoxin in a predictable manner. Up to 105 mg podophyllotoxin on a dry weight basis was accumulated by *L. album* transformed with *A. rhizogenes* (Chashmi et al. 2011). Other researchers also reported that *L. album* hairy root cultures biosynthesise up to 0.6% podophyllotoxin on a dry weight basis (Baldi et al. 2008a; Farkya and Bisaria 2008). Transformed roots also offer the possibility for elicitation, further increasing the levels of various

[5] We recommend two good reviews provided by Yousefzadi et al. 2010a and Malik et al. 2014.

lignans, as observed by Ionkova (Ionkova 2009), and Bahabadi and colleagues (Bahabadi et al. 2014). Large-scale production of podophyllotoxin was reported by Nippon Oil Ltd. (Japan) and ROOTec bioactives Ltd. (Switzerland), which was recently acquired by the start-up Green2Chem S.A. (Belgium).

The medicinal plant research also benefited from the advances in the plant molecular biology techniques, which enabled scrutinising non-model plants with large and non-sequenced genomes, like *Podophyllum*. There has been a significant change in the way the biosynthetic pathway is discovered: the stepwise experiments, where the proteins were purified and sequenced were replaced by transcriptomics and homology-based cloning approach, which speeded up the process (De Luca et al. 2012). In the recent years, interrogating the transcriptome and proteome of elicited cell cultures has been used to advance the knowledge of the lignan biosynthetic pathway and its regulatory elements (Bhattacharyya et al. 2012; Bhattacharyya et al. 2016).

The reconstitution of the biosynthetic pathway to etoposide aglycone enabled the development of Platform 3, which involved metabolomiic engineering. Several studies already showed that it is possible to engineer the upstream lignan biosynthesis. The whole set of reactions to etoposide aglycone, though, will require further knowledge about the branch points and potential sinks of the pathway. Already successful was the *A. tumefaciens* transient transformation of *N. tabacum*, where Lau and Sattely combinatorially expressed the whole lignan biosynthetic pathway. Although the yield was very low, the study opens the door for speedy research to engineer the pathway in host systems (Lau and Sattely 2015).

Several studies explored the RNAi technologies, successfully down-regulating the expression of *PLR* in *L. corymbulosum* (Bayindir et al. 2008) and *F. koreana* (Kim et al. 2009; Murata et al. 2015). Although these studies targeted the upstream biosynthetic pathway, they showed that the technique can be used to prove biosynthetic steps by suppressing candidate genes (Bayindir et al. 2008) and to engineer the production of uncharacteristic lignans in host plants (sesamin in *F. koreana*, Kim et al. 2009, Murata et al. 2015). In 2015, Murata and colleagues created triple transgenic *F. koreana* cell line, which produced high amounts of sesamin at the expanse of pinoresinol. The transgenic plants had two down-regulated transcripts to abolish the pinoresinol glycosylation and lariciresinol/secoisolariciresinol formation. To redirect the lignan to another metabolic pathway, the authors introduced *CYP81Q1*, which is responsible for the formation of sesamin from pinoresinol. Single- and double-RNAi lines in this experiment were unsuccessful, suggesting increased cytotoxicity generated by higher aglycone concentration (Murata et al. 2015).

Conclusions

The combinatorial biochemistry is most likely the next playground in the podophyllotoxin research. Many questions are still unanswered, but the recent discoveries showed that it can be possible to have the entire biosynthetic pathway expressed in alternative systems. The interest in uncovering the biosynthetic pathway of podophyllotoxin has intertwined both fundamental and commercial purposes. Finding reliable and sustainable sources of the etoposide precursors is an exciting avenue that can have a positive impact for the wild populations of the only commercial source of podophyllotoxin, *P. hexandrum*, and provides alternative systems to biosynthesise the lignan on economical scale.

References

Airi S, Rawal RS, Dhar U, Purohit AN (1997) Population studies on *Podophyllum hexandrum* Royle—a dwindling medicinal plant of the Himalaya. Plant Genet Resour Newslett 110:20–34

Bahabadi SE, Sharifi M, Safaie N, Murata J, Yamagaki T, Satake H (2011) Increased lignan biosynthesis in the suspension cultures of *Linum album* by fungal extracts. Plant Biotechnol Rep 5(4):367–373

Bahabadi SE, Sharifi M, Behmanesh M, Safaie N, Murata J, Araki R, Yamagaki T, Satake H (2012) Time-course changes in fungal elicitor-induced lignan synthesis and expression of the relevant genes in cell cultures of *Linum album*. J Plant Physiol 169(5):487–491

Bahabadi SE, Sharifi M, Chashmi NA, Murata J, Satake H (2014) Significant enhancement of lignan accumulation in hairy root cultures of *Linum album* using biotic elicitors. Acta Physiol Plant 36:3325–3331

Baldi A, Srivastava AK, Bisaria VS (2008a) Improved podophyllotoxin production by transformed cultures of *Linum album*. Biotechnol J 3(9–10):1256–1263

Baldi A, Jain A, Gupta N, Srivastava AK, Bisaria VS (2008b) Co-culture of arbuscular mycorrhiza-like fungi (*Piriformospora indica* and *Sebacina vermifera*) with plant cells of *Linum album* for enhanced production of podophyllotoxin: a first report. Biotechnol Lett 30:1671–1677

Bayindir U, Alfermann AW, Fuss E (2008) Hinokonin biosynthesis in *Linum corymbulosum* Reichenb. Plant J 55:810–820

Bhadula SK, Singh A, Lata H, Kunyal CP, Purohit AN (1996) Genetic resources of *Podophyllum hexandrum* Royle, an endangered medicinal species from Garhwal Himalaya, India. Plant Genet Resour Newslett 106:26–29

Bhattacharyya D, Sinha R, Ghanta S, Chakraborty A, Hazra S, Chattopadhyay S (2012) Proteins differentially expressed in elicited cell suspension culture of *Podophyllum hexandrum* with enhanced podophyllotoxin content. Proteome Sci 10:34

Bhattacharyya D, Sinha R, Hazra S, Datta R, Chattopadhyay S (2013) De novo transcriptome analysis using 454 pyrosequencing of the Himalayan Mayapple, *Podophyllum hexandrum*. BMC Genomics 14:748

Bhattacharyya D, Hazra S, Banerjee A, Datta R, Kumar D, Chakrabarti S, Chattopadhyay S (2016) Transcriptome-wide identification and characterization of CAD isoforms specific for podophyllotoxin biosynthesis from *Podophyllum hexandrum*. Plant Mol Biol 92(1–2):1–23

Biswas S, Hazra S, Chattopadhyay S (2016) Identification of conserved miRNAs and their putative target genes in *Podophyllum hexandrum* (Himalayan Mayapple). Plant Gene 6:82–89

Botta B, Monache GD, Misiti D, Vitali A, Zappia G (2001) Aryltetralin lignans: chemistry, pharmacology and biotransformations. Curr Med Chem 8:1363–1381

Broomhead AJ, Rahman MMA, Dewick PM, Jackson DE, Lucas JA (1991) Matairesinol as precursor of *Podophyllum* lignans. Phytochemistry 30:1489–1492

Bruschi M, Orlandi M, Rindone M, Rindone B, Saliu F, Suarez-Bertoa R, Tollpa EL, Zoia L (2010) Podophyllotoxin and antitumor synthetic aryltetralines. Toward a biomimetc preparation. In: Mukherjee A (ed) Biomimetics learning from nature. InTech, Rijeka

Capranico G, Tinelli S, Austin CA, Fisher ML, Zunino F (1992) Different patterns of gene expression of topoisomerase II isoforms in differentiated tissues during murine development. Biochim Biophys Acta 1132:43–48

Chashmi NA, Sharifi M, Yousefzadi M, Behmanesh M, Palazon J (2011) The production of cytotoxic lignans by hairy root cultures of *Linum album*. World Acad Sci Eng Technol 80:401–402

Chattopadhyay S, Srivastava AK, Bhojwani SS, Bisaria VS (2001) Development of suspension culture of *Podophyllum hexandrum* for the production of podophyllotoxin. Biotechnol Lett 23:2063–2066

Chattopadhyay S, Srivastava AK, Bhojwani SS, Bisaria VS (2002) Production of podophyllotoxin by plant cell culture of *Podophyllum hexandrum* in bioreactor. J Biosci Bioeng 93:215–220

Chattopadhyay S, Bisaria VS, Bhojwani SS, Srivastava AK (2003) Enhanced production of podophyllotoxin by fed batch cultivation of *Podophyllum hexandrum*. Can J Chem Eng 81:1011–1018

Chu A, Dinkova A, Davin LB, Bedgar DL, Lewis NG (1993) Stereospecificity of (+)-pinoresinol and (+)-lariciresinol reductases from *Forsythia intermedia*. J Biol Chem 268:27026–27033

Corbin C, Renouard S, Lopez T, Lamblin F, Laine E, Hano C (2013) Identification and characterization of *cis*-acting elements involved in the regulation of ABA- and/or GA-mediated *LuPLR1* gene expression and lignan biosynthesis in flax (*Linum usitatissimum* L.) cell cultures. J Plant Physiol 170:516–522

Dalisay DS, Kim KW, Lee C, Yang H, Rubel O, Bowen BP, Davin LB, Lewis NG (2015) Secoisolariciresinol diglucoside formation and cyanogenic glucosides in flax seed: MALDI mass spectrometry imaging. J Nat Prod 78:1231–1242

Damayanthi Y, Lown JW (1998) Podophyllotoxins: current status and recent developments. Curr Med Chem 5:205–252

Davin LB, Lewis NG (1995) Lignin and lignan biochemical pathways in plants: an unprecedented discovery in phenolic coupling. An Acad Bras Cienc Suppl 3(67):363–378

Davin LB, Lewis NG (2003) An historical perspective on lignan biosynthesis: monolignol, allylphenol and hydroxycinnamic acid coupling and downstream metabolism. Phytochem Rev 2:257–288

Davin LB, Wang HW, Crowell AL, Bedgar DL, Martin DM, Sarkanen S, Lewis NG (1997) Stereoselective bimolecular phenoxy radical coupling by an auxiliary (dirigent) protein without an active center. Science 275:362–366

De Luca V, Salim V, Atsumi SM, Yu F (2012) Mining the biodiversity of plants: a revolution in the making. Science 336:1658–1661

Dinkova-Kostova AT, Gang DR, Davin LB, Bedgar DL, Chu A, Lewis NG (1996) (+)-Pinoresinol/(+)-lariciresinol reductase from *Forsythia intermedia*: protein purification, cDNA cloning, heterologous expression and comparison to isoflavone reductase. J Biol Chem 271:29473–29482

Dorleans A, Gigant B, Ravelli RBG, Mailliet P, Mikol V, Knossow M (2009) Variations in the colchicine-binding domain provide insight into the structural switch of tubulin. Proc Natl Acad Sci U S A 106(33):13775–13779

Drwal MN, Marinello J, Manzo SG, Wakelin LPG, Capranico G, Griffith R (2014) Novel DNA topoisomerase IIα inhibitors from combined ligand- and structure-based virtual screening. PLoS One 9(12):e114904

Farkya S, Bisaria VS (2008) Exogenous hormones affecting morphology and biosynthetic potential of hairy root line (LYR2i) of *Linum album*. J Biosci Bioeng 2:140–146

Federolf K, Alfermann WA, Fuss E (2007) Aryltetralin-lignan formation in two different cell suspension cultures of *Linum album*: deoxypodophyllotoxin 6-hydroxylase, a key enzyme for the formation of 6-methoxypodophyllotoxin. Phytochemistry 68:1397–1406

Fujii Y (1991) *Podophyllum* spp.: in vitro regeneration and the production of podophyllotoxin. In: Bajaj YPS (ed) Biotechnology in agriculture and forestry. Springer Verlag, Berlin/Heidelberg

Fujita M, Gang DR, Davin LB, Lewis NG (1999) Recombinant pinoresinol/lariciresinol reductases from western red cedar (*Thuja plicata*) catalyze opposite enantiospecific conversions. J Biol Chem 274:618–627

Fuss E (2003) Lignans in plant cell and organ cultures: an overview. Phytochem Rev 2:307–320
Giri A, Narasu ML (2000) Production of podophyllotoxin from *Podophyllum hexandrum*: a potential natural product for clinically useful anticancer drugs. Cytotechnology 34:17–26
Gordaliza M, Garcia PA, Miguel del Corral JM, Castro MA, Gomez-Zurita MA (2004) Podophyllotoxin: distribution, sources, applications and new cytotoxic derivatives. Toxicon 44:441–459
Halls SC, Lewis NG (2002) Secondary and quaternary structures of the (+)-pinoresinol-forming dirigent protein. Biochemistry 41:9455–9461
Hano C, Martin I, Fliniaux O, Legrand B, Gutierrez L, Arroo RRJ, Mesnard F, Lamblin F, Laine E (2006) Pinoresinol–lariciresinol reductase gene expression and secoisolariciresinol diglucoside accumulation in developing flax (*Linum usitatissimum*) seeds. Planta 224:1291–1301
Hartwell JL, Schrecker AW (1951) Components of podophyllin. V. The constitution of podophyllotoxin. J Am Chem Soc 73:2909–2916
Henges A (1999) Biosynthese und Kompartimentierung von Lignanen in Zellkulturen von *Linum album*. PhD Thesis, University of Düsseldorf, Germany
Humphreys JM, Chapple C (2002) Rewriting the lignin roadmap. Curr Opin Plant Biol 5:224–229
Huzil JT, Winter P, Johnson L, Weis AL, Bakos T, Banerjee A, Luduena RF, Damaraju S, Tuszynski JA (2010) Computational design and biological testing of highly cytotoxin colchicine ring a modifications. Chem Biol Drug Des 75:541–550
Ionkov T, Ionkova I, Sasheva P (2012) Strategy to control the production of rare anticancer substances in the endangered plant *Linum linearifolium* in bioreactor. Ind J Fund Appl Life Sci 2(1):170–176
Ionkova I (2009) Effect of methyl jasmonate on production of aryltetralin lignans in hairy root cultures of *Linum tauricum*. Pharmacogn Res 1(3):102–105
Ionkova I (2010) Podophyllotoxin and related lignans: biotechnological production by in vitro plant cell cultures. In: Arora R (ed) Medicinal plant biotechnology. C.A.B. International, Oxfordshire
Ionkova I, Fuss E (2009) Influence of different strains of *Agrobacterium rhizogenes* on induction of hairy roots and lignan production in *Linum tauricum* ssp. *tauricum*. Pharmacogn Res 5:14–18
Ionkova I, Antonova I, Momekov G, Fuss E (2010) Production of podophyllotoxin in *Linum linearifolium* in vitro cultures. Pharmacogn Res 6(23):180–185
Kadkade PG (1981) Formation of podophyllotoxin by *Podophyllum peltatum* tissue cultures. Naturwissenschaften 68:481–482
Kadkade PG (1982) Growth and podophyllotoxin production in callus tissues of *Podophyllum peltatum*. Plant Sci Lett 25:107–115
Kaplan IW (1942) Condylomata acuminata. New Orleans Med Surg J 94:388–390
Katayama T, Davin LB, Chu A, Lewis NG (1993) Novel benzylic ether reductions in lignan biogenesis in *Forsythia intermedia*. Phytochemistry 33:581–591
Kim MK, Jeon JH, Fujita M, Davin LB, Lewis NG (2002) The western red cedar (*Thuja plicata*) 8-8′ DIRIGENT family displays diverse expression patterns and conserved monolignol coupling specificity. Plant Mol Biol 49:199–214
Kim HJ, Ono E, Morimoto K, Yamagaki T, Okazawa A, Kobayashi A, Satake H (2009) Metabolic engineering of lignan biosynthesis in *Forsythia* cell culture. Plant Cell Physiol 50:2200–2209
Kim KW, Moinuddin SGA, Atwell KM, Costa MA, Davin LB, Lewis NG (2012) Opposite stereoselectivities of dirigent proteins in *Arabidopsis* and *Schizandra* species. J Biol Chem 287:33957–33972
Kim KW, Smith CA, Daily MD, Cort JR, Davin LB, Lewis NG (2015) Trimeric structure of (+)-pinoresinol forming dirigent protein at 1.95 Å resolution with three isolated active sites. J Biol Chem 290:1308–1318
King LS, Sullivan M (1946) The similarity of the effect of podophyllin and colchicine and their use in the treatment of condylomata acuminata. Science 104:244–245

Kumar P, Jaiswal V, Pal T, Singh J, Chauhan RS (2016) Comparative whole-transcriptome analysis in Podophyllum species identifies key transcription factors contributing to biosynthesis of podophyllotoxin in *P. hexandrum*. Protoplasma 254(1):217–228. doi:10.1007/s00709-015-0938-7

Lau E, Sattely ES (2015) Six enzymes from mayapple that complete the biosynthetic pathway to the etoposide aglycone. Science 349:1224

Loike JD, Horwitz SB, Grollman AP (1974) Effects of VP 16-213 on HeLa cells. Pharmacologist 16:209

Loike JD, Brewer CF, Sternlich H, Gensler WJ, Horwitz SB (1978) Structure-activity study of the inhibition of microtubule assembly in vitro by podophyllotoxin and its congeners. Cancer Res 38:2688–2693

Long BH, Minocha A (1983) Inhibition of topoisomerase II by VP-16 (etoposide), VM-26 (teniposide) and structure congeners as explanation for in vivo DNA breakage and cytotoxicitry. Proc Am Ass Cancer Res 24:321

Malik S, Biba O, Gruz J, Arroo RRJ, Strnad M (2014) Biotechnological approaches for producing aryltetralin lignans from Linum species. Phytochem Rev 13:893–913

Marques JV, Kim KW, Lee C, Costa MA, May GD, Crow JA, Davin LB, Lewis NG (2013) Next generation sequencing in predicting gene function in podophyllotoxin biosynthesis. J Biol Chem 288:466–479

Min T, Kasahara H, Bedgar DL, Youn B, Lawrence PK, Gang DR, Halls SC, Park HJ, Hilsenbeck JL, Davin LB, Lewis NG, Kang C (2003) Crystal structures of pinoresinol-lariciresinol and phenylcoumaran benzylic ether reductases and their relationship to isoflavone reductases. J Biol Chem 278:50714–50723

Mohagheghzadeh A, Hemmati S, Mehregan I, Alfermann AW (2003) *Linum persicum*: lignans and placement in Linaceae. Phytochem Rev 2(3):363–369

Mohagheghzadeh A, Hemmati S, Alfermann AW (2006) Quantification of aryltetralin lignans in *Linum album* organs and in vitro cultures. Iran J Pharm Sci 2(1):47–56

Moraes RM, Lata H, Bedir E, Maqbool M, Cushman K (2002) The American mayapple and its potential for podophyllotoxin production. In: Janick J, Whipkey A (eds) Trends in new crops and new uses. ASHS, Alexandria

Moss GP (2000) Nomenclature of lignans and neolignans. Pure Appl Chem 72(8):1493–1523

Muranaka T, Miyata M, Kazutaka I, Tachibana S (1998) Production of podophyllotoxin in *Juniperus chinensis* callus cultures treated with oligosaccharides and a biogenetic precursor. Phytochemistry 49:491–496

Murata J, Matsumoto E, Morimoto K, Koyama T, Satake H (2015) Generation of triple-transgenic *Forsythia* cell cultures as a platform for the efficient, stable, and sustainable production of lignans. PLoS One 10(12):e0144519

Nitiss JL (2009) Targeting DNA topoisomerase II in cancer chemotherapy. Nat Rev Cancer 9:338–350

Pare PW, Wang HB, Davin LB, Lewis NG (1994) (+)-pinoresinol synthase: a stereoselective oxidase catalysing 8,8-lignan formation in *Forsythia intermedia*. Tetrahedron Lett 35:4731–4734

Pickel B, Schaller A (2013) Dirigent proteins: molecular characteristics and potential biotechnological applications. Appl Microbiol Biotechnol 97:8427–8438

Pickel B, Constantin MA, Pfannstiel J, Conrad J, Beifuss U, Schaller A (2010) An enantiocomplementary dirigent protein for the enantioselective laccase-catalyzed oxidative coupling of phenols. Angew Chem Int Ed 49:202–204

Podwyssotzki V (1880) Pharmacologische studien uber *Podophyllum peltatum*. Arch Exp Pathol Pharmakol 13:29–52

Portyanko AS, Akalovich ST, Doroshenko TM (2015) Beta-III tubulin as a potential target for blocking invasive growth of malignant epithelial tumors. Eksp Klin Farmakol 78(8):25–28

Ragamustari SK, Nakatsubo T, Hattori T, Ono E, Kitamura Y, Suzuki S, Yamamura M, Umezawa T (2013) A novel O-methyltransferase involved in the first methylation step of yatein biosynthesis from matairesinol in *Antriscus sylvestris*. Plant Biotechnol 30:375–384

Ravanbakhsh S, Gajewski M, Greiner R, Tuszynski JA (2013) Determination of the optimal tubulin isotype target as a method for the development of individualized cancer chemotherapy. Theor Biol Med Modell 10:29

Ravelli RBG, Gigant B, Curmi PA, Jourdain I, Lachkar S, Sobel A, Knossow M (2004) Insight into tubulin regulation from a complex with colchicine and a stathmin-like domain. Nature 428:198–202

Renouard S, Corbin C, Lopez T, Montguillon J, Gutierrez L, Lamblin F, Laine F, Hano C (2012) Abscisic acid regulates pinoresinol–lariciresinol reductase gene expression and secoisolariciresinol diglucoside accumulation in developing flax (*Linum usitatissimum*) seeds. Planta 235:85–98

Sakakibara N, Suzuki S, Umezawa T, Shimada M (2003) Biosynthesis of yatein in *Anthriscus sylvestris*. Org Biomol Chem 1:2474–2485

Sasheva P, Ionkova I (2015) Podophyllotoxin production by cell cultures of *Linum thracicum* ssp. *thracicum* Degen elicited with methyl jasmonate and salicylic acid. C R Acad Bulg Sci 68(7):883–888

Sasheva P, Letkarska G, Ionkova I (2013) Biotechnological production of podophyllotoxin and podophyllotoxin-related lignans in cultures of *Linum thracicum* Degen. C R Acad Bulg Sci 66(10):1445–1450

Sasheva P, Ionkova I, Stoilova N (2015) Methyl jasmonate induces enhanced podophyllotoxin production in cell cultures of thracian flax (*Linum thracicum* ssp. *thracicum*). Nat Prod Commun 10(7):1225–1228

Satake H, Ono E, Murata J (2013) Recent advances in metabolic engineering of lignan biosynthesis pathways for the production of transgenic plant-based foods and supplements. J Agric Food Chem 61:11721–11729

Shams-Ardakani M, Hemmati S, Mohagheghzadeh A (2005) Effect of elicitors on the enhancement of podophyllotoxin biosynthesis in suspension cultures of *Linum album*. DARU 13(2):56–60

Smollny T, Wichers H, de Rijk T, van Zwam A, Shahsavari A, Alfermann AW (1992) Formation of lignans in suspension cultures of *Linum album*. Planta Med 58:A622–A624

Smollny T, Wichers H, Kalenberg S, Shahsavari A, Petersen M, Alfermann AW (1998) Accumulation of podophyllotoxin and related lignans in cell suspension cultures of *Linum album*. Phytochemistry 48:575–579

Stahelin HF, von Wartburg A (1991) The chemical and biological route from podophyllotoxin glucoside to etoposide: ninth cain memorial award lecture. Cancer Res 51:5–15

Stanton RA, Gernet KM, Nettles JH, Aneja R (2011) Drugs that target dynamic microtubules: a new molecular perspective. Med Res Rev 31(3):443–481

Suzuki S, Umezawa T (2007) Biosynthesis of lignans and norlignans. J Wood Sci 53:273–284

Suzuki S, Umezawa T, Shimada M (2002a) Stereochemical diversity in lignan biosynthesis of *Arctium lappa*. Biosci Biotechnol Biochem 66:1262–1269

Suzuki S, Sakakibara N, Umezawa T, Shimada M (2002b) Survey and enzymatic formation of lignans of *Anthriscus sylvestris*. J Wood Sci 48:536–541

Tahsili J, Sharifi M, Safaie N, Esmaelizadeh-Bahabadi S, Behmanesh M (2014) Induction of lignans and phenolic compounds in cell culture of *Linum album* by culture filtrate of *Fusarium graminearum*. J Plant Interact 9:412–417

Ter Haar E, Rosenkranz HS, Hamel E, Day BW (1996) Computational and molecular modeling evaluation of the structural basis for tubulin polymerisation inhibition by colchicine site agents. Bioorg Med Chem 4:1659–1671

Toyoda E, Kagaya S, Cowell IG, Kurosawa A, Kamoshita K, Nishikawa K, Iizumi S, Koyama H, Austin CA, Adachi N (2008) NK314, a topoisomerase II inhibitor that specifically targets the alpha isoform. J Biol Chem 283:23711–23720

Tseng CY, Mane JY, Winter P, Johnson L, Huzil T, Izbicka E, Luduena RF, Tuszynski JA (2010) Quantitative analysis of the effect of tubulin isotype expression on sensitivity of cancer cell lines to a set of novel colchicine derivatives. Mol Cancer 9:131

Umezawa T (2003) Phylogenetic distribution of lignan producing plants. Wood Res 90:27–110

Umezawa T, Davin LB, Lewis NG (1991) Formation of lignans(−)-secoisolariciresinol and (−)-matairesinol with *Forsythia intermedia* cell-free extracts. J Biol Chem 266:10210–10217

van Furden B, Humburg A, Fuss E (2005) Influence of methyl jasmonate on podophyllotoxin and 6-methoxypodophyllotoxin accumulation in *Linum album*. Plant Cell Rep 24:312–317

van Uden W, Pras N, Visser JF, Malingre TM (1989) Detection and identification of podophyllotoxin produced by cell cultures from *Podophyllum hexandrum* Royle. Plant Cell Rep 8:165–168

van Uden W, Pras N, Malingre TM (1990) The accumulation of podophyllotoxin-b-D-glucoside by cell suspension cultures derived from the conifer *Callitris dummondii*. Plant Cell Rep 9:257–260

Vasilev N, Ionkova I (2006) Isolation and structure elucidation of aryltetralin lignans from *Linum tauricum* ssp. *bulgaricum*. Pharmacogn Mag 2(7):172–177

Wankhede DP, Biswas DK, Rajkumar S, Sinha AK (2013) Expressed sequence tags and molecular cloning and characterization of gene encoding pinoresinol/lariciresinol reductase from *Podophyllum hexandrum*. Protoplasma 250(6):1239–1249

Wendorff TJ, Schmidt BH, Heslop P, Austin CA, Berger JM (2012) The structure of DNA-bound human topoisomerase II alpha: conformational mechanisms for coordinating inter-subunit interactions with DNA cleavage. J Mol Biol 424:109–124

Weng JK, Chapple C (2010) The origin and evolution of lignin biosynthesis. New Phytol 187:273–285

Wu CC, Li TK, Farh L, Lin LY, Lin TS, Yu YJ, Yen TJ, Chiang CW, Chan NL (2011) Structural basis of type II topoisomerase inhibition by the anticancer drug etoposide. Science 333:459–462

Xia ZQ, Costa MA, Proctor J, Davin LB, Lewis NG (2000) Dirigent-mediated podophyllotoxin biosynthesis in *Linum flavum* and *Podophyllum peltatum*. Phytochemistry 55:537–549

Xia ZQ, Costa MA, Pelissier HC, Davin LB, Lewis NG (2001) Secoisolariciresinol dehydrogenase purification, cloning and functional expression: implications for human health protection. J Biol Chem 276:12614–12623

Youn B, Moinuddin SGA, Davin LB, Lewis NG, Kang CH (2004) Crystal structures of apo-form and binary/ternary complexes of *Podophyllum* secoisolariciresinol dehydrogenase, an enzyme involved in formation of health-protecting and plant defense lignans. J Biol Chem 280:12917–12926

Yousefzadi M, Sharifi M, Behmanesh M, Moyano E, Bonfill M, Cusido RM, Palazon J (2010a) Podophyllotoxin: current approaches to its biotechnological production and future challenges. Eng Life Sci 10(4):281–292

Yousefzadi M, Sharifi M, Chashmi NA, Behmanesh M, Ghasempour A (2010b) Optimization of podophyllotoxin extraction method from *Linum album* cell cultures. Pharm Biol 48(12):1421–1425

Yousefzadi M, Sharifi M, Behmanesh M, Ghasempour A, Moyano E, Palazon J (2010c) Salicylic acid improves podophyllotoxin production in cell cultures of *Linum album* by increasing the expression of genes related with its biosynthesis. Biotechnol Lett 32(11):1739–1743

Yousefzadi M, Sharifi M, Behmanesh M, Ghasempour A, Elisabeth M, Palazon J (2012) The effect of light on gene expression and podophyllotoxin biosynthesis in *Linum album* cell culture. Plant Physiol Biochem 56:41–46

Chapter 9
Hairy Root Culture for the Production of Useful Secondary Metabolites

Jyothi Abraham and T. Dennis Thomas

Abstract Hairy root is induced when *Agrobacterium rhizogenes,* a naturally occurring gram-negative soil bacterium, infects a plant that is susceptible to infection. During the infection process of the *A. rhizogenes* to a plant, parts of its plasmid DNA (T-DNA) in the Ri plasmid will be transferred and integrated into the nuclear genome of the host plant. This transformation process eventually resulted in a valuable by-product called as hairy root. During infection of *A. rhizogenes* the plant oncogenes (i.e. *rol* A, *rol* B, *rol* C and *rol* D) present in bacteria are incorporated into the plant genomes and cause tumour formation resulting in hairy root disease. The characteristic feature of hairy root is that they have the capability to grow fast on basal media, genetic and biosynthetic stability and ability to produce elevated amount of several useful secondary metabolites. This chapter reviews the recent reports on hairy root culture and secondary metabolite production in various plant species.

Keywords *Agrobacterium rhizogenes* • Elicitors • Ri plasmid • Rol genes • Secondary metabolites

Introduction

Plant tissue culture offers a number of useful techniques for various purposes including the improvement of crop species. The numerous in vitro techniques employed for improvement of plants include micropropagation (Thomas and Hoshino 2010; Cheruvathur et al. 2012), somatic embryogenesis (Kumar and Thomas 2012; Thomas and Jacob 2004), protoplast isolation and culture (Rahmani et al. 2016; Thomas 2009), androgenesis (Cimò et al. 2016; Kurtar et al. 2016),

J. Abraham
Research Department of Botany, St. Thomas College, Palai, Arunapuram (P.O.), Kottayam, Kerala 686574, India

T.D. Thomas (✉)
Department of Plant Science, Riverside Transit Campus, Central University of Kerala, Padannakkad, Kasaragod, Kerala 671314, India
e-mail: den_thuruthiyil@yahoo.com

S. Malik (ed.), *Biotechnology and Production of Anti-Cancer Compounds*,
DOI 10.1007/978-3-319-53880-8_9

gynogenesis (Alan et al. 2016; Thomas et al. 1999) and endosperm culture (Thomas and Chaturvedi 2008). One of the major reasons for culturing plant cell, tissue or organ is its capability to synthesize secondary metabolites which are present in whole plants (Tamer and Mavituna 1996). However, this protocol is met with many limitations such as instability of cultures and low productivity of secondary metabolites (Bonhomme et al. 2000). These limitations could be avoided by using hairy root culture for the production of useful secondary metabolites.

Roots of many plants yield a number of secondary metabolites useful for various purposes. In a similar fashion these phytochemicals could be synthesized in hairy roots and thereby save the natural population of medicinal plants from extinction. A diseased condition commonly called 'hairy root' usually appears when *Agrobacterium rhizogenes,* a naturally occurring gram-negative soil bacterium, infects a plant that is susceptible to infection. While the bacteria infect a plant, parts of its plasmid DNA (T-DNA) in the Ri plasmid will be migrated and incorporated into the nuclear genome of the host plant. This transformation process ultimately gives rise to a valuable by-product known as hairy root (Hu and Du 2006). Plant oncogenes, i.e. *rol* A, *rol* B, *rol* C and *rol* D, are present in *A. rhizogenes* plasmids. During infection of *A. rhizogenes* these genes are incorporated into the plant genomes and cause tumour formation resulting in hairy root disease (Bulgakov 2008). Moreover these *rol* genes are supposed to stimulate the production of secondary metabolites in transgenic plant tissues. The specialty of hairy roots includes its ability to grow fast on basal media, genetic and biosynthetic stability and ability to produce elevated amount of several useful secondary metabolites (Cheruvathur et al. 2015). Recently, there is an increase in use of hairy technology for the isolation of useful secondary metabolites (Table 9.1).

Hairy root system formed as a result of genetic transformation is considered as a stable biotechnological system. In order to enhance the natural product formation in hairy root culture elicitors from various sources have been routinely employed. Recently, this methodology has been routinely used in stimulating the production of several classes of active plant secondary metabolites especially alkaloids and flavonoids. The working of the elicitors is unique and can activate a series of stress or defence responses so that specific genes responsible for secondary metabolite biosynthesis can be triggered. Although the crucial role of elicitors in enhancing the production of useful secondary metabolites and examining their biosynthesis is well explained, such studies have been applied to only limited number of species. This chapter is an attempt to review the recent reports on hairy root culture and secondary metabolite production in various plant species.

Rhinacanthus nasutus (L.) Kurz

Rhinacanthus nasutus (Family: Acanthaceae) is distributed widely in some areas of subcontinent, in the region of Southeast Asia and China (Farnsworth and Bunyapraphatsara 1992). This shade loving plant is usually seen wild in the

Table 9.1 Some recent reports on normal and hairy root culture research in medicinal plants

Name of the plant	Explant source	Type of medium, growth regulators/additives	Culture response/secondary metabolite	References
Phyllanthus odonladenius	Seeds	½MS, GA_3	Hairy root culture/not assessed	Nakweti et al. (2015)
Gentiana dinarica	L	½MS	Hairy root culture/xanthone	Vinterhalter et al. (2015)
Isatis tinctoria	P	LB	Hairy root culture/flavonoids	Gai et al. (2015)
Plumbago rosea	TR	MS	Hairy root culture	Satheeshkumar et al. (2014)
Capsicum spp.	C, H, Ra	½MS	Hairy root culture/not assessed	Md Setamam et al. (2014)
Linum mucronatum	H, R	LB	Hairy root culture/podophyllotoxin, 6-methoxy podophyllotoxin	Samadi et al. (2014)
Portulaca oleracea	R, S, C	½MS	Hairy root culture/dopamine	Moghadam et al. (2014)
Gentiana scabra	L	B_5 + NAA	Hairy root culture/iridoids, secoiridoids	Huang et al. (2014)
Valeriana officinalis	R, H, L	MS	Hairy root culture/valerenic acid	Torkamani et al. (2014b)
Catharanthus roseus	L	½B_5	Hairy root culture/indol alkaloids	Thakore et al. (2013)
Artemisia annua	Seed, S	MS + NAA	Hairy root culture/artemisinin	Patra et al. (2013)
Gloriosa superba	Ca, Co	MS + 2, 4-D + KIN	Hairy root culture/colchicine, colchicoside	Glorybai and Agastian (2013)
Inula helenium	L, S	MS	Hairy root culture/insulin, helenin, sesquiterpene lactones	Shirazi et al. (2013)
Arnica montana	L, CN	MS + B_5 + SH	Hairy root culture	Petrova et al. (2013)
Silybum marianum	L	MS	Hairy root culture/flavonolignans	Rahnama et al. (2013)
Hyoscyamus niger	P	MS + IBA	Normal root culture/hyoscyamine, scopolamine	Hong et al. (2012)
Coleus forskohlii	L	MS + IBA	Hairy root culture/forskolin	Reddy et al. (2012)
Nicotiana tabacum	L	MS	Hairy root culture/nicolin	Zhao et al. (2013)
Atropa belladonna	Seed, L	½MS	Hairy root culture/tropane	Yang et al. (2011)
Camptotheca acuminate	S	MS	Hairy root culture/camptothecin	Ni et al. (2011)
Coleus forskohlii	S, L	MS + B_5 vitamins	Hairy root culture/forskolin	Maheswari et al. (2011)
Salvia miltiorrhiza	L	B_5 + ABA + TDZ + BAP	Hairy root culture/tanshinone	Gupta et al. (2011)

(continued)

Table 9.1 (continued)

Name of the plant	Explant source	Type of medium, growth regulators/additives	Culture response/secondary metabolite	References
Pogostemon cablin	L	½MS + BA + NAA	Hairy root culture	Ping et al. (2011)
Catharanthus roseus	L	½B_5	Hairy root culture/catharanthine, vincristine	Wang et al. (2010)
Psoralea drupacea	Ca, IN, RCa	MS + BAP + 2,4-D + Methyl Jasmonate	Hairy root culture/bakuchiol	Lystvan et al. (2010)
Echinacea spp.	S	½B_5	Hairy root culture/alkamide	Romero et al. (2009)
Plumbago rosea	N	MS + BAP + KIN	Hairy root culture/plumbagin	Yogananth and Basu (2009)
Plumbago zeylanica	S, L	MS + NAA + IBA	Hairy root culture/not assessed	Sivanesan and Jeong (2009)
Stephania suberosa	L	½MS	Hairy root culture/dicentrine	Putalun et al. (2009)
Silybum marianum	Seeds, C	MS	Hairy root culture/flavonolignan	Rahnama et al. (2008)
Rauvolfia serpentine	L	MS + NAA	Hairy root culture/vomilenine, reserpine	Madhusudanan et al. (2008)
Genista tinctoria	Seed, H	SH + ABA	Hairy root culture/isoliquiritigenin	Łuczkiewicz and Kokotkiewicz (2005)

C cotyledon, *Ca* callus, *Co* corms, *CN* cotyledonary node, *H* hypocotyl, *L* leaf, *N* node, *P* petiole, *Ra* radical, *R* root, *S* stem, *TR* tuberous root roadside bushes as a perennial shrub commonly called as Rangchita. *R. nasutus* is a tiny slender shrub measuring about 60–76 cm in height. The requirement for the optimum growth of this plant includes 1000–1200 mm rainfall and 20–28 °C temperature. Vigorous growth of the plant was observed during rainy season whereas most of the aerial parts dry up and root portion remains intact during summer. Since *R. nasutus* is highly susceptible to water logging, water stagnation for a period of 1–2 days severely damages the plant (Das 2006). This medicinal plant has been employed in the treatment and prevention of several diseases as folklore medicine. Various parts of *R. nasutus* have been used in traditional medicine for the treatment of diseases like eczema, pulmonary tuberculosis, herpes, hepatitis, diabetes, hypertension and several skin diseases (Siripong et al. 2006). Its potential role in the treatment of diseases such as cancer, liver disorders, skin diseases, peptic ulcers, helminthiasis, scurvy, inflammation and obesity has been well documented (Suja et al. 2003). The leaves and roots of *R. nasutus* produce several important bioactive molecules.

Cheruvathur et al. (2015) isolated three natural naphthoquinone esters such as rhinacanthin (RC)-C, RC-D, and RC-N, from *R. nasutus* hairy root cultures. Following transformation with *Agrobacterium rhizogenes* (MTCC Strain No. 532) the leaves, stems, and cotyledons induced hairy roots. Maximum (73%) hairy root induction frequency was obtained with cotyledons. Murashige and Skoog (MS; Murashige and Skoog 1962), ½ MS, Schenk and Hildebrandt (Schenk and Hildebrandt 1972) and Woody Plant Medium (WPM; Lloyd and McCown 1980) were used to study their effects on hairy root growth and RC production. The hairy roots were analysed for its transgenic nature by Polymerase Chain Reaction (PCR) and Southern hybridization analysis. The results indicated that optimum hairy root biomass (0.89 g/flask dry weight [DW]) and RC contents (RC-C, 3.8 mg/g DW; RC-D, 0.43 mg/g DW; and RC-N, 0.18 mg/g DW) were obtained on MS medium after 6 weeks of culture. These results were further enhanced [root biomass (1.41 g DW/flask) and RC contents (RC-C, 4.4 mg/g DW; RC-D, 0.69 mg/g DW; and RC-N, 0.21 mg/g DW)] by using 4% sucrose. This study revealed that hairy root culture may offer an effective and reliable alternative method for the production of anticancer compound RC.

In another report Cheruvathur and Thomas (2014) studied the role of auxins and elicitors on hairy root growth and rhinacanthin (RC) accumulation in *R. nasutus*. Two auxins, i.e. indole-3-butyric acid (IBA) and naphthalene acetic acid (NAA), and two elicitors, i.e. methyl jasmonate (MJ) and salicylic acid (SA), were employed in this study. MS medium was used throughout the experiment. Although IBA and NAA at 2.5 μM concentration gave the optimum biomass after 4 weeks, the RC production was highest after 6 weeks. The presence of MJ and SA decreased the hairy root biomass. However, the presence of MJ and SA significantly enhanced the RC induction in hairy roots harvested 7 days after elicitation as compared to control. MJ was more prominent than SA for RC induction. Maximum RC content (6.3 mg/g DW RC-C; 1.1 mg/g DW RC-D and 0.61 mg/g DW RC-N) was obtained when treated with 10 μM MJ which was 1.7-, 2.5- and 3.5-fold higher RC-C, RC-D and RC-N, respectively, than the control.

Stevia rebaudiana Bert.

The genus Stevia consists of about 154 species of which only two species produce the sweet steviol glycosides (Robinson 1930; Soejarto et al. 1982, 1983). *S. rebaudiana* belongs to the family Asteraceae and is a tiny perennial shrub that grows up to a height of about 65 cm with sessile, oppositely arranged lanceolate to oblanceolate leaves, serrated above the middle. The speciality of trichome on the leaves of this plant is that it appear in two distinct sizes, i.e. one large (4–5 μm) and one small (2.5 μm) in size (Shaffert and Chetobar 1994). The tiny (7–15 mm), white flowers appear on the irregular cyme. Several feathery pappus hairs appear on the seed which is an achene (Robinson 1930). The bioactive plant secondary metabolite chlorogenic acid and its derivatives (CADs) have several health benefits. Hairy root cultures were induced from *S. rebaudiana* for the production of CADs by Fu et al. (2015). Leaves were used for the induction of hairy roots after coculturing with *Agrobacterium rhizogenes* (C58C1) which was verified by PCR detection of *rol* B and *rol* C genes. Liquid chromatography–mass spectrometry (LC-MS or HPLC-MS) and High Performance Liquid Chromatography (HPLC) analysis confirmed the presence of some major CADs compounds such as chlorogenic acid (3-caffeoylquinic acid, 3-CQA), 3, 5-dicaffeoylquinic acid (3, 5-CQA), and 4, 5-dicaffeoylquinic acid (4, 5-CQA) in hairy roots. Rapidly growing 8 hairy root lines were selected for CADs estimation and the results showed that T3 line has the maximum yield. Moreover the CADs production was highest on B_5 medium (Gamborg et al. 1968) supplemented with 40 g/L sucrose than other media. The total content of CADs reached 105.58 mg/g and total yield was 234.40 mg/100 mL under this optimum condition.

Astragalus membranaceus (Fisch.) Bge.

A. membranaceus, a perennial Fabaceae member, is usually distributed in the northern, north-eastern and north-western parts of China (Ma et al. 2000). This plant is commonly called yellow leader. It is extensively used in Chinese traditional medicine for thousands of years to treat various ailments. The root is the most important part since it is utilized to prepare a tonic which can enhance metabolism and digestion, to enhance the immune system, and to trigger the curing of wounds and injuries. This plant is effectively utilized against cancer, anaemia, diabetes, hepatitis, liver and heart diseases (Wagner et al. 1997). It is also employed as an immunostimulant, antiperspirant, diuretic and as a supplementary medicine during cancer therapy (Zheng 2005). The antiviral, anti-inflammatory and antibacterial properties of *A. membranaceus* have been reported (Wagner et al. 1997). This plant contains several useful antioxidants that can prevent cell damage caused by free radicals (Zheng 2005). The primary metabolic components in *A. membranaceus* roots include astragalosides (ASTs), calycosin and calycosin-7-*O*-β-D-glucoside (CG).

Jiao et al. (2015) standardized a technique for the simultaneous determination of six astragalosides (ASTs) and five isoflavonoids in *A. membranaceus* hairy root cultures (AMHRCs) by direct analysis approach, namely, high speed homogenization coupled with cavitation-accelerated extraction (HSH-CAE) followed by liquid chromatography-tandem mass spectrometry (LC-MS/MS). The advantage of using the present sample preparation procedure (HSH-CAE) over soxhlet extraction (SE) and ultrasound-assisted extraction (UAE) methods is that it offers significant improvement with regard to simplicity in operation (elimination of biomass drying and grinding), enhanced yield and green aspects in terms of saving energy cost, high efficiency and minimizing the generation of waste. Further, the standardized LC-MS/MS protocol offered linearity with correlation coefficients beyond 0.9991, limit of detections (LODs) under 1.77 ng/mL, relative standard deviations (RSDs) below 6.01% and recoveries above 96.84%. Moreover, the standardized HSH-CAE-LC-MS/MS protocol was effectively utilized for screening high-productive AMHRCs. In short, this investigation has significant importance since it can be utilized for the direct estimation of secondary metabolite profiles from fresh plant in vitro cultures, which was essential for enhancing the quality control of plant cell/organ cultures.

Park et al. (2015) investigated the expression levels of genes related to the biosynthetic pathways of ASTs, calycosin and CG. This study also compares the variations between hairy roots, adventitious roots and seedling roots (SRs) using quantitative real-time polymerase chain reaction (qRT-PCR). The results of qRT-PCR study disclosed that the transcription level of genes involved in the AST biosynthetic pathway was lowest in adventitious roots and a comparable pattern was also noticed in hairy roots and SRs (Park et al. 2015). It was also noticed that the majority of genes responsible for the synthesis of calycosin and CG displayed the highest expression levels in seedling roots. The results obtained by using HPLC indicated that the expression level of the genes correlated with the content of ASTs, calycosin and CG in the three different types of roots. Seedling roots showed rich source of ASTs. CG accumulation was higher than calycosin accumulation in adventitious roots and hairy roots, whereas the reverse was true in seedling roots. In total 40 metabolites were detected and identified using gas chromatography–time-of-flight mass spectrometry (GC-TOF-MS). The differences among seedling roots, adventitious roots and hairy roots were documented by using principal component analysis (PCA). From this study it is clear that hairy roots were different from adventitious roots and seedling roots with reference to the high quantity of sugars and clusters derived from closely similar biochemical pathways. Similarly, adventitious roots had higher concentration of a precursor for the phenyl propanoid biosynthetic pathway namely phenylalanine, as well as CG. Higher amount of TCA cycle intermediates including succinic acid and citric acid was observed in seedling roots than in adventitious roots and hairy roots.

Coleus forskohlii Briq.

Coleus forskohlii [syn: *Coleus barbatus* (Andr.) Benth.] belonging to the mint family Lamiaceae is a plant of Indian origin (Valdes et al. 1987). This perennial plant is distributed over tropical and subtropical regions of India, Pakistan, Sri Lanka, East Africa and Brazil at 600–1800 m elevation. It is cultivated in various parts of India including Gujarat, Maharashtra, Rajasthan, Karnataka and Tamil Nadu.

C. forskohlii has been cultivated in an area of more than 2500 hectares for harvesting its tuberous roots. The tuberous roots of this plant have been utilized traditionally as condiments in pickles and for ayurvedic medicinal purposes (Ammon and Muller 1985). Children suffering from constipation will be provided with the root juice of this plant (Singh et al. 2011).

Agrobacterium rhizogenes (MTCC 2364 strain) mediated transformation leading to hairy root production has been reported in *C. forskohlii* by Pandey et al. (2014). The explants used for hairy root induction were nodal stem part and mature leaves. Of these two explants nodal stem part gave rise to highest response with huge amount of hairy root production. Initially the shoots were emerged from the nodal part (having a size of 3.0–4.0 cm long) and later hairy roots were induced from the base of the node within 12 days of culture. When these shoots were excised and subcultured on MS basal medium new hairy roots were induced within 5 days. Further, the transformed nature of hairy roots was confirmed by PCR with *rol A* gene primer. From this study it was confirmed that hairy roots contain the highest amount of forskolin in comparison to any other plant parts.

Valeriana officinalis L.

Valeriana officinalis is a plant native to Asia and Europe (Larsen 1986). The root or rhizome of this plant is used in traditional medicine as a mild sedative and tranquilizer in many countries. It is employed as a mild sedative and sleeping aid in India. *V. officinalis* has depressant activities on the central nervous system. This plant is used as a sedative in agitated states and a stimulant in fatigue (Fernández et al. 2006). The important phytochemicals responsible for its clinical use include valepotriates and their breakdown products valerenal, valeranone, baldrinals and valerenic acid and some other constituents in the essential oil (Nishiya et al. 1992; Bos et al. 1996). Valerenic acid (VA) is another pharmacologically active sesquiterpene seen in the root and rhizome of this plant. *V. officinalis* produces only small amounts of secondary metabolites naturally. Therefore, induction of hairy roots and elicitation are important protocols for the large-scale commercial production of secondary metabolites.

Hairy roots were induced by *A. rhizogenes* (Strain ATCC 15834) mediated genetic transformation in *V. officinalis* by Parizi et al. (2014). Various media such as MS, B_5 (1.0× and 0.5× strength), N6 and a modified MS without phytohormones were employed to compare the effect of growth of *V. officinalis* hairy roots. The effect of various NH_4^+ to NO_3^- ratios in MS medium was also investigated. The results of these experiments were evaluated after 21 days of culture in relation to hairy root growth. The highest hairy root induction and growth was observed on B_5 and ½B_5 media. For highest biomass yield and growth rates of hairy root culture MS medium fortified with a 20:20 ratio (mM) of NH_4^+ to NO_3^- was optimum. This study confirmed the positive role of media type, composition and the ratio of different nitrogen sources on the growth rate and biomass yield of *V. officinalis* hairy roots.

In another study, Zebarjadi et al. (2011) reported hairy root induction in *V. officinalis* under in vitro conditions. The leaf, stem and root of this plant were co-cultivated with *A. rhizogenes* (strains AR15834 and LBA 9402) for hairy root induction. Of the two bacterial strains used for hairy root induction, the strain LBA9402 was better than AR15834. On MS medium supplemented with 20 mg/L rifampicin and 250 mg/L cefotaxime without any plant growth regulators gave the highest growth of the hairy roots. This study revealed that bacterial strains play an important role in rapid growth of the hairy roots of *V. officinalis*.

Torkamani et al. (2014a) employed *A. rhizogenes* wild-type strain 'A13' for inducing hairy roots in *V. officinalis*. In this study, the effect of various elicitors such as *Fusarium graminearum* extract (FE), methyl jasmonate (MJ) and salicylic acid (SA) on VA production in the selected hairy root line 'LeVa-C4' was investigated. Various concentrations of elicitors were treated after 23 days of culture at exposure time of 3 and 7 days. The results indicated that FE (1%) and MJ (100 μM) were highly favourable for VA production at 7 days after elicitation, to a level of 12.31- and 6-fold higher than that of non-elicited controls, respectively. It was also noticed that FE did not play any negative role on biomass yield of hairy roots. The application of SA did not significantly influence the VA production. From this study it is confirmed that elicitation of hairy root culture is an efficient way to promote VA biosynthesis in *V. officinalis* and *F. graminearum* extract and MJ was indeed a potent inducer of VA biosynthesis.

Atropa komarovii L.

Atropa belongs to the family Solanaceae and is a perennial herbaceous plant. It is the most significant commercial source of the pharmaceutical tropane alkaloids. There are four different species of the genus atropa and are distributed in the Mediterranean region, South Europe and Asia. *A. belladonna* has long been utilized as a reputed drug in Europe and is still considered as the most important and indispensable drug of the plant origin (Parvaz et al. 2006). However, *A. komarovii*, (Kopet-Dagh Mountains) the species found in central Asia, is considered as equivalent to *A. belladonna* particularly in chemical constituents. The drugs extracted from atropa include scopolamine and hyoscyamine which are considered as an important stimulant to the sympathetic nervous system.

Banihashemi et al. (2015) reported that *A. komarovii* was infected by the soil gram-negative bacterium *A. rhizogenes* will be able to produce the neoplastic disease with the formation of hairy roots at the site of the infection (Banihashemi et al. 2015). In this study, hairy roots were induced directly from the wounds of leaves. The leaves were collected from 15-day-old in vitro grown seedlings of *A. komarovii*. For hairy root induction leaves were cultured on MS medium and the *A. rhizogenes* strain used was ATCC15834. About 70% of cultured leaves showed positive reaction to bacterium. On hormone free MS medium the hairy roots demonstrated rapid, plagiotropic growth and remained highly branched. PCR method was employed for the amplification of *rol* B gene region. Thus the integration of T-DNA fragment of Ri plasmid from *A. rhizogenes* to plant genome was confirmed. For optimum hairy root induction MS basal medium with 500 mg/L cefotaxime in dark was essential. The induction of hairy roots was observed within 3 weeks of culture.

Artemisia annua L.

Artemisia annua is an Asteraceae member also called sweet sagewort or annual wormwood. It is the most common kind of wormwood that is cultivated throughout the world. It is considered as a natural pesticide and has been effectively utilized for the production of antimalarial and antibacterial agents. The value of *A. annua* as a herbal medicine was first utilized by the Chinese and later several pharmaceutical companies tried to exploit this plant for the manufacture of artemisinin for artemisinin-based combination therapies (ACTs) in the treatment of malaria. It has been reported that ACTs have rapid resolution to fever and parasitaemia, low toxicity and are well tolerated (Dangash et al. 2015).

Commonly artemisinin has been extracted from the shoot of *A. annua* plant growing in the natural habitat. The seasonal availability of this plant is one of the important limitations which check the supply of this plant. However, the supply of this drug is considerably lower than its requirement. Biotechnological tools could be utilized effectively for the ample supply of this drug and mass scale hairy root culture in bioreactors could be considered as one such efficient tool. Patra and Srivastava (2014) developed an efficient tool for the production of in vitro hairy roots in *A. annua* by *A. rhizogenes* mediated genetic transformation. They have optimized the various conditions for the hairy root growth and thereby enhance the biomass and artemisinin accumulation. The optimum result was obtained when the following conditions were maintained—temperature (25 °C), size of inoculum (1 g/L DW), rotational speed (70 rpm), age of inoculum (8 days) and medium to vessel volume ratio (0.18). Further, a stirred tank bioreactor was employed for the scale-up of the hairy roots and for the mass scale artemisinin production. Stirred tank bioreactor cultivation of hairy roots not only resulted in maximum biomass accumulation of 6.3 g/L dry weight (37.50 g fresh weight) but also artemisinin content of 0.32 mg/g by using standardized media after 25 days of batch cultivation.

Rauwolfia Serpentina (L.) Benth

Rauvolfia serpentina belongs to the family Apocynaceae. It is commonly known as 'Sarpagandha'. This woody perennial shrub is an endangered medicinal plant found in India, Indonesia, Malaysia, Nepal, Bangladesh, China, Pakistan, Sri Lanka and Vietnam (Dey and De 2010; Susila et al. 2013). In Ayurvedic preparations, the roots of this plant are an important ingredient. The laxative, diuretic, thermogenic, bitter, sedative and acrid properties of the root of *R. serpentina* are well known (Prakash 2001). The use of this plan as medicine for anxiety, insomnia, high blood pressure and epilepsy has been reported (Jain et al. 2003). It is highly used for the treatment of hypertension, wounds, dyspepsia, strangury, fever, rectifying the opacities of the cornea and antidote to snake venom (Prakash 2001; Jain et al. 2003; Panwar et al. 2011).

An effective protocol has been standardized for the production of *A. rhizogenes* transformed hairy roots were reported by Shetty et al. (2014). The extraction and analysis of secondary metabolites were further carried out by thin layer chromatography (TLC). This study indicates that there are several factors such as media composition, genotype and culture conditions which play a crucial role in hairy root induction. The leaf explants were used for the induction of hairy roots and were infected with *A. rhizogenes* for the efficient production of secondary metabolites. Large-scale production of profusely branched hairy roots was obtained from leaf cultures within 1 month after culture.

Gmelina arborea Roxb.

Gmelina arborea commonly known as Gambhari belongs to the Verbenaceae family which consists of about 2600 species, of which 107 species are found in India (Dutta 1964). It is distributed in almost all parts of India including North-West Himalaya, Deccan Peninsula, Chittagong region and Western Ghats. *G. arborea* is a deciduous plant that grows to a medium-sized tree attaining about 15–20 m height (Bhawan 2011). It is usually planted in gardens as an ornamental tree and as an avenue tree. One of the most important ingredients of Dashamoola and Brihathpanchamoola includes *G. arborea* (Murthy 2012). The medicinally useful parts of this plant include leaf, root, flower, bark and fruit.

For induction of hairy roots in *G. arborea*, seedling explants were employed. The explants were infected with a wild-type *A. rhizogene* (strain ATTCC 15834). Subsequently the hairy roots were induced in 32% explants (Dhakulkar et al. 2005). Of the different hairy root lines, eight transformed lines were selected and established in liquid medium for growth assessment and secondary metabolite extraction. PCR protocol using *rol* B primer was used for the confirmation of the transgenic nature of hairy roots and subsequently by Southern analysis of the PCR products.

Salvia miltiorrhiza Bunge

Salvia miltiorrhiza Bunge is a well-known Chinese herb of the family Lamiaceae. Its rhizome is the medicinally important part and is an ancient drug in Chinese traditional medicine (Duke and Avensu 1985). The various medicinal uses of this plant include the treatment of hypertension, menstrual disorders, coronary heart disease, viral hepatitis and miscarriage (Chang and But 1986). Danshen, a famous Chinese herbal medicine in which *S. miltiorrhiza* is an ingredient, has been used for the treatment of cardiovascular diseases (Zhou et al. 2006; Cheng 2007; Jiang et al. 2005). A number of studies revealed its wide range of pharmacological activities including antimicrobial, anticancer, antioxidant, cardiovascular protective effect, anti-inflammatory and neuroprotective (Cheng 2007; Jiang et al. 2005).

Tanshinone is an important constituent of *S. miltiorrhiza* which forms the most potent diterpene diketones used to cure many illnesses. Gupta et al. (2011) standardized a protocol for the production of tanshinone by hairy root culture. The influence of various plant growth regulators (PGR) on the biomass growth of hairy roots and production of three tanshinone constituents (i.e. cryptotanshinone, tanshinone I and IIA) were investigated. Leaf explants of *S. miltiorrhiza* when inoculated with *A. rhizogenes* strain BCRC15010 produced 78% hairy roots. The optimum hairy root growing lines were obtained when the hairy roots were cultured on B_5 liquid medium under dark conditions. For confirmation of the transgenic nature of hairy roots, PCR protocol using *rol* B and C gene specific primers was employed. The addition of various plant growth regulators like auxins, cytokinins and abscisic acid (ABA) considerably enhanced the production of cryptotanshinone, tanshinone I and II. The addition of 1.0 mg/L ABA and thidiazuron (TDZ) enhanced the production of Tanshinone I and cryptotanshinone to 5- and 7.5-fold higher respectively compared to the mother plant. The presence of TDZ, ABA and 6- benzyl adenine (BA) in the medium enhanced the cryptotanshinone content to 6.3-, 5.0- and 3.75-fold respectively compared to field grown plants. This protocol is an effective method for the enhanced production of tanshinone from hairy root culture in *S. miltiorrhiza.*

In another report, a hairy root induction method was developed by Hao et al. (2012) from leaf explants of Chinese variety of *S. miltiorrhiza. A. rhizogenes* strain CC10060 was employed for this study. Two main salvianolic acids, i.e. salvianolic acid B and salvianolic acid A, were isolated from the hairy roots of *S. miltiorrhiza* by using HPLC. The influence of a number of additives such as ABA and polyamines [putrescine (Put), spermidine (Spd) and spermine (Spe)] on salvianolic acids production from hairy roots was also studied. The presence of ABA in the medium enhanced the phenylalanine ammonia-lyase (PAL) activity and increased the salvianolic acid B and salvianolic acid A contents to 1.8, 2.0 and 3.3 times after 80 μ mol/L ABA treatment for 10 days, 12 days and 10 days, respectively. Among the three polyamines employed for secondary metabolite production from hairy roots, Put was the most effective. Comparatively, the presence of Put and Spd in the medium enhanced the salvianolic acids better than that of Spe. Put at 50 mg/L produced the maximum level of hairy root growth (13.23 g/L culture), salvianolic acid B (12.13 mg/g DW) and salvianolic acid A (3.95 mg/g DW). The results were further improved by adding a mixture of Put and Spd (50 mg/L each) for 10 days, 12 days and 20 days and the PAL activity, salvianolic acid B and salvianolic acid A production were about 1.82, 2.05 and 3.45 times more than that of control, respectively. Hao et al. (2012) concluded that ABA and PAs can enhance the two salvianolic acid production in hairy root cultures of *S. miltiorrhiza.*

Kai et al. (2011) reported the introduction of genes encoding 3-hydroxy-3-methylglutaryl CoA reductase (HMGR), 1-deoxy-D-xylulose-5-phosphate synthase

(DXS) and geranylgeranyl diphosphate synthase (GGPPS) responsible for tanshinone biosynthesis in *S. miltiorrhiza* hairy roots by *Agrobacterium*-mediated gene transfer technology. An active diterpene, tanshinone is commonly used in the treatment of cardiovascular illness. It was observed that overexpression of SmGGPPS and/or SmHMGR as well as SmDXS in transgenic hairy root lines can profoundly increase the biosynthesis of tanshinone to greater quantity than the control. The effect of SmDXS was much more influential than SmHMGR in tanshinone production. However, SmGGPPS plays a crucial role in tanshinone accumulation than the upstream enzyme SmHMGR or SmDXS in *S. miltiorrhiza*. The highest accumulation of tanshinone (about 2.727 mg/g dw) was obtained in line HG9 was observed due to the co-expression of SmHMGR and SmGGPPS, which was about 4.74-fold greater than the control (0.475 mg/g dw). Further, these hairy root lines showed higher antioxidant activity as compared to control. According to Kai et al. (2011) this is the first report on improvement of tanshinone accumulation and antioxidant activity attained through metabolic engineering of hairy roots by push-pull strategy in *S. miltiorrhiza*.

Cannabis sativa L.

The Cannabaceae family member *Cannabis sativa* is commonly known as hemp. This annual multipurpose herb has been commercially cultivated for its fibre from the bark. It is also employed in oil extraction, paper industry, drug preparation and various medicinal purposes (Ranalli et al. 1999). The world production of hemp has come down considerably due to the competitive presence of other fibre producing plants such as sisal, jute and cotton (Rode et al. 2005). Recently, there is an increased interest in hemp cultivation due to its wide ranging adaptability to various agro-ecological climates, more yield as compared to other fibre yielding plants and the extraction of various medicinal components such as tetrahydrocannabinol (Struik et al. 2000).

Farag and Kayser (2015) developed a protocol for the hairy root culture in *C. sativa* and estimated the content of cannabinoid, an antitumour phytochemical. The hairy roots were induced from calli cultured in vitro on B_5 medium fortified with 4.0 mg/L NAA in darkness at 25 °C. The shake flask culture of hairy roots enhanced the growth and biomass periodically during 35 days of growth cycle. The liquid chromatography method was adopted for the estimation of contents of cannabinoid and it was confirmed by mass spectrometry. The estimation of cannabinoid contents showed a quantity slightly below 2.0 μg/g dry weight.

Berberis aristata DC.

Berberis aristata commonly known as Daruharidra is an important medicinal shrub of the family Berberidaceae. Berberine is an important isoquinoline alkaloid obtained from the root and stem bark of this plant as well as a close relative of this plant *Berberis chitria*. Therefore, *B. Chitria* is used as a substitute of *B. aristata*. (Srivastava et al. 2006). One of the most important uses of berberine is that it is used against cholera (Dutta and Panse 1962). It is also employed to cure acute diarrhoea (Lahiri and Dutta 1967), eye ailments, rheumatism, diabetes, malarial fever, jaundice, fever, ear diseases and skin disease (Srivastava et al. 2006; Kirtikar and Basu 1975). It is also used as a tonic (Chopra et al. 1958). In China Berberine has been effectively used as a drug to treat diabetes from time immemorial (Yin et al. 2008).

An effective way of inducing hairy roots was developed in *B. aristata* by Brijwal and Tamta (2015). In this study two *A. rhizogenes* strains were utilized, i.e. MTCC 532 and 2364. Of the two strains used in this study, strain 532 was found more vigorous than strain 2364 in terms of hairy root induction. For hairy root induction, in vitro callus was found more responding (61.11 ± 1.60% transformation frequency) than leaves (42.59 ± 0.92% transformation frequencies) and nodal segments (34.25 ± 0.92% transformation frequencies) of in vitro grown microshoots. Transformed nature of the roots was confirmed by analysing the presence of *rol* A and *rol* B genes during amplification. By using acetosyringone (100 μM) during co-cultivation period (48 h), the transformation frequency of callus was further increased from 61 to 72% on MS semisolid medium. This may be the first step towards the extraction and production of berberine from hairy root culture, thereby reducing the risk of overharvesting of this endangered species from its natural habitat.

Withania somnifera (L.) Dunal

Withania somnifera (Family Solanaceae) is a well-known medicinal plant used for a variety of ailments and is commonly called as Indian ashwagandha. One important feature of this plant is that it can grow in most arid and nutrient deficient soils. *W. somnifera* is one of the best known and maximum explored Ayurvedic plants and holds Ayurvedic traditions parallel to Ginseng in Chinese therapies. Because of this reason *W. somnifera* is generally considered as 'Indian Ginseng' in Ayurvedic world. Another noted feature of this plant is that a tonic made from this plant has several properties such as anti-inflammatory, antiarthritic and aphrodisiac (Baba et al. 2013). This plant is normally distributed in the dry arid areas of Bangladesh, Pakistan, India and China where it has been grown wildly for ages.

A. rhizogenes-mediated genetic transformation for hairy root induction was reported in *W. somnifera* by Saravanakumar et al. (2012). Three strains of A. *rhizogenes* (ATCC 15834, R1000 and K599) were used for this study. Various explants like leaf, petiole and internodal segments were employed for hairy root induction. Of the three strains used, only two strains (ATCC 15834 and R1000) were successful in inducing hairy roots. When petiole explants infected with R1000 produced the highest frequency (64%) of hairy root induction, it gave rise to five distinct morphotypes (callus (fragile), callus (hard), callus + hairy roots, hairy roots and callusing roots). The presence of acetosyringone further increased the transformation frequency up to 93.2%. PCR analysis of the transformed roots confirmed the presence of *rol C* gene. For obtaining highest root biomass accumulation half-strength MS medium was optimum than other media such as MS full strength, B_5 full strength and B_5 half strength. The accumulation of the phytochemical withaferin-A was quantified (72.3 mg/g dw) by HPLC analysis of hairy roots. From this study it is clear that *W. somnifera* hairy root culture was influenced by several factors such as type of *Agrobacterium* strains, explant types and presence of acetosyringone.

Gossypium hirsutum L.

Gossypium hirsutum is one of the important crops which yield commercial fibre in the world. This plant belongs to the family Malvaceae. The seeds of this plant are used for making oil and food for animals (Mishra et al. 2003; Aragao et al. 2005).

A protocol was developed for gossypol production in *G. hirsutum* by hairy root culture by Esyanti and Sahroni (2014). *A. rhizogenes* ATCC-15834 was employed for the induction of hairy roots. For hairy root culture two media, i.e. full-strength LS and half-strength LS, were used. Of the two media used, the highest dry weight of roots was observed on half-strength LS 20 days after culture. After 16 days the optimum gossypol content in roots which was directly related with the growth was observed irrespective of treatments whereas the amount of gossypol that was secreted into culture medium was obtained 20 days after treatment. Half-strength LS medium was the best for hairy root induction and gossypol production. Therefore, it was concluded that half-strength LS medium plays a crucial role for the growth and gossypol production in hairy root culture of *G. hirsutum.*

Isatis tinctoria L.

The perennial medicinal herb *Isatis tinctoria* (family Brassicaceae) is native to the steppe and desert zones of the Central Asia to Eastern Siberia, Western Asia, South West Asia (Shu 2001) and Europe (Tan 2002). It was a source of indigo and cultivated mainly in Europe from the twelfth to the seventeenth century (Hurry 1930; Clark et al. 1993). It is used as a dye yielding plant from its leaves in various parts of Europe including England, France, Germany and Italy. It is also cultivated in Turkey for extracting an economically important blue dye which is used for the carpet industry.

Hairy roots were induced in *I. tinctoria* by Gai et al. (2015) as an alternative source of flavonoids (FL) production. The transformed nature of the hairy roots was confirmed by PCR amplification of *rol* B, *rol* C and aux1 genes. The hairy root line no. 5 was found most productive and efficient. Eight bioactive FL components namely kaempferol, rutin, isorhamnetin, neohesperidin, liquiritigenin, buddleoside, quercetin, and isoliquiritigenin were isolated and quantified by LCMS/MS protocols. Total FL content reached the highest (438.10 μg/g dry weight) under optimal conditions as compared to normal roots (341.73 μg/g DW) when 24-day-old hairy roots were used for extraction. Hence, the hairy roots showed a significantly superior FL production than normal field grown roots. Furthermore, the hairy roots showed significantly higher antioxidant activity with lower IC50 values (0.41 and 0.39 mg/ml) than field grown roots (0.56 and 0.48 mg/mL).

Silybum marianum L.

Silybum marianum is commonly called by the name blessed milk thistle, Mediterranean milk thistle or variegated thistle and belongs to the family Asteraceae. *S. marianum* is supposed to be originated from the Mediterranean basin and is now cosmopolitan in distribution (Vaknin et al. 2008). Since ancient times, this plant had been used as a traditional medicine especially for liver diseases and as a liver cancer therapy in Europe and Asia (Khalili et al. 2009; Zhu et al. 2014).

A unique flavonoid named silymarin which has reported hepatoprotective properties was isolated from the hairy root culture of *S. marianum.* This protocol has been considered as a viable option for extracting silymarin without destroying the natural plant populations. In this study Hasanloo et al. (2015) investigated the interaction of *Trichoderma* strains and *S. marianum* hairy root culture interaction on flavonoid silymarin production. The influence of two *Trichoderma* Strains (KHB and G46–7) at various concentrations (0, 0.5, 1, 2 and 4 mg/50 mL culture) and six different exposure times (0, 24, 48, 72, 96 and 120 h) has been investigated on flavonolignans production. The quantification of flavonoids was done by HPLC protocol. The important role of two *Trichoderma* strains (KHB and G46–7) during elicitation was studied on flavonoid accumulation and the activation of cell defence system in *S. marianum* hairy root cultures. The maximum result of silymarin accumulation (0.45 and 0.33 mg/g DW) was observed on media elicited with 0.5 mg/50 mL cultures of *T. harzianum* strains KHB and G46–3 respectively after 120 h. Feeding time experiments showed that the optimum content of the flavonoid silymarin was obtained 120 and 72 h after treatment in media containing 0.5 mg/50 mL cultures of KHB and G46–3, respectively. These results clearly indicated that *S. marianum* treated by KHB strain showed significant increase in silymarin production. In hairy roots the H_2O_2 content was lower as compared to the treated cultures. The activities of peroxidase and ascorbate peroxidase were significantly higher reaching a peak after 72 h in treated hairy roots. This study by Hasanloo et al. (2015) demonstrated the promotive role of some *Trichoderma* strains as elicitors for enhancing silymarin accumulation in hairy root cultures of *S. marianum.*

Cichorium intybus L.

The Asteraceae member *Cichorium intybus* is commonly known as chicory. This erect hairy glandular biennial herb has tuberous taproot system. Although this plant is biennial, its vegetative growth takes place during the first year followed by flowering and completion of life cycle during the second year since it is an absolute long day plant (Varotto et al. 1997).

Malarz et al. (2002) developed a root transformation technique in *C. intybus* for the production of sesquiterpene lactones of guaiane and germacrane type. The cultures were raised on modified MS medium. The highest dry weight (0.38 g per flask) of the roots reached after 30 days of culture which was about 10 times higher than initial weight. The biomass increase was the highest from 15 to 25 days of culture. The various phytochemicals isolated from the transformed hairy roots include lactucopicrin, 8-desoxylactucin and three sesquiterpene lactone glycosides: crepidiaside B, sonchuside A and ixerisoside D. The yield of 8-desoxylactucin varied from 0.03 to 0.18 mg/g fresh weight, in 5 and 25 days old cultures, respectively.

Salvia officinalis L.

The genus *Salvia* comprises about 700 species and is one of the most widespread members of the family Lamiaceae. *Salvia officinalis* is mostly distributed in the Mediterranean Basin, in Central and South America and in South East Africa. This plant is mainly cultivated for culinary and medicinal purposes.

The hairy roots were induced from shoot explants of *S. officinalis* by Grzegorczyk et al. (2006). The shoots were infected with *A. rhizogenes* strains ATCC 15834 and A4 which resulted in the induction of hairy roots in 57% and 37% of the explants, respectively. On WP liquid medium under light and dark conditions, seven lines of hairy roots were identified. PCR using *rol* B and *rol* C specific primers confirmed the transformed nature of hairy roots. Cultures of transformed hairy roots of *S. officinalis* demonstrated variations in biomass and rosmarinic acid production. This type of bacterial strain plays a crucial role in biomass and rosmarinic acid production. The growth and rosmarinic acid production of strain ATCC 15834 induced significantly greater than the A4-induced lines. The highest production of rosmarinic acid (about 45 mg/g of dry weight) was obtained by using hairy root line 1 (HR-1) at the end of the culture period (45–50 days). This level of rosmarinic acid was significantly greater than untransformed root culture (19 mg/g of dry wt).

Picrorhiza kurroa Royle Ex Benth.

Picrorhiza kurroa is an endemic medicinal herb belonging to the family Scrophulariaceae. This plant is distributed widely all through the higher altitudes of alpine Himalayas from west to east, between 3000 and 4500 m above sea level. The dried rhizomes and roots of *P. kurroa* yield a number of useful phytochemicals such as picroside I and II, glycosides and kutkoside. It has wide medicinal properties like hepatoprotective, stomachic, diuretic, cathartic, antiperiodic and cholagogues (Jain 1984).

A. rhizogenes induced fast-growing transformed hairy root cultures were obtained in *P. kurroa* by Verma et al. (2015). This protocol is considered as a viable and convenient option for extraction and production of iridoid glycosides. In this study the influence of various nutrient media such as WP, B_5, MS, Nitsch and Nitsch (NN; Nitsch and Nitsch 1969) and sucrose concentrations (1–8%) on the root biomass and glycoside production of selected clones (14-P) of *P. kurroa* hairy roots were investigated. For optimum hairy root biomass yield (Growth index = 32.72 ± 0.44) full-strength B_5 medium was the best subsequently by the NN medium of the same strength (Growth index = 22.9 ± 0.43). B_5 medium was the best for secondary metabolite production (i.e. kutkoside and picroside I) from hairy roots and was 1.1 and 1.3 times higher levels of the kutkoside and picroside I, respectively, than the same in the full-strength B_5 medium. Of the various sucrose concentrations used, the highest biomass accumulation and optimum picroliv content were obtained at 4% level in basal medium. On MS medium the presence of RT vitamin and thiamine-HCl considerably influenced the secondary metabolite production. However, RT vitamin and thiamine-HCl did not have any influence on other media. The pH of the medium considerably affects the biomass and secondary metabolite production and the yield was highest at pH 6.0 and lowest at pH 3.0. In order to enhance the productivity of picroliv further, a bioreactor was employed and a 27-fold (324 FW) higher growth rate was obtained and a corresponding increase in picroliv production was also noticed.

Lobelia inflata L.

Lobelia inflata (Family—Lobeliaceae) contains several useful secondary metabolites such as di-substituted and mono-substituted piperidine alkaloids. More than 20 alkaloids were reported from this plant (Kaczmarek 1961). The most important among the alkaloids is lobeline which has a promotary role on the respiratory centre.

The biomass assessment and alkaloid production of *L. inflata* hairy roots were carried out by Balvanyos et al. (2001). The *A. rhizogenes* strain R 1601 was used for inducing hairy roots. B_5 medium supplemented with various plant growth regulators [kinetin (KIN), indole −3- acetic acid (IAA) or NAA] were used for hairy root induction. KIN has a negative effect and thereby decreases the biomass production as well as secondary metabolite production. However, the presence of IAA or NAA increased the biomass production and lobeline production. These hormones also have some significant morphological influence on growth. It also enhanced the lateral root formation in hairy roots. The presence of IAA or NAA in the medium significantly enhanced the biomass formation and lobeline production of hairy roots. When 0.2 mg/L IAA is added to the medium, the highest amount of lobeline was obtained.

Psoralea corylifolia Linn.

Psoralea corylifolia, also called as Babchi, is an endangered medicinal plant. This plant belongs to the family Fabaceae. It is an important plant mentioned well in Chinese and Indian folkloric medicine as a diuretic laxative, anthelmintic, aphrodisiac and diaphoretic in febrile conditions (Anonymous 1989). The medicinal use of seeds of this plant is well known and has been used in the treatment of leprosy, leucoderma, psoriasis and skin inflammation (Anonymous 1989).

Baskaran and Jayabalan (2009) investigated the possibility of inducing hairy roots and adventitious roots from hypocotyl explants of *P. corylifolia*. The purpose of this study was to evaluate the psoralen content in hairy roots and adventitious roots grown either in suspension cultures or on agar solidified medium. The psoralen content was higher (3.0 mg/g DW) in suspension culture of hairy roots than hairy roots grown on semisolid media and in solid and suspension-grown adventitious roots.

Fagopyrum tataricum Gaertn

Fagopyrum tataricum (tartary buckwheat) is a plant belonging to the family Polygonaceae. It is grown as a crop in some areas of Northern India, Nepal, Bhutan, mountainous regions of southwest China (Sichuan) and a minor part of northwest Europe (Bonafaccia et al. 2003; Xuan and Tsuzuki 2004). *F. tataricum* is an exceptional nutrient rich and medicinal plant. This plant contains rutin and other phenolic compounds than common buckwheat (Fabjan et al. 2003). Moreover, *F. tataricum* has several pharmacological and biological properties such as antioxidant activities (Liu et al. 2008), antidiabetic (Yao et al. 2008) and anticancer (Guo et al. 2007a).

Sterile young stems were used to induce hairy roots in *F. tataricum* by inoculating with *A. rhizogenes* (Park et al. 2011). Two morphological phenotypes, i.e. thin and thick phenotypes, of hairy roots were obtained when cultured on MS basal medium. However, the higher growth rate was observed with thin phenotype than thick phenotype. Similarly, the thin phenotype produced significantly higher phenolic compound than that of the thick phenotype. The thin phenotype produced almost double the quantity of epigallocatechin along with more than 51.5% caffeic acid, 40% rutin and 65% chlorogenic acid compared to the thick phenotype 21 days after culture. From this study it is clear that selection of the optimal morphological phenotype of hairy roots of tartary buckwheat is absolutely essential for maximum phenolic compound production.

Calendula officinalis L.

The Asteraceae annual herb *Calendula officinalis* is cultivated for ornamental as well as medicinal purposes. This plant is commonly called as marigold. The economic importance of this plant includes its use as a raw material in pharmaceutical, cosmetic and food processing industries. Food and Drug Administration (FDA) approved the flower extract of this plant as an additive to food since it has 'generally recognized as safe' (GRAS) status. The flower extract of *C. officinalis* is added in food as a natural colourant. Marigold extract also has very high antioxidant properties since it contains both carotenoids and flavonoids (Piccaglia and Venturi 1998; Guinot et al. 2008). The extract of this plant is employed in about 200 cosmetic formulations, mainly in creams and shampoos (Andersen et al. 2010). Since this plant has high therapeutical property, it is effectively utilized in folk medicine for centuries.

An efficient hairy root induction protocol was standardized by Dlugosz et al. (2013) using cotyledons and hypocotyls explants of *C. officinalis* infected with *A. rhizogenes* strain ATCC 15834. The same strain containing pCAMBIA 1381Z vector with β-glucuronidase reporter gene under control of promoter of NIK (Nematode Induced Kinase) gene is also used for inducing hairy roots. Of the two explants utilized for hairy induction, 33.8% cotyledon and 66.6% hypocotyl explants responded positively. A total number of nine transformed lines and eight control lines were established as long-term culture. *C. officinalis* hairy roots mainly showed the ability to synthesize oleanolic acid (97%) as glycosides. The various lines showed high positive correlation between dry/fresh weight and oleanolic acid concentration in tissue.

Angelica gigas Nakai

Angelica gigas commonly called as Korean angelica is a member of the family Umbelliferae. This perennial herb is an important indigenous medicinal plant in Korea. The root of *A. gigas* is mostly used for medicinal purposes especially in the treatment of injuries, migraine, arthritis, anaemia, abdominal pain, and female afflictions and it has been also recommended for health-promoting effects (Chi and Kim 1970; Choi et al. 2003; Sarker and Nahar 2004). Although there are several phytochemical compounds present in *A. gigas*, the major compounds include pyranocoumarin compounds decursin and decursinol angelate which have major anticancer, neuroprotective and anti-androgen-receptor signalling properties (Guo et al. 2007b; Kang et al. 2005; Yim et al. 2005).

The effect of various media, strength of media and influence of different sucrose concentrations on hairy induction, optimization of the biomass and pyranocoumarin production has been studied in *A. gigas* by Xu et al. (2009). Of the various media employed for hairy root growth and decursin production, SH medium was the best. On this medium the root growth was 9.26 g dry weight/L and decursin production reached 2.69 mg/g dry weight. However, on half-strength SH medium the hairy root growth was 43% higher, decursin production increased to twofold and decursinol angelate production was 93% more as compared to the double strength SH medium which showed the lowest response. The production of decursinol angelate was a little more on MS medium as compared to SH medium. Sucrose also plays a critical role in hairy root biomass and secondary metabolite production. Sucrose concentration at 40 g/L produced the highest hairy root growth and pyranocoumarins production. An increasing concentration of sucrose above 40 g/L decreased the response.

Przewalskia tangutica Maxim

The Solanaceae member *Przewalskia tangutica* distributed in the Tibet plateau of China is an important and rare medicinal plant. The root of this plant contains some vital tropane alkaloids such as hyoscyamine and scopolamine (Xiao et al. 1973). These alkaloids are important and are widely employed as parasympatholytics that competitively antagonize acetylcholine with significant marketable demands (Zhang et al. 2007a). Due to the serious destruction of wild resources the population of *P. tangutica* has come down and there is an increasing demand from medicinal plant industry.

Lan and Quan (2010) demonstrated that hairy root culture is possible in *P. tangutica* by using bacteria free seedling explants. Acetosyringone induced *A. rhizogenes* strain A4 was employed for the genetic transformation of seedling derived leaves for inducing hairy roots. Hundred percentage hairy root induction was noticed from the wounded sites of leaf explants. The fast growing independent hairy root lines were developed and maintained on MS medium. The presence of rooting genes (*rol* B and *rol* C) integrated into the genome of *P. tangutica* was confirmed by genomic DNA PCR analysis. However, high amount of hyoscyamine and scopolamine was extracted (as compared to wild-type normal roots) only after 5 weeks of culture in liquid MS medium. The best line which produced the highest amount (0.68 mg/g DW) of scopolamine and hyoscyamine (1.13 mg/g DW) was isolated and selected. The authors claimed that this is the first report on hairy root culture and tropane alkaloid production in *P. tangutica.*

Rubia akane Nakai

Rubia akane is a perennial climber mostly distributed in East Asian countries. This Rubiaceae member produces anthraquinone pigment and is recognized as a vegetable and red dye yielding plant in Korea. These anthraquinones form important group of natural product and are commercially used for the production of dyes, such as alizarin and purpurin and also used as pharmacological compounds (Shin 1989; Singh et al. 2004). The pharmacological and biological activities of the pigment anthraquinones include antimicrobial (Lenta et al. 2007; Xiang et al. 2008), anticancer (Zhang et al. 2007b; Son et al. 2008), antifungal (Singh et al. 2006), antioxidant (Galindo et al. 2008) and antimalarial (Bringmann et al. 2008) properties.

A protocol for hairy root culture of *R. akane* was developed by Park and Lee (2009). They used various media combinations fortified with several concentrations of auxins to obtain optimum hairy root growth and anthraquinones production. Of the various media used, full-strength SH medium exhibited maximum levels of root growth (12.2 g/L). Similarly, on the same medium the highest yield of alizarin (4.5 mg/g DW) and purpurin production (5.5 mg/g DW) was noticed as compared to *R. akane* hairy roots grown in half- and full-strength B_5 and MS medium, and half-strength SH medium. The most successful auxin which produced the highest response in terms of growth as well as alizarin and purpurin production from hairy roots was NAA. NAA at 0.5 mg/L concentration yielded the maximum amount of dry weight (14.2 g/L) and the optimum quantity of both alizarin (5.9 mg/g DW) and purpurin (7.2 mg/g DW). In short, for hairy root culture in *R. akane* SH medium fortified with NAA (0.5 mg/L) could be a valuable option for high biomass and anthraquinone production.

Pueraria candollei Wall. Ex Benth.

Pueraria candollei commonly known as White Kwao Krua is an indigenous herb mostly distributed in Thailand. This plant belongs to the family Leguminosae and has long been employed in traditional system of medicine for rejuvenation. The extract of the tuber of this plant showed estrogenic effects in a dose-dependent manner on the reproductive system (Cherdshewasart et al. 2004, 2007). *P. candollei* exhibited antioxidant capacity (Cherdshewasart and Sutjit 2008) and capacity to avert bone loss (Urasopan et al. 2008). The chief phytochemicals isolated from the tuber of *P. candollei* comprise isoflavonoids, including genistin, daidzein, puerarin, daidzin and genistein (Chansakaow et al. 2000).

By using *A. rhizogenes* (strain ATCC15834) Udomsuka et al. (2009) standardized a protocol for the induction of hairy roots from *P. candollei*. The hairy roots were induced on half-strength MS medium. The maximum yield of isoflavonoids was found to be 36.48 ± 4.09 mg/g DW which include 0.76 ± 0.03 mg/g DW daidzein, 0.76 ± 0.03 mg/g DW genistein, 3.39 ± 0.20 mg/g DW puerarin, 29.91 ± 3.74 mg/g DW daidzin and 1.65 ± 0.09 mg/g DW genistin. The hairy root culture yielded a 5.18-fold higher isoflavonoid production than natural tuber. The concentration of sugar significantly influenced the growth, biomass and isoflavonoid production in *P. candollei.* The sucrose level at 5% (w/v) yielded the highest growth and isoflavonoid accumulation in hairy roots. Although half-strength MS medium resulted in high biomass production of hairy roots, isoflavonoid production was highest on wood plant medium.

References

Alan AR, Toprak FC, Kaska A (2016) Production and evaluation of gynogenic leek (*Allium ampeloprasum* L.) plants. Plant Cell Tissue Organ Cult 125:249–259

Ammon HPT, Muller AB (1985) Forskolin: from an ayurvedic remedy to a modern agent. Planta Med 46:473–477

Andersen F, Bergfeld W, Belsito D, Hill R, Klaassen C, Liebler D, Marks J, Shank R, Slaga T, Snyder P (2010) Final report of the cosmetic ingredient review expert panel amended safety assessment of *Calendula officinalis*-derived cosmetic ingredients. Int J Toxicol 29:21–243

Anonymous (1989) The wealth of India: a dictionary of Indian raw materials and industrial products, raw materials, vol 8, Publication and information directorate, CSIR, New Delhi, pp 296–297

Aragao FJL, Vianna GR, Carvalheira SBRC, Rech EL (2005) Germ line genetic transformation in cotton (*Gossypium hirsutum* L.) by selection of transgenic meristematic cells with a herbicide molecule. Plant Sci 168:1227–1233

Baba IA, Alia A, Saxena RC, Itoo A, Kumar S, Ahmad M (2013) In vitro propagation of *Withania somnífera* (L.) Dunal (Ashwagandha) an endangered medicinal plant. Int J Pharm Sci Invent 2:6–11

Balvanyos I, Kursinszki L, Szoke E (2001) The effect of growth regulators on the biomass formation and lobeline production of *Lobelia inflate* L. hairy root cultures. Plant Growth Regul 34:339–345

Banihashemi O, Nejad RAK, Yassa N, Najafi F (2015) Induction of hairy roots in *Atropa komarovii* using *Agrobacterium rhizogenes*. Ind J Fund Appl Life Sci 5:2014–2020

Baskaran P, Jayabalan N (2009) Psoralen production in hairy roots and adventitious roots cultures of *Psoralea coryfolia*. Biotechnol Lett 31:1073–1077

Bhawan VR (2011) Medicinal plants unit Indian council of medical research. Quality standards of Indian medicinal plants, vol VI. Indian Council of Medical Research, New Delhi, p 151

Bonafaccia G, Gambelli L, Fabjan N, Kreft I (2003) Trace elements in flour and bran from common and tartary buckwheat. Food Chem 83:1–5

Bonhomme V, Laurain-Matter D, Lacoux J, Fliniaux MA, Jacquin-Dubreuil A (2000) Tropan alkaloid production by hairy roots of *Atropa belladonna* obtained after transformation with *Agrobacterium rhizogenes* 15834 and *Agrobacterium tumefaciens* containing *rol* A, B, C genes only. J Biotechol 81:151–158

Bos R, Woerdenbag HJ, Hendriks H, Zwaving JH, De SPAM, Tittel G, Wikstroem HV, Scheffer JJC (1996) Analytical aspects of phytotherapeutic valerian preparations. Phytochem Anal 7:143–151

Brijwal L, Tamta S (2015) *Agrobacterium rhizogenes* mediated hairy root induction in endangered *Berberis aristata* DC. Springer Plus 4:443

Bringmann G, Mutanyatta-Comar J, Knauer M, Abegaz BM (2008) Knipholone and related 4-phenylanthraquinones: structurally, pharmacologically and biosynthetically remarkable natural products. Nat Prod Rep 25:696–718

Bulgakov VP (2008) Functions of *rol* genes in plant secondary metabolism. Biotechnol Adv 26:318–324

Chang HM, But P (1986) Pharmacology and applications of Chinese materia medica. World Sci 1:255–268

Chansakaow S, Ishikawa T, Sekine K, Okeda M, Higuchi Y, Kudo M, Chaichantipyuth C (2000) Isoflavonoids from *Pueraria mirifica* and their estrogenic activity. Planta Med 66:572–575

Cheng TO (2007) Cardiovascular effects of Danshen. Int J Cardiol 121:9–22

Cherdshewasart W, Sutjit W (2008) Correlation of antioxidant activity and major isoflavonoid contents of the phytoestrogen-rich *Pueraria mirifica* and *Pueraria lobata* tubers. Phytomedicine 15:38–43

Cherdshewasart W, Cheewasopit W, Picha P (2004) The differential anti-proliferation effect of white (*Pueraria mirifica*), red (*Butea superba*), and black (*Mucuna collettii*) Kwao Krua plants on the growth of MCF-7 cells. J Ethnopharmacol 93:255–260

Cherdshewasart W, Panriansaen R, Picha P (2007) Pretreatment with phytoestrogen-rich plant decreases breast tumor incidence and exhibits lower profile of mammary ER alpha and ER beta. Maturitas 58:174–181

Cheruvathur MK, Thomas TD (2014) Effect of plant growth regulators and elicitors on rhinacanthin accumulation in hairy root cultures of *Rhinacanthus nasutus* (L.) Kurz. Plant Cell Tissue Organ Cult 118:169–177

Cheruvathur MK, Britto J, Thomas TD (2012) Pulvinus: an ideal explant for plant regeneration in *Caesalpinia bonduc* (L.) Roxb., an important ethnomedicinal woody climber. Acta Physiol Plant 34:693–699

Cheruvathur MK, Jose B, Thomas TD (2015) Rhinacanthin production from hairy root cultures of *Rhinacanthus nasutus* (L.) Kurz. In Vitro Cell Dev Biol Plant 51:420–427

Chi HJ, Kim HS (1970) Studies on the components of Umbelliferae plants in Korea: pharmacological study of decursin, decursinol and nodakenin. Korean J Pharmacol 1:25–32

Choi SS, Han KJ, Lee HK, Han EJ, Suh HW (2003) Antinociceptive profiles of crude extract from roots of *Angelica gigas* NAKAI in various pain models. Biol Pharm Bull 26:1283–1288

Chopra RN, Chopra IC, Handa KL, Kapur LD (1958) Indigenous drugs of India, vol 2. Academic, Calcutta, pp 508–674

Cimò G, Marchese A, Germanà MA (2016) Microspore embryogenesis induced through in vitro anther culture of almond (*Prunus dulcis* Mill.) Plant Cell Tissue Organ Cult. doi:10.1007/s11240-016-1086-2

Clark RJH, Cooksey CJ, Daniels MAM, Withnall R (1993) Indigo, woad and tyrian purple: important vat dyes from antiquity to the present. Endeavour 17:191–199

Dangash A, Ram M, Niranjan R, Bharillya A, Misra H (2015) In vitro selection and hormonal regulation in cell culture of *Artemisia annua* L. Plant. JSM Cell Dev Biol 3:1013

Das N (2006) Propagation prospects of dye yielding plant *Rhinacanthus nasutus* (Linn.) Kurz. Nat Prod Rad 5:42–43

Dey A, De JN (2010) Ethnobotanical aspects of *Rauvolfia serpentina* (L). Benth. ex Kurz. in India, Nepal and Bangladesh. J Med Plant Res 5:144–150

Dhakulkar S, Bhargava S, Ganapathi TR, Bapat VA (2005) Induction of hairy roots in *Gmelina arborea* Roxb. Using *Agrobacterium rhizogenes*. Paper presented in the National Seminar on 'Assessment, Utilization and Conservation of Biodiversity' held at Department of Botany, University of Pune, Pune, India

Dlugosz M, Wiktorowska E, Eisniewska A, Paczkowski C (2013) Production of oleanolic acid glycosides by hairy root established cultures of *Calendula officinalis* L. Acta Biochim Pol 60:467–473

Duke JA, Avensu ES (1985) Medicinal plants of China, vol 2. Reference, Algonac, MI, p 381

Dutta AC (1964) Botany for degree students, 6th edn. Oxford University Press, New Delhi, pp 583–585

Dutta NK, Panse MV (1962) Usefulness of berberine (an alkaloid from *Berberis aristata*) in the treatment of cholera. Indian J Med Res 50:732–735

Esyanti RR, Sahroni A (2014) Root growth and Gossypol content in *Gossypium hirsutum* L. root and hairy root cultures. Global Illuminator 1:223–234

Fabjan N, Rode J, Kosir IJ, Wang Z, Zhang Z, Kreft I (2003) Tartary buckwheat (*Fagopyrum tataricum* gaertn.) as a source of dietary rutin and quercitrin. J Agric Food Chem 51:6452–6455

Farag S, Kayser O (2015) Cannabinoids production by hairy root cultures of *Cannabis sativa* L. Am J Plant Sci 6:1874–1884

Farnsworth NR, Bunyapraphatsara N (1992) Thai medicinal plant: recommended for primary health care system. Prachachon Company, Bangkok

Fernández SP, WasowskiC PAC, Marder M (2006) Central nervous system depressant action of flavonoid glycosides. Eur J Pharmacol 539:168–176

Fu X, Yin ZP, Chen JG, Shangguan XC, Wang X, Zhang QF, Peng DY (2015) Production of chlorogenic acid and its derivatives in hairy root cultures of *Stevia rebaudiana*. J Agric Food Chem 63:262–268

Gai QY, Jiao J, Luo M, Wei ZF, Zu YG, Ma W (2015) Establishment of hairy root cultures by *Agrobacterium rhizogenes* mediated transformation of *Isatis tinctoria* L. for the efficient production of flavonoids and evaluation of antioxidant activities. PLoS One 10:e0119022

Galindo F, Kabir N, Gavrilovic J, Russell DA (2008) Spectroscopic studies of 1,2-diaminoanthraquinone (DAQ) as a fluorescent probe for the imaging of nitric oxide in living cells. Photochem Photobiol Sci 7:126–130

Gamborg O, Miller RA, Ojima K (1968) Plant cell cultures: I. Nutrient requirements of suspension cultures of soybean root cells. Plant Exp Cell Res 50:151–158

Glorybai AL, Agastian P (2013) *Agrobacterium rhizogenes* mediated hairy root induction for increased colchicine content in *Gloriosa superba* L. J Acad Ind Res 2:68–78

Grzegorczyk I, Krolicka A, Wysokinska H (2006) Establishment of *Salvia officinalis* L. Hairy root cultures for the production of rosmarinic acid. Z Naturforsch 61:351–356

Guinot P, Gargadennec A, Valette G, Fruchier A, Andary C (2008) Primary flavonoids in marigold dye: extraction, structure and involvement in the dyeing process. Phytochem Anal 19:46–51

Guo X, Zhu K, Zhang H, Yao H (2007a) Purification and characterization of the antitumor protein from Chinese tartary buckwheat (*Fagopyrum tataricum* gaertn.) water soluble extracts. J Agric Food Chem 55:6958–6961

Guo J, Jiang C, Wang Z, Lee HJ, Hu H, Malewicz B, Lee HJ, Lee JH, Baek NI, Jeong JH, Kim DK, Kang KS, Kim SH, Lu J (2007b) A novel class of pyranocoumarin anti-androgen receptor signaling compounds. Mol Cancer Ther 6:907–917

Gupta SK, Liu RB, Liae SY, Chan HS, Tsay HS (2011) Enhanced tanshinone production in hairy roots of *Salvia miltiorrhiza* Bunge under the influence of plant growth regulators in liquid culture. Bot Stud 52:435–443

Hao G, Ji H, Li Y, Shi R, Wang J, Feng L, Huang L (2012) Exogenous ABA and polyamines enhanced salvianolic acids contents in hairy root cultures of *Salvia miltiorrhiza* Bge. *F. alba.* Plant Omics J 5:446–452

Hasanloo T, Eskandari S, Kowsari M (2015) *Trichoderma* strains- *Silybum marianum* hairy root cultures interactions. Res J Pharmacol 2:33–46

Hong MLK, Bhatt A, Ping NS, Keng CL (2012) Detection of elicitation effect on *Hyoscyamus niger* L. root cultures for the root growth and production of tropane alkaloids. Rom Biotechol Lett 17:7340–7352

Hu ZB, Du M (2006) Hairy root and its application in plant genetic engineering. J Integr Plant Biol 48:121–127

Huang SH, Vishwakarma RK, Lee TT, Chan HS, Tsay HS (2014) Establishment of hairy root lines and analysis of iridoids and secoiridoids in the medicinal plant *Gentiana scabra.* Bot Stud Int J 55:17

Hurry JB (1930) The woad plant and its dye. Oxford University Press, Oxford

Jain SK (1984) Conservation of Tropical Plant Resources. Botanical Survey of India, Howrah, pp 110–117

Jain V, Singh D, Saraf S, Saraf S (2003) In vitro micropropagation of *Rauwolfia serpentina* through multiple shoot generation. Anc Sci Life 23:1–5

Jiang RW, Lau KM, Hon PM, Mak TCW, Woo KS, Fung KP (2005) Chemistry and biological activities of caffeic acid derivatives from *Salvia miltiorrhiza.* Curr Med Chem 12:237–246

Jiao J, Gai QY, Fu YJ, Ma W, Yao LP, Feng C, Xia XX (2015) Optimization of *Astragalus membranaceus* hairy root induction and culture conditions for augmentation production of astragalosides. Plant Cell Tissue Organ Cult 120:1117–1130

Kaczmarek F (1961) Badania hemiczneniektory chgatunkowrodzaju Lobelia. Pharm Biul IRL 26:89–156

Kai G, Xu H, Zhou C, Liao P, Xiao J, Luo X, You L, Zhang L (2011) Metabolic engineering tanshinone biosynthetic pathway in *Salvia miltiorrhiza* hairy root cultures. Metab Eng 13:319–327

Kang SY, Lee KY, Sung SH, Kim YC (2005) Four new neuroprotective dihydropyrano coumarins from *Angelica gigas.* J Nat Prod 68:5659

Khalili M, Hasanloo T, Tabar SKK, Rahnama H (2009) Influence of exogenous salicylic acid on flavonolignans and lipoxygenase activity in the hairy root cultures of *Silybum marianum.* Cell Biol Int 33:988–994

Kirtikar KR, Basu BD (1975) Indian medicinal plants, vol 3. Bishen Singh Mahendra Pal Singh, Dehradun, pp 2327–2329

Kumar KG, Thomas TD (2012) High frequency somatic embryogenesis and synthetic seed production in *Clitoria ternatea* Linn. Plant Cell Tissue Organ Cult 110:141–151

Kurtar ES, Ahmet Balkaya A, Kandemir D (2016) Evaluation of haploidization efficiency in winter squash (*Cucurbita maxima* Duch.) and pumpkin (*Cucurbita moschata* Duch.) through anther culture. Plant Cell Tissue Organ Cult. doi:10.1007/s11240–016–1074-6

Lahiri SC, Dutta NK (1967) Berberine and chloramphenicol in the treatment of cholera and severe diarrhea. J Indian Med Assoc 48:1–11

Lan X, Quan H (2010) Hairy root culture of *Przewalskia tangutica* for enhanced production of pharmaceutical tropane alkaloids. J Med Plants Res 4:1477–1481

Larsen BB (1986) A taxonomic revision of Phyllactis and Valeriana sect. Bracteata (Valerianaceae). Nord J Bot 6:427–446

Lenta BN, Weniger B, Antheaume C, Noungoue DT, Ngouela S, Assob JC, Vonthron-Sénécheau C, Fokou PA, Devkota KP, Tsamo E, Sewald N (2007) Anthraquinones from the stem bark of *Stereospermum zenkeri* with antimicrobial activity. Phytochemistry 68:1595–1599

Liu CL, Chen YS, Yang JH, Chiang BH (2008) Antioxidant activity of tartary (*Fagopyrum tataricum* L. gaertn.) and common (*Fagopyrum esculentum* Moench) buckwheat sprouts. J Agric Food Chem 56:173–178

Lloyd G, McCown B (1980) Commercially feasible micropropagation of mountain laurel, *Kalmia latifolia*, by use of shoot-tip culture. Int Plant Prop Soc Proc 30:421

Łuczkiewicz M, Kokotkiewicz A (2005) *Genista tinctoria* hairy root cultures for selective production of isoliquiritigenin. Z Naturforsch 60:867–875

Lystvan K, Belokurova V, Sheludko Y, Ingham JL, Prykhodko V, Kishchenko O, Paton E, Kuchuk M (2010) Production of bakuchiol by in vitro systems of *Psoralea drupacea* Bge. Plant Cell Tissue Organ Cult 101:99–103

Ma XQ, Duan JA, Zhu DY, Dong TTX, Tsim KWK (2000) Species identification of Radix Astragali (Huang-qi) by DNA sequence of its 5S-rRNA spacer domain. Phytochemistry 54:363–368

Madhusudanan KP, Banerjee S, Khanuja SPS, Chattopadhyay SK (2008) Analysis of hairy root culture of *Rauvolfia serpentine* using direct analysis in real time mass spectrometric technique. Biomed Chromatogr 22:596–600

Maheswari R, Selvamurugan C, Jayabarath J, Prabha AL (2011) Hairy root culture of an important medicinal plant: *Coleus forskohlii* UMA. Int J Agri 3:82–89

Malarz J, Stojakowska A, Kisiel W (2002) Sesquiterpene lactones in a hairy root culture of *Cichorium intybus*. Z Naturforsch 57:994–997

Md Setamam N, Sidik NJ, Rahman ZA, Zain CRCM (2014) Induction of hairy roots by various strains of *Agrobacterium rhizogenes* in different types of *Capsicum* species explants. BMC Res Notes 7:414

Mishra R, Wang H, Yadav NR, Wilkins TA (2003) Development of highly regenerable elite Acala cotton (*Gossypium hirsutum* cv. Maxxa)—a step towards genotype-independent regeneration. Plant Cell Tissue Organ Cult 73:21–35

Moghadam YA, Piri KH, Bahramnejad BH, Ghiasvand T (2014) Dopamine production in hairy root cultures of *Portulaca oleracea* (Purslane) using *Agrobacterium rhizogenes*. J Agric Sci Technol 16:409–420

Murashige T, Skoog F (1962) A revised medium for rapid growth and bioassays with tobacco tissue cultures. Physiol Plant 15:473–497

Murthy KRS (2012) Illustrated Sushrutha Samhitha, vol 1. Chaukambha Orientalia, Varanasi, p 274

Nakweti KR, Vaissayre V, Ndofunsu DA, Ndiku LS, Bonneau J, Franche C (2015) Hairy roots production in *Phyllanthus odontadenius* Mull. Arg. by seedlings transformed with *Agrobacterium rhizogenes* A4RS/pHKN29. African J Plant Sci 9:50–55

Ni X, Wen S, Wang W, Wang X, Xu H, Kai G (2011) Enhancement of camptothecin production in *Camptotheca acuminate* hairy roots by over expressing ORCA3 gene. J Appl Pharm Sci 1:85–88

Nishiya K, Kimura T, Takeya K, Itokawa H (1992) Sesquiterpenoids and iridoid glycosides from *Valeriana fauriei*. Phytochemistry 31:3511–3514

Nitsch JP, Nitsch C (1969) Haploid plants from pollen grains. Science 163:85–87

Pandey R, Krishnasamy V, Kumaravadivel N, Rajamani K (2014) Establishment of hairy root culture and production of secondary metabolites in Coleus (*Coleus forskohlii*). J Med Plants Res 8:58–62

Panwar GS, Attitalla IH, Guru SK (2011) An efficient in vitro clonal propagation and estimation of reserpine content in different plant parts of *Rauwolfia serpentina*. Am Eurasian J Sci Res 6:217–222

Parizi AP, Farsi M, Nematzadeh GA, Mirshamsi A (2014) Impact of different culture media on hairy roots growth of *Valeriana officinalis* L. Acta Agric Slovenica 103:299–305
Park SU, Lee SY (2009) Anthraquinone production by hairy root culture of *Rubia akane* Nakai: Influence of media and auxin treatment. Sci Res Essay 4:690–693
Park N, Li X, Uddin MDR, Park SU (2011) Phenolic compound production by different morphological phenotypes in hairy root cultures of *Fagopyrum tataricum* gaertn. Arch Biol Sci Belgrade 63:193–198
Park YJ, Thwe AA, Li X, Kim YJ, Kim JK, Arasu MV, Al-Dhabi NA, Park S (2015) Triterpene and flavonoid biosynthesis and metabolic profiling of hairy roots, adventitious roots, and seedling roots of *Astragalus membranaceus*. J Agric Food Chem 63:8862–8869
Parvaz A, Irshad A, Nawchoo BA, Wafai C (2006) Improvement of sexual destination in *Atropa acuminta* Royle (Solanaceae): A critically endangered medicinal plant of Northwest Himalaya. Hortic Sci 44:233–236
Patra N, Srivastava AK (2014) Enhanced production of artemisinin by hairy root cultivation of *Artemisia annua* in a modified stirred tank reactor. Appl Biochem Biotechnol 174:2209–2222
Patra N, Srivastava AK, Sharma S (2013) Study of various factors for enhancement of artemisinin in *Artemisia annua* hairy roots. Int J Chem Eng Appl 4:157–160
Petrova M, Zayova E, Vlahova M (2013) Induction of hairy roots in *Arnica montana* L. by *Agrobacterium rhizogenes*. Cent Eur J Biol 8:470–479
Piccaglia R, Venturi G (1998) Dye plants: a renewable source of natural colours. Agro Food Ind Hi-Tech 9:27–30
Ping SH, Yue LY, Shan ST, Eric TPK (2011) Induction of hairy roots and plant regeneration from the medicinal plant *Pogostemon cablin*. Plant Cell Tissue Organ Cult 107:251–260
Prakash P (2001) Indian medicinal plants, Forgotten Healers. A Guide to Ayurvedic Herbal Medicine, Chaukhamba Sanskrit Pratishthan, Delhi, pp 229–230
Putalun W, Yusakul G, Dicentrine DP (2009) Production from a hairy roots culture of *Stephania suberosa*. Z Naturforsch 64:692–696
Rahmani M, Pijut PM, Shabanian N (2016) Protoplast isolation and genetically true-to-type plant regeneration from leaf- and callus-derived protoplasts of *Albizia julibrissin*. Plant Cell Tissue Organ Cult. doi:10.1007/s11240-016-1072-8
Rahnama H, Hasanloo T, Shams MR, Sepehrifar R (2008) Silymarin production by hairy root culture of *Silybum marianum* (L.) Gaertn. Iran J Biotechnol 6:113–118
Rahnama H, Razi Z, Dadgar MN, Hasanloo T (2013) Enhanced production of flavonolignans in hairy root cultures of *Silybum marianum* by over expression of chalcone synthase gene. J Plant Biochem Biotechnol 22:138–143
Ranalli P, Candilo MD, Mandolino G, Grassi G, Carboni A (1999) Hemp for sustainable agricultural systems. Agro Ind 10:33–38
Reddy CS, Praveena CH, Veeresham C (2012) Strategies to improve the production of forskolin from hairy root cultures of *Coleus forskohlii* Briq. Int J Pharm Sci Nanotech 5:1720–1727
Robinson BL (1930) Contributions from the Gray Herbarium of Harvard University. The Gray Herbarium of Harvard University, Cambridge, MA
Rode J, Chol KI, Saal B, Flachowsky H, Krises U, Weber WE (2005) Sex-linked SSR markers in hemp. Plant Breed 124:167–170
Romero FR, Delate K, Kraus GA, Solco AK, Murphy PA, Hannapel DJ (2009) Alkamide production from hairy root cultures of *Echinacea*. In Vitro Cell Dev Biol Plant 45:599
Samadi A, Jafari M, Nejhad NM, Hossenian F (2014) Podophyllotoxin and 6-methoxy podophyllotoxin production in hairy root cultures of *Linum mucronatum* ssp. *Mucronatum*. Pharmacogn Mag 10:154–160
Saravanakumar A, Aslam A, Shajahan A (2012) Development and optimization of hairy root culture systems in *Withania somnifera* (L.) Dunal for withaferin- A production. Afr J Biotechnol 11:16412–16420
Sarker SD, Nahar L (2004) Natural medicine: the genus *Angelica*. Curr Med Chem 11:1479–1500
Satheeshkumar K, Jose B, Dhanya BP, Krishnan PN (2014) Prospects of *Plumbago rosea* L. hairy root culture in traditional preparations: a phytochemical comparison with tuberous roots. Plant Roots 8:13–23

Schenk RU, Hildebrandt AC (1972) Medium and techniques for induction and growth of monocotyledonous and dicotyledonous plant cell cultures. Can J For Res 50:199–204
Shaffert EE, Chetobar AA (1994) Structure, topography and ontogeny of *Stevia rebaudiana*. Bot Z 79:38–48
Shetty MR, Harisha GA, Jayanth Y, Kumar AH (2014) Production of secondary metabolites from *in vitro* cultures of *Rauwolfia serpentina* (L.) Benth. Int J Sci Technol Res 2:844–852
Shin SH (1989) Studies on the production of anthraquinone derivatives by tissue culture of *Rubia* species. Arch Pharm Res 12:99–102
Shirazi Z, Piri K, Asl AM, Hasanloo T (2013) Establishment of *Inula helenium* hairy root culture with the use of *Agrobacterium rhizogenes*. Int J Sci Basic Appl Res 4:1034–1038
Shu SL (2001) Isatis. In: Taiyan Z, Lianli Z, Lianli L, Guang Y, Al-Shehbaz IA (eds) Flora of China, vol 8. Missouri Botanical Garden, Missouri
Singh R, Geetanjali, Chauhan SM (2004) 9, 10-Anthraquinones and other biologically active compounds from the genus *Rubia*. Chem Biodivers 1:1241–1246
Singh DN, Verma N, Raghuwanshi S, Shukla PK, Kulshreshtha DK (2006) Antifungal anthraquinones from *Saprosma fragrans*. Bioorg Med Chem Lett 16:4512–4514
Singh R, Gangwar SP, Singh D, Singh R, Pandey R, Kalra A (2011) Medicinal plant *Coleus forskohlii* Briq.: disease and management. Med Plants 3:1–7
Siripong P, Yahuafai J, Shimizu K, Ichikawa K, Yonezawa S, Asai T (2006) Antitumor activity of liposomal naphthoquinone esters isolated from Thai medicinal plant: *Rhinacanthus nasutus* Kurz. Biol Pharm Bull 29:2279–2283
Sivanesan I, Jeong BR (2009) Induction and establishment of adventitious and hairy root cultures of *Plumbago zeylanica* L. Afr J Biotechnol 8:5294–5300
Soejarto DD, Kinghorn AD, Farnsworth NR (1982) Potential sweetening agents of plant origin. III. Organoleptic evaluation of stevia leaf herbarium samples for sweetness. J Nat Prod 45:590–599
Soejarto DD, Compadre CM, Medon PJ, Kamath SK, Kinghorn AD (1983) Potential sweetening agents of plant origin. II. Field search for sweet-tasting Stevia species. Econ Bot 37:71–79
Son JK, Jung SJ, Jung JH, Fang Z, Lee CS, Seo CS, Moon DC, Min BS, Kim MR, Woo MH (2008) Anticancer constituents from the roots of *Rubia cordifolia* L. Chem Pharm Bull 56:213–216
Srivastava SK, Rawat AKS, Srivastava M, Mehrotra S (2006) Pharmacognostic evaluation of the roots of *Berberis chitria* Lindl. Nat Prod Sci 12:19–23
Struik PC, Amaducci S, Bullard MJ, Stutterheim NC, Venturi G, Cromack HTH (2000) Agronomy of fibre hemp (*Cannabis sativa* L.) in Europe. Ind Crop Prod 11:107–118
Suja SR, Lath PG, Pushpangadan P, Rajasekharan S (2003) Evaluation of hepatoprotective effects of *Rhinacanthus nasutus* root extracts. Trop Bot Garden Res Doc 4:151–157
Susila T, Reddy GS, Jyothsna D (2013) Standardization of protocol for in vitro propagation of an endangered medicinal plant *Rauwolfia serpentine* Benth. J Med Plants Res 7:2150–2153
Tamer IM, Mavituna F (1996) Protease from callus and cell suspension cultures of *Onopordum turcicum*. Biotechnol Lett 18:361–366
Tan K (2002) Isatis. In: Strid A, Tan K (eds) Flora Hellenica, vol 2. Gantner Verlag, Ruggell, pp 126–127
Thakore D, Srivastava AK, Sinha AK (2013) Yield enhancement strategies for enhancement of indole alkaloids in hairy root cultures of *Catharanthus roseus*. Int J Chem Eng Appl 4:153–156
Thomas TD (2009) Isolation, callus formation and plantlet regeneration from mesophyll protoplasts of *Tylophora indica* (Burm. f.) Merrill: an important medicinal plant. In Vitro Cell Dev Biol Plant 45:591–598
Thomas TD, Chaturvedi R (2008) Endosperm culture: a novel method for triploid plant production. Plant Cell Tissue Organ Cult 93:1–14
Thomas TD, Hoshino Y (2010) In vitro propagation for the conservation of a rare medicinal plant *Justicia gendarussa* Burm. f. by nodal explants and shoot regeneration from callus. Acta Physiol Plant 32:943–950
Thomas TD, Jacob A (2004) Direct somatic embryogenesis of *Curculigo orchioides* Gaertn., an endangered medicinal herb. J Plant Biotechol 6:193–197

Thomas TD, Bhatnagar AK, Razdan MK, Bhojwani SS (1999) A reproducible protocol for the production of gynogenic haploids of mulberry, *Morus alba* L. Euphytica 110:169–173
Torkamani MRD, Abbaspour N, Jafari M, Samadi A (2014a) Elicitation of Valerenic acid in the hairy root cultures of *Valeriana officinalis* L (Valerianaceae). Trop J Pharm Res 13:943–949
Torkamani MRD, Jafari M, Abbaspour N, Heidary R, Safaie N (2014b) Enhanced production of valerenic acid in hairy root culture of *Valeriana officinalis* by elicitation. Cent Eur J Biol 9:853
Udomsuka L, Jarukamjorna K, Tanakab H, Putaluna W (2009) Isoflavonoid production in a hairy roots culture of *Pueraria candollei*. Z Naturforsch 64:687–691
Urasopan N, Hamada Y, Cherdshewasart W, Malaivijitnond S (2008) Preventive effects of *Pueraria mirifica* on bone loss in ovariectomized rats. Maturitas 59:137–148
Vaknin Y, Hadas R, Schafferman D, Murkhovsky L, Bashan N (2008) The potential of milk thistle (*Silybum marianum* L.), an Israeli native, as a source of edible sprouts rich in antioxidants. Int J Food Sci Nutr 59:339–346
Valdes LJ, Mislankars SG, Paul AG (1987) *Coleus barbatus* (Lamiaceae) and the potential new drug forskolin (Coleonol). Econ Bot 41:474–483
Varotto S, Lucchin M, Parrini P (1997) Plant regeneration from protoplast of Italian red chicory (*Cichorium intybus*). J Genet Breed 15:17–22
Verma PC, Singh H, Negi AS, Saxena G, Rahman L, Banerjee S (2015) Yield enhancement strategies for the production of picroliv from hairy root culture of *Picrorhiza kurroa* Royle ex Benth. Plant Signal Behav 10:e1023976
Vinterhalter B, Milosevic DK, Jankovic T, Plejevljakusic D, Ninkovic S, Smigocki A, Vinterhalter D (2015) *Gentiana dinarica* Beck. hairy root cultures and evaluation of factors affecting growth and xanthone production. Plant Cell Tissue Organ Cult 121:667–679
Wagner H, Bauer R, Xiao PG, Chen JM, Michler G (1997) Chinese drug monographs and analysis: radix astragali (Huangqi). Verlag Wald, Germany, pp 1–17
Wang CT, Liu H, Gao XS, Zhang HX (2010) Overexpression of G10H and ORCA3 in the hairy roots of *Catharanthus roseus* improves catharanthine production. Plant Cell Rep 29:887–894
Xiang W, Song QS, Zhang HJ, Guo SP (2008) Antimicrobial anthraquinones from *Morinda angustifolia*. Fitoterapia 79:501–504
Xiao PG, Xia GC, He LY (1973) The occurrence of some important tropane alkaloids in Chinese solanaceous plants. J Integr Plant Biol 15:187–194
Xu H, Park JJ, Kim YK, Park N, Lee SY, Park SU (2009) Optimization of growth and pyranocoumarins production in hairy root culture of *Angelica gigas* Nakai. J Med Plants Res 3:978–981
Xuan TD, Tsuzuki E (2004) Allelopathic plants: buckwheat (*Fagopyrum* spp.) Allelopath J 13:137–148
Yang C, Chen M, Zeng L, Zhang L, Liu X, Tang K, Liao Z (2011) Improvement of tropane alkaloids production in hairy root cultures of *Atropa belladonna* by overexpressing pmt and h6h genes. Plant Omics J 4:29–33
Yao Y, Shan F, Bian J, Chen F, Wang M, Ren G (2008) D-chiro-Inositol enriched tartary buckwheat bran extract lowers the blood glucose level in kk-Ay mice. J Agric Food Chem 56:10027–10031
Yim D, Singh RP, Agarwal C, Lee S, Chi H, Agarwal R (2005) A novel anticancer agent, decursin, induces G1 arrest and apoptosis in human prostate carcinoma cells. Cancer Res 65:1035–1044
Yin J, Gao Z, Liu D, Liu Z, Ye J (2008) Berberine improves glucose metabolism through induction of glycolysis. Am J Physiol Endocrinol Metab 294:148–156
Yogananth LN, Basu MJ (2009) TLC method for the determination of plumbagin in hairy root culture of *Plumbago rosea*. Global J Biotechol Biochem 4:66–69
Zebarjadi AR, Najafi SH, Ghasempour HR, Motamedi J (2011) Establishment of a practical tissue culture for producing hairy roots of *Valeriana officinalis* L. via *Agrobacterium rhizogenes*. J Med Plants Res 5:4984–4992
Zhang JY, Wu HY, Xia XK, Liang YJ, Yan YY, She ZG, Lin YC, Fu LW (2007a) Anthracenedione derivative 1403P-3 induces apoptosis in KB and KBv200 cells via reactive oxygen species-independent mitochondrial pathway and death receptor pathway. Cancer Biol Ther 6:1413–1421
Zhang L, Yang B, Lu BB, Kai GY, Wang ZN, Xia YY, Ding RX, Zhang HM, Sun XF, Chen WS, Tang KX (2007b) Tropane alkaloids production in transgenic *Hyoscyamus niger* hairy root

cultures over expressing Putrescine N-methyl transferase is methyl jasmonate dependent. Planta 225:887–896

Zhao B, Agblevor FA, Ritesh KC, Jelesko JG (2013) Enhanced production of the alkaloid nicotine in hairy root cultures in *Nicotiana tabacum* L. Plant Cell Tissue Organ Cult 113:121–129

Zheng XY (2005) Pharmacopoeia of the Peoples Republic of China, vol 1, Chinese edn. Chemical Industry Press, Beijing, pp 212–213

Zhou LG, Wu JY (2006) Development and application of medicinal plant tissue cultures for production of drugs and herbal medicinals in China. Nat Prod Rep 23:789–810

Zhu SY, Dong Y, Tu J, Zhou Y, Zhou XH, Xu B (2014) *Silybum marianum* oil attenuates oxidative stress and ameliorates mitochondrial dysfunction in mice treated with D-galactose. Pharmacogn Mag 10:S92–S99

Chapter 10
Edible Mushrooms and Their In Vitro Culture as a Source of Anticancer Compounds

Bożena Muszyńska, Katarzyna Kała, and Katarzyna Sułkowska-Ziaja

Abstract Cancer diseases are one of the main causes of death in the world. The experiences from Asian and Eastern European countries show that mushrooms can play an important role in the prevention and treatment of cancers. An example would be *Piptoporus betulinus,* traditionally used in the Czech Republic in colorectal cancer treatment, while fruiting bodies of *Inonotus obliquus* were used in folk medicine in Eastern Europe to treat cancer diseases since the sixteenth or seventeenth century. It is commonly known that substances with immunostimulating, antioxidant, and revitalizing properties naturally occurring in food, including edible mushrooms, can be an important element in anticancer prevention. Due to seasonal occurrence of mushrooms' fruiting bodies in the natural state, an access to many of them is difficult, and the problem is also their correct identification; thus it is obvious that the easiest way to obtain them is edible mushrooms from commercial cultivations, for example *Agaricus bisporus*, *Pleurotus ostreatus*, or *Lentinula edodes*. The search for an effective material for the prevention and treatment of cancer diseases resulted in an undertaking of an introduction of edible mushrooms mycelial cultures in vitro. Thus, the last three decades is the period of the most intensive studies on substances isolated from mushrooms, both fruiting bodies and mycelial cultures in vitro.

Keywords In vitro culture • Anticancer potential • β-glucans • Lectins • Basidiomycota • *Agaricus bisporus* • *Lentinula edodes* • *Pleurotus ostreatus*

Abbreviations

5-FU	5-Fluorouracil
AAL	Lectin from *A. aegerita*
ABL	Lectin from *A. bisporus*
ACF	Aberrant crypt foci

B. Muszyńska (✉) • K. Kała • K. Sułkowska-Ziaja
Department of Pharmaceutical Botany, Jagiellonian University Medical College, Medyczna 9, Krakow 30-688, Poland
e-mail: muchon@poczta.fm

S. Malik (ed.), *Biotechnology and Production of Anti-Cancer Compounds*,
DOI 10.1007/978-3-319-53880-8_10

Cis-DPP	*Cis*-dichlorodiammine platinum
CLA	Conjugated linoleic acid
CTL	Cytotoxic T lymphocyte
DF	Dietary fiber
DMBA	7,12-Dimethylbenz[a]anthracene
ergone	Ergosta-4,6,8(14),22-tetraen-3-one
FIPs	Fungal immunomodulating proteins
G-CSF	Granulocyte colony-stimulating factor
NK	Natural killer
RIP	Ribosome inactivating proteins
S-GAP-P	Chemically sulfated polysaccharide
SOD	Superoxide dismutase
TML1 and TML2	Lectins from *T. mongolicum*
TPA	12-*O*-tetradecanoylphorbol-13-acetate
VEGF	Vascular endothelial growth factor

Introduction

The reasons for many diseases development, including cancers, are inter alia free radicals. People who are subjected to strong, prolonged stress or improper nutrition are exposed to the formation of free radicals and thus an increased oxidative stress. These changes are also the result of the natural aging process. Although the human body has mechanisms protecting and maintaining internal balance through the production of, inter alia, superoxide dismutase (SOD) or catalase, which are the factors neutralizing free radicals, the support for defense reaction of the organism is needed (Elmastas et al. 2007; Hu et al. 2015; Maseko et al. 2014; Tsai et al. 2008). Currently, cancers, next to cardiovascular diseases, are the main causes of deaths in the world, and the available treatment methods, except that not always effective, often involve a number of side effects and burden already weakened organism. Thus, the interest of the scientific communities is focused on natural and less burdening methods of treatment, and first of all on the prevention based on a properly balanced diet. Health properties of edible mushrooms are analyzed for years especially in traditional Eastern medicine and as part of folk medicine of the West. Edible mushrooms are increasingly popular due to numerous biological activities, such as an antioxidant, anticancer, immunostimulating, anti-atherosclerotic, neuroprotective, anti-inflammatory, anti-allergic, antibacterial, antiviral, or hypoglycemic activity. The range of these activities is due to bioactive components present in the fruiting bodies, such as proteins, e.g., fungal immunomodulating proteins (FIPS), lectins, glycoproteins, polysaccharides, phenolic compounds, indole compounds, terpenoids, or lipids (including ergosterol and its derivatives) (Moro et al. 2012; Muszyńska et al. 2013; Patel and Goyal 2012; Roupas et al. 2012; Zong et al. 2012). It is commonly known that anticancer compounds naturally present in foods, including edible mushrooms, can be an important element in prevention of cancer. Due to the seasonal occurrence of mushrooms' fruiting bodies in the natural state, an access to many of them is difficult, and the problem is also their correct identification; thus it is obvious that the easiest way to obtain them are edible mushrooms from commercial crops. The search for an effective material for the prevention and treatment of cancer diseases resulted in an undertaking of an introduction of edible mushrooms mycelial cultures in vitro. Thus, the last three decades is the period of the most intensive studies on substances isolated from mushrooms, both fruiting bodies and mycelial cultures in vitro.

Edible mushrooms can be used in cancer treatment, or as additives to conventional therapy, or even as substances combating the side effects of cancer therapy (Patel and Goyal 2012; Zong et al. 2012). Extracts and substances isolated from mushrooms demonstrate numerous mechanisms of anticancer activity; for example, they may act through the inhibition of the kinase, and thus the cell cycle, or inhibition of angiogenesis, and they are also inducers of reactive oxygen species, antimitotic agents, or inhibitors of topoisomerases, finally stimulating the cancer cells for apoptosis (Kosanić et al. 2016; Patel and Goyal 2012). The effective anticancer therapy needs the use of processes comprehensively acting with respect to cancers, since the cells that mutated to cancer cells have the unique ability for proliferation and spreading, forming metastases, and just metastases are mainly responsible for such high mortality (Zong et al. 2012).

According to the present state of knowledge on edible mushrooms and their health-promoting potential, they seem to be an interesting alternative to improve the quality of life concurrently reducing the risk of civilization diseases, and thus can be included to the group of nutraceuticals (Roupas et al. 2012; Singdevsachan et al. 2016). At present, there are many pharmaceutical companies specializing in the production of anticancer formulations, including extracts, or dietary supplements in various forms. Due to the progress of the research methods, the studies can be carried out both in vitro and in vivo. An interest in the subject of edible mushrooms and constantly growing number of studies suggest that this material administered directly or in the form of extracts is well tolerated and safe for the human body. The full credibility of the results and evaluation of the actual effect on the health need large amounts of clinical studies. In these studies, mushroom extracts derived from different species or edible mushrooms or their in vitro cultures biomass are provided to cancer patients in a double-blind samples next to placebo (Patel and Goyal 2012; Roupas et al. 2012; Singdevsachan et al. 2016).

Anticancer Polysaccharides

Significance of edible mushrooms as effective agents in the prevention and treatment of cancer diseases was confirmed in clinical studies in recent years. The main compounds responsible for this effect are polysaccharides and their connections to peptides (proteoglycans) or steroids (Kosanić et al. 2016; Meng et al. 2016; Ruthes et al. 2015; Singdevsachan et al. 2016). Mushroom polysaccharides are one of the most known groups of mushroom compounds with anticancer activities (Patel and Goyal 2012). Edible mushrooms are also a source of dietary fiber (DF) with polysaccharide structure and anticancer activity. The content of DF depends on mushrooms structure and the stage of their morphogenesis (Cheung 2013). Different species of mushrooms produce various kinds of polysaccharides, which can be both soluble and insoluble in water. The molecules of simple sugars are combined in polysaccharides by glycoside bonds. Some of them are built with residues of only one type of saccharides—homopolysaccharides, while some of them belong to the heteropolysaccharides and are composed of residues of various monosaccharides. The most important simple sugars included in the polysaccharides are glucose, mannose, galactose, xylose, arabinose, fucose, uronic acid, and glucuronic acid, forming linear or branched structures. These compounds can be used as prebiotics and in the treatment of cancer or viral diseases, e.g., AIDS. It is assumed that polysaccharides isolated from edible mushrooms activate the immune response in vitro and in vivo, acting as biological stimulants. The most important anticancer polysaccharides are made up of compounds with a structure similar to those which are components of the cell wall of mushrooms. They include polysaccharide–protein complexes, cellulose, $(1 \rightarrow 3)$-α-glucans, and $(1 \rightarrow 3)$, $(1 \rightarrow 6)$-β-glucans. In turn, heteropolysaccharides which exhibit an antiproliferative effect with respect to cancer cells turned out to be substances that inhibit the cancers development (carcinostatics), especially after intraperitoneal or oral administration (Singdevsachan et al. 2016; Zhu et al. 2015; Zong et al. 2012).

Particularly important is mushroom polysaccharides significance in the modulation of immune system function, and thus the potential inhibitory effect on cancers. One of the first clinically described activities of mushroom polysaccharides in the treatment of cancer originated in 1957 and was made by Byerrum et al. (1957; Meng et al. 2016). The mechanism of this effect on the immune system confirmed in subsequent studies involves stimulation of immune system cells, including T lymphocytes and cytotoxic T lymphocytes (CTL), B lymphocytes, granulocytes (eosinophils and neutrophils), natural killers (NK), or macrophages (Meng et al. 2016; Roupas et al. 2012; Singdevsachan et al. 2016; Zhang et al. 2007). This mechanism is particularly characteristic for β-1,3-glucans, but numerous studies also suggest that β-glucans may enhance the specific cellular response by enhancing the secretion of IL-6, IL-8, IL-12, and IFN-ɣ from neutrophils, macrophages, and NK cells (Meng et al. 2016; Singdevsachan et al. 2016). Moreover, β-glucans found in the fruiting bodies of edible mushrooms may constitute the factors that stimulate new effector cells contributing to the formation of, inter alia, antibodies directed against cancer antigens, which is less popular than the classic cytotoxic effect caused by chemotherapy (Singdevsachan et al. 2016).

The ability to bind other molecules (proteins, steroids) resulting in an increased anticancer activity is also important in order to achieve an anticancer effect. In clinical practice, mushroom polysaccharides are usually used as part of polytherapy next to the standard treatment using chemotherapy or radiotherapy. Most among applied polysaccharides are glucans, as well as homo- or heteroglycans which, when combined with other protein molecules, can be transformed into glycoproteins, glycopeptides, or proteoglycans. Also the conformation of polysaccharide chain is crucial to induce a therapeutic effect. The most active polysaccharides are usually the complexes with proteins of a molecular weight of 10,000 kDa. It was examined that the human macrophages contain polysaccharide receptor highly specific for glucose and mannose molecules, hence the anticancer activity of polysaccharides with these particular groups (Meng et al. 2016; Patel and Goyal 2012; Roupas et al. 2012; Tian et al. 2016; Zhang et al. 2007).

Among the edible mushrooms rich in polysaccharides with the structure described above, anticancer effect is exhibited, inter alia, by the following species: *Armillaria mellea* (honey Mushrooms), *Lactarius deliciosus* (saffron milk cap), *Macrolepiota procera* (parasol mushroom), *Lentinus edodes* (Shiitake), *Grifola frondosa* (dancing mushroom), *Tremella fuciformis* (snow fungus), *Hericium erinaceus* (pom pom mushroom), *Agaricus bisporus* (white mushrooms), *Agaricus blazei* (almond mushroom), *Agaricus campestris* (field mushroom), *Cantharellus cibarius* (Chanterelle), *Flammulina velutipes* (winter mushroom), *Pleurotus ostreatus* (oyster mushrooms), *Sparassis crispa* (cauliflower mushrooms), *Boletus edulis* (King bolete), and *Boletus badius* (Bay bolete). In modern medicine, mainly Eastern one, but all over the world, *L. edodes* is used as an agent enhancing the strength of the organism, but the extract from this species is also used as an anticancer substance. The therapeutic effect is primarily due to the polysaccharide—lentinan (Fig. 10.1), whose action causes a reduction in tumor size even by 90% (Kosanić

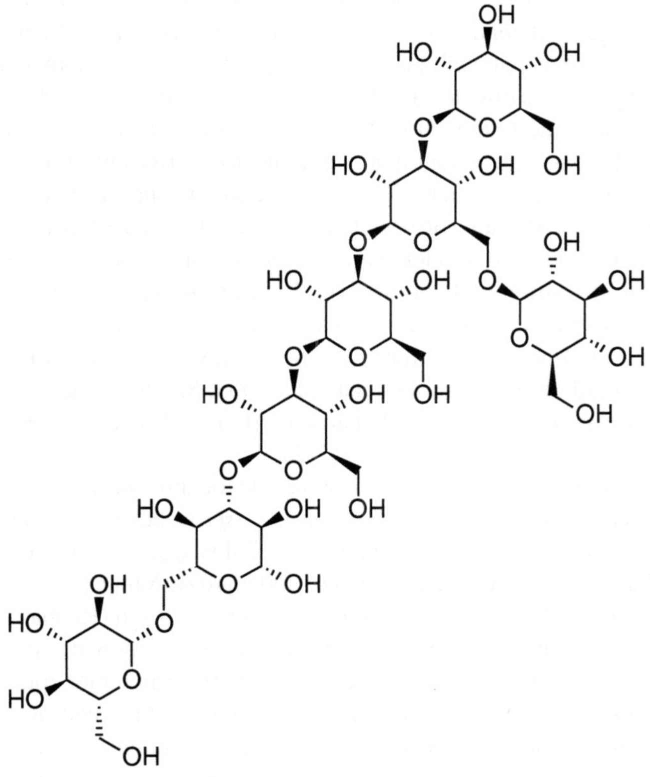

Fig. 10.1 Lentinan

et al. 2016; Meng et al. 2016; Patel and Goyal 2012; Roupas et al. 2012; Tian et al. 2016; Zhang et al. 2007).

In terms of chemical structure, lentinan is a β (1 → 3) glucan with β branches (1 → 6), with a molecular weight from 400 to 800 kDa. This polysaccharide is most commonly used to treat solid tumors of stomach, lung, breast, colon, and malignant leukemia. It works by an activation of the immune system and causes restoration of normal defense reaction of the body, thereby causing an anticancer effect. It induces a humoral immune response of the body in a manner involving restoration of reduced activity of T helper cells in host cells occupied by the cancer. Lentinan does not exhibit cytotoxic activity, practically does not demonstrate any side effects, only the ones such as local irritation after injection or sporadically observed fever and vomiting, but usually these are only the episodes, and what the most important, it is generally well tolerated by the patient's organism. It extends an average time of the survival of patients with cancer in whom it is used. The studies on the mechanism of lentinan action prove that it is dependent on the thymus and involves the strengthening of T helper precursor cells and macrophages response, and thus some of the cytokines produced by lymphocytes after the recognition of cancer cells. The induction of interferon (IFN-ɣ) is also important for this activity. Lentinan is the most commonly used in the treatment of gastric cancer, as an adjunct to conventional treatment, including surgical removal of the tumor, chemotherapy, or radiotherapy. This is a kind of synergistic effect, improving the general condition of the patient. There were also effective attempts of the treatment of hepatocellular carcinoma, colorectal, and pancreatic cancer, while reducing the side effects, and thereby improving the quality of life. Lentinan is also effective in the treatment of leukemia, as a factor inhibiting lymphocytes proliferation. It is also important that it exhibits a selective antiproliferative effect with respect to skin cancer cells (CH72), concurrently not affecting the healthy keratinocytes (C50). It is also proved that lentinan inhibits the formation of metastases in mice liver in case of adenocarcinoma-26. This activity is possible due to the activation of Kupffer cells in the liver. It was also proved that the use of lentinan as an adjuvant can enhance the quality of life for patients with cancer diseases, since it eliminates (like other mushroom polysaccharides) the side effects of chemo- and radiotherapy (Lindequist et al. 2005; Mantovani et al. 2008; Meng et al. 2016; Muszyńska et al. 2013; Patel and Goyal 2012; Roupas et al. 2012; Singdevsachan et al. 2016; Zong et al. 2012).

Another example of polysaccharide responsible for anticancer activity is β-glucan most prevalent in the edible mushroom of *S. crispa* species. Clinical trials were conducted, in which powdered fruiting bodies of this species were orally administered to patients with cancer in an amount of 300 mg/day. A major improvement was noted in case of many patients compared to the control group (Roupas et al. 2012). *L. deliciosus* and *M. procera* are edible mushroom species showing similar anticancer potential in vitro. This activity was examined for human tumor epithelial cancer cells (HeLa), human colon cancer (LS174), and human lung cancer (A549). The IC_{50} values ranged from 19.01 to 74.01 μg mL^{-1} for *L. deliciosus* extract, and from 25.55 to 68.49 μg mL^{-1} for *M. procera* extract depending on the type of cell lines.

However, the activity of these extracts was not as effective as in case of *cis*-dichlorodiammine platinum (*Cis*-DPP). Stronger anticancer effect with respect to A549 and LS174 cell lines was demonstrated for *M. procera* species, while higher activity with respect to HeLa was noted for *L. deliciosus*. However, it was not proved whether this activity is selective for malignant cells, or also includes other types of cancer cells. It is possible that the presence of the CD is one of the factors affecting the cytotoxic effect with respect to the examined cell lines, but these are further studies confirming this activity. In the case of *M. procera*, inhibitory activity with respect to tumor metastasis of colon 26-M3.1 was demonstrated for aqueous extracts (Kosanić et al. 2016).

Two polysaccharide factions which are α-glucans were isolated from the fruiting bodies of *Armillaria mellea* in the Centro de Investigaciones Biológicas in Madrid. The major fraction consisted of linear chains of α-(1,3)- and α-(1,4)-glucan linked to the protein, while the second fraction contained α-(1,3)-glucan. Also the fraction containing β-glucan with peptide was obtained. The authors demonstrated anticancer activity of this fraction. Glucose molecules linked with β-(1,3) and β-(1,6) bonds were detected in the study on the chemical structure of peptide-glucan sugar portion (Muszyńska et al. 2011).

A. blazei is a species commonly used in cancer prevention, since it demonstrates immunomodulatory and antimutagenic properties. Polysaccharide fractions with high content of β-glucan derived from this species are shown to be effective in the treatment of prostate cancer, both dependent and independent on androgens. Induction of prostate cancer cells apoptosis was directly related to the activation of caspase-3 as a pro-apoptotic factor. Polysaccharide fraction obtained from this species proved to be effective in cancers treatment in vivo, in the absence of cytotoxic action (examined in vitro). In addition, in the therapy using 5-fluorouracil (5-FU), this species protected against leucopenia, as a side effect of the treatment using this substance. There were some attempts to improve β-glucan derived from this species by modifications involving the incorporation of sulfate groups improving the solubility (Mantovani et al. 2008; Patel and Goyal 2012; Roupas et al. 2012; Zong et al. 2012). In turn, *A. campestris* and glycoprotein fraction obtained from this species showed an anticancer activity against sarcoma 180 in ICR mice. Importantly, the protein moiety of glycoprotein contributing to the anticancer effect consists of 17 amino acids (Singdevsachan et al. 2016).

Also another glucan—pleuran (β-1,3-d-glucan)—derived from *P. ostreatus* species is worth paying attention (Fig. 10.2).

Its activity slowing down the formation of precancerous Aberrant Crypt Foci (ACF) of the colon of Wistar rats was demonstrated. This action involved an inhibition of cancer cells proliferation and an induction of apoptosis. Additionally, the new polysaccharide (POPS-1) obtained from hot aqueous extracts of this species showed considerably reduced toxicity compared to commonly used 5-FU. Diet rich in dried fruiting bodies of *P. ostreatus* reduced toxicity in mice treated with cyclophosphamide, and reduced the pathological lesions resulting from the emergence of colon cancer induced by dimethylhydrazine in rats. The mechanism of action resulted from a strong antioxidant capacity demonstrated by these fruiting bodies

Fig. 10.2 Pleuran

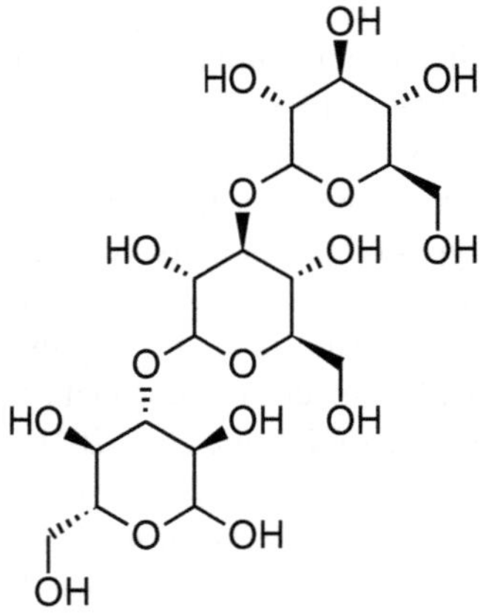

and DF content. In turn, the aqueous extracts obtained at a high temperature from fruiting bodies of *H. erinaceus* proved to be rich in β-glucan, the administration of which contributed to the reduction in tumor mass in mice with induced colon cancer. Reduction in tumor size resulted from an induction of tumor necrosis factor and NK cells secretion, as well as macrophages activation and an inhibition of angiogenesis (Lindequist et al. 2005; Meng et al. 2016; Patel and Goyal 2012; Roupas et al. 2012; Ruthes et al. 2015; Singdevsachan et al. 2016).

G. frondosa species, commonly known as Maitake, is also a source of β-glucan, including an active fraction D and the fraction MD obtained by its further purification (both improving the effectiveness of therapy using cisplatin treatment). Fraction MD acts by enhancing the activity inhibiting the spread of cancers, and reducing nephrotoxicity and immunosuppression induced by cisplatin treatment. In the combined therapy with cyclophosphamide, this fraction acts inducing the production of granulocyte colony-stimulating factor (G-CSF), which helps to stimulate granulopoiesis and neutrophils mobilization in mice treated with cyclophosphamide. Also heteropolysaccharide MZF demonstrating an activity inhibiting tumor growth in vivo by stimulation of cellular immunity was isolated from this species. An application of whole powdered fruiting bodies with an addition of isolated fraction MD was used in the patients with cancer diseases in the stages II-IV of cancer development. The use of such a combination improved general condition of the patient, and even caused a regression of cancers in 68.8% of patients diagnosed with breast cancer, 58.3% of patients with liver cancer, and 62.5% treated for lung cancer. MD fraction alone is believed to be mainly responsible for the stimulation of NK cells activity, and thus inhibition of mentioned cancers development (Lindequist et al. 2005; Cheung 2013; Patel and Goyal 2012; Roupas et al. 2012; Zong et al. 2012). Water-insoluble polysaccharides, like, e.g., chemically sulfated polysaccharide (S-GAP-P), were also recognized to be effective in the treatment of gastric cancer in humans (cancer cells SGC-7901). *G. frondosa* appeared to be effective not only in the combination therapy next to 5-FU, but also in monotherapy with S-GAP-P where it induced an apoptosis of cancer cells SGC-7901. This effect was dose dependent. In addition, the combined use of 5-FU and S-GAP-P enhanced the anticancer effect.

β-glucan from *G. frondosa* also exhibited cytotoxic effect on human prostate cancer cells (PC-3 cells). Induction of apoptosis was confirmed in vitro for adrogeno-independent tumor (Meng et al. 2016; Patel and Goyal 2012; Roupas et al. 2012; Ruthes et al. 2015; Singdevsachan et al. 2016). It was also demonstrated in the studies that polysaccharide isolated from another species—*Pleurotus pulmonarius*—is effective in liver cancer treatment, both in vitro and in vivo. This activity involved an inhibition of cancer development via inhibiting effect on VEGF-induced (vascular endothelial growth factor) PI3K/AKT signaling pathway (Cheung 2013). Anticancer effect in *B. edulis* species was investigated in mice. This activity was based on an inhibition of cancer cell Sarcoma 180.

The structure of mushroom polysaccharides is an important issue for their anticancer activity. Polysaccharides include β-glucans, glycans, heteroglycans, and hetero-β-glucans. The medical use mainly concerns the polysaccharides in the form of a triple helix, e.g., already mentioned lentinan, although their mechanism of action due to the structure is not fully understood. However, the polysaccharides in other forms, e.g., linear like (1 → 3)-β-*d*-glucan, soluble in water and isolated from *Auricularia auricula* species, are also characterized by a strong anticancer potential (Meng et al. 2016; Ruthes et al. 2015; Singdevsachan et al. 2016; Zhang et al. 2007). Mushroom polysaccharides, depending on the structure, also exhibit immunomodulatory activity, strengthening and accelerating the organism's defensive response, which produce an anticancer effect. This effect is more common than the typical cytotoxic activity (Meng et al. 2016; Ruthes et al. 2015; Singdevsachan et al. 2016; Zhang et al. 2007). The result of mushroom polysaccharides application can thus be prevention of cancers, and in case of already detected cancers, cell cycle inhibition and apoptosis induction in cancer cells, affecting its growth and development (growth reduction even up to 50%), as well as an inhibition of metastases formation, prolonging patient's survival (Zhang et al. 2007; Zong et al. 2012). Preventive effect was especially observed in the group of farmers involved in the commercial cultivation of edible mushrooms, including *F. velutipes* and *A blazei*. Naturally, these mushrooms species were a popular element of their daily diet. As a result of mushrooms consumption, cancer incidence rates were set at about 40% lower level than in the general population. The studies on mice were also conducted in order to confirm these observations. They were regularly fed with those mushroom species, and then the cancer cells were implanted. It was clearly observed that the murine tumors did not develop in the examined group, in contrast to the control one, which confirms the potent preventive anticancer activity of polysaccharides from *F. velutipes* and *A. blazei* species (Zhang et al. 2007).

Mushroom Lectins

In addition to polysaccharides, mushrooms are also rich in active proteins which can be isolated from them. Among the bioactive anticancer proteins, attention should be especially paid to ribosome inactivating proteins (RIP), FIP lectins, and ribonucleases (Roupas et al. 2012). Lectins are a kind of storage proteins needed for proper growth of mushroom fruiting bodies. They are also a protective factor for the mushrooms themselves, protecting them from toxins from the environment, e.g., bacteria, pesticides, and even viruses (Singh et al. 2015; Varrot et al. 2013). Lectins contained in the edible mushrooms were investigated also due to their anticancer activity. This is related to their immunomodulatory potential realized by the stimulation of immune system cells maturation. The mixture of lectins (lectin A and lectin B) isolated from *Agaricus bisporus* species (ABL) demonstrated an antiproliferative effect on endothelial cancer cells, without a direct cytotoxic effect. In vitro studies using human colonic carcinoma cells HT-29 demonstrated that lectins isolated from *A. bisporus* have the ability to induce apoptosis in just these cells, without the cytotoxic effect. It is suggested that this effect may be related to an increase in caspase-3 activity (Carrizo et al. 2005; Roupas et al. 2012; Singh et al. 2015; Wang et al. 1998). The exact relationship between the structural elements and anticancer effect induced in both mouse and human cells was determined for the lectin from *Agrocybe aegerita* species (AAL). This is important, since it may be the first step for the design of specific molecules used later in the therapy. AAL acts inducing an apoptosis in cancer cells. Dimeric organization of lectin molecule is needed to achieve this activity. The presence of glucose and galactose in carbohydrate functional moiety is also necessary. Irrespective of these structural properties, also another hydrophobic pocket responsible for the induction of apoptosis was identified (Li et al. 2008; Singh et al. 2015; Xu et al. 2011).

Also lectins from edible mushrooms such as *G. frondosa*, *F. velutipes*, *Tricholoma mongilicum*, *Volvariella volvacea*, *Pleurotus citrinopileatus*, and *P. ostreatus* proved to have immunomodulatory potential and anticancer, antiproliferative, and cytotoxic properties. It was possible to isolate heterodimeric lectin which exhibits the potential inhibiting cancer cells growth and development from *V. volvacea* species. However, the inhibitory potential is mild. Importantly, this lectin prolongs the survival in mice with sarcoma 180. An extension of survival is dependent on the dose. *G. frondosa* species was studied due to the cytotoxic effects with respect to cancer cells HeLa. This activity was due to *N*-acetylgalactosamine-specific lectin (Li et al. 2008; Singh et al. 2015; Wang et al. 1998). On the other hand, *P. citrinopileatus* species is rich in homodimeric lectin inhibiting cancer growth (murine sarcoma 180) already within 20 days with intraperitoneal administration in a daily dose of 5 mg kg^{-1} body weight. The sources of this lectin were fresh fruiting bodies. An anticancer activity is similar to that exhibited by *P. ostreatus* (Li et al. 2008; Patel and Goyal 2012). Lectin isolated from *P. ostreatus* species demonstrated an anticancer activity with respect to murine hepatoma H-22 and sarcoma S-180, reducing the number of cancer cells (Li et al. 2008; Ng 2004; Singh et al. 2015). This lectin inhibits the growth of both cancers by 75% and 88%, respectively. To achieve this, lectin dose of 1.5 mg kg^{-1} body weight for 20 days in a form of intraperitoneal injection was necessary. This treatment resulted in weight loss in mice compared to the control group; however, the time of their survival was prolonged significantly (Ng 2004).

Lectins from *T. mongolicum* species (TML1 and TML2) inhibit the growth of sarcoma 180. These are the lectins of a molecular weight of about 36–38 kDa. Both

TML1 and TML2 inhibit the growth of these cells, while TML2 proved to be more effective. TML1 and TML2 are the compounds of the dimeric structure. They also inhibit the proliferation of mouse cells, including mastocytoma P815 and monocyte-macrophage PU5–1.8 in vitro. In this case, higher effectiveness was proven for TML1. The activity of both lectins is further based on macrophages stimulation to the production of nitrite ions (NO_2^-). Both lectins exhibit an anticancer effect stimulating the immune system in vivo, while an antiproliferative effect is only observed in vitro (Ng 2004; Singh et al. 2015; Wang et al. 1996; Wang et al. 1998).

Other Anticancer Compounds

Currently, there are a number of studies showing a close relationship between the consumption of given mushroom species and the risk of breast cancer. This effect was especially observed in postmenopausal women. This anticancer effect is caused by an inhibition of aromatase, which is important in case of hormone-dependent cancers, especially breast cancer. The studies confirming this effect were conducted in vitro, but also in subsequent steps on animals and finally on humans. Special attention should be paid to already mentioned *A. bisporus* species, which is one of the most popular species of edible mushrooms. The studies suggest that conjugated linoleic acid (CLA) may be the compound inhibiting aromatase activity (Chen et al. 2006; Kanaya et al. 2011; Roupas et al. 2012). *A. bisporus* additionally constitutes not only the source of linoleic acid, but palmitic and linolenic acids as well, which also protect against breast cancer development. This action also includes an inhibition of aromatase activity and estrogens biosynthesis. This is important, as it was proved that abnormal expression of aromatase, enzyme responsible for the synthesis of estrogens in breast cancer cells, may play a key role in the growth and development of cancer cells of this type (Chen et al. 2006; Patel and Goyal 2012). In the studies previously carried out by Chen et al., the extracts of *A. bisporus* were more effective in an inhibition of human placental aromatase activity than a few compared plant extracts. The method of cancer therapy using this mushroom species is an alternative to the conventional treatment, as well as one of the possibilities of anticancer prevention (Chen et al. 2006).

The inhibitory effect of the extracts of *A. bisporus* fruiting bodies was confirmed not only with respect to the development of breast, but also prostate cancer. The compounds responsible for this activity, except β-glucans, are phenolic compounds, ergothioneine, arginine, lectins, and the aforementioned fatty acids. Unsaturated fatty acid content was calculated to be 79.7% of total fatty acids found in this species. This is important because the inefficient metabolism, and especially the absence of certain fatty acids in the diet, can cause the development of many diseases, including cancers (Ahmad et al. 2013; Novaes et al. 2011; Öztürk et al. 2011; Patel and Goyal 2012; Yilmaz et al. 2006).

Attention should be also paid to the closed, young fruiting bodies of *Coprinus comatus*, which are edible. The studies demonstrated that they can be also effective in the treatment of estrogen independent breast cancer. An inhibitory effect was also demonstrated with respect to prostate cancer cells (LNCaP cells). Immunomodulatory protein (FIP-fve), which is characterized by an anticancer activity realized by T-cells activation, was isolated from *F. velutipes* species. This activity was tested on a murine model of hepatoma. Aqueous extract appeared to be rich in another anticancer substance—flammulin. Also the presence of stable hemagglutinin, responsible for the inhibition of cancer cells proliferation in leukemia (L1210 cells), was a significant issue. The aqueous extracts both of *C. comatus* and *F. velutipes* are found to be effective in the treatment of an estrogen-dependent and estrogen-independent breast cancer in vitro. Anticancer effect was dose-dependent for both the extracts, and caused by an induction of altered cells apoptosis. *Calvatia utriformis* species is also effective in the treatment of this type of cancers due to the presence of ubiquitin-like peptide. It can be thus concluded that an increase in the consumption of certain edible mushrooms can significantly contribute to the reduced risk of breast cancer in women both before and after menopause (Patel and Goyal 2012; Roupas et al. 2012).

The presence of arginine in *A. bisporus* species should also be emphasized, since it is a substance used in the supplementation of cancer patients. It not only improves the condition of the body and exhibits a beneficial effect on the immune system, but also prolongs the life of cancer patients. This activity results from the delay in tumor growth and metastases emergence (Novaes et al. 2011). Also ergothioneine—a strong antioxidant isolated from *A. bisporus*—is beneficial in cancers therapy due to its potent antimutagenic properties and chemo- and radioprotective activity (Chen et al. 2012; Muszyńska et al. 2013) (Fig. 10.3).

An anticancer effect was also demonstrated for ergosterol (5,7,22-ergostatrien-3β-ol) present not only in *A. bisporus* but also in other species of *Agaricales* taxon (Fig. 10.4).

Fig. 10.3 Ergothioneine

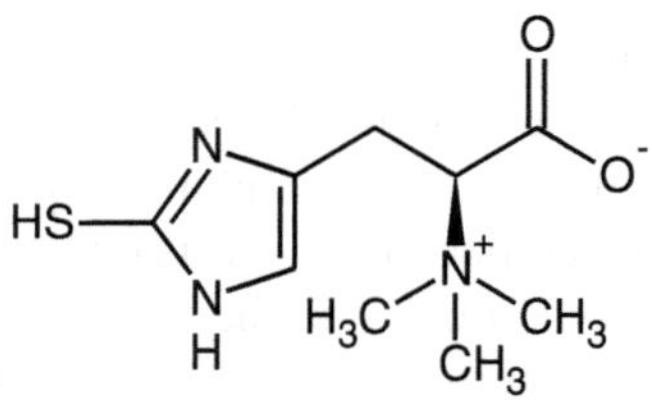

Fig. 10.4 Ergosterol

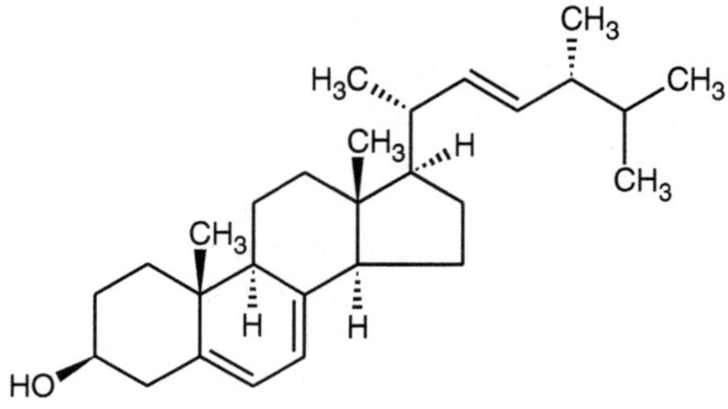

It is found in most species of the mushrooms of *Basidiomycota* class. Ergosterol constitutes 83–89% of the total amount of mushroom sterols. It inhibits the process of angiogenesis, which accompanies solid tumors preventing tumor growth, and prevents the migration and proliferation of cancer cells. Such effect was found in cancer cell lines in vitro cultures, as well as in studies on rats conducted in vivo (Novaes et al. 2011; Roupas et al. 2012; Shao et al. 2010; Yuan et al. 2008). The content of ergosterol in cultivated mushroom species such as *A. bisporus*, *L. edodes*, or *P. ostreatus* is in the range of 3.7–5.1 mg g^{-1} dry weight, while in case of the wild species of *Cantharellus tubaeformis*, *Cantharellus cibarius*, and *B. edulis,* this amount is slightly lower, in the range of 1.4–4.0 mg g^{-1} dry weight (Teichmann et al. 2007). Also the protective effect of ergosterol on lymphocytes level in patients subjected to chemotherapy was confirmed. This therapy is safe and well tolerated. Also ergocalciferol (vitamin D_2) present in the edible mushrooms, formed from ergosterol as a result of UV irradiation with a wavelength of 280–320 nm, is one of the preventive factors in cancer therapy (Novaes et al. 2011; Roupas et al. 2012; Shao et al. 2010). It demonstrates antimutagenic properties, with low toxicity, causing only a very small hypercalcemic effect. Vitamin D_2, in particular its hydroxylated form, is considered to be a potential anticancer agent used in therapy of, inter alia, melanoma. This effect results from the inhibition of keratinocytes differentiation in vivo, inducing thus protection from photo damage after the local administration. In the studies on melanoma, also ergosterol and dihydroergosterol inhibited DNA synthesis due to the local metabolism of ergosterol causing an expression of cytochrome P-450scc by melanoma cells. Similarly, 17α- and 24-dihydroxyergosterol inhibited the proliferation of cell lines of human epidermal keratinocytes. These results confirm an antiproliferative activity, and thus an anticancer effect of ergosterol metabolites to the cells, not only animal but also human ones (Słomiński et al. 2005). The studies also indicate the existence of the relationship between the suitable levels of vitamin D_2, also of mushroom origin, and a reduced risk of prostate, ovarian, breast, and colon cancer (Shao et al. 2010). Some mushroom species are a rich source of ergocalciferol, except *A. bisporus*; they include *G. frondosa*, *Morchella* spp., or *L. edodes* (Phillips et al. 2011).

Derivatives of ergosterol, including ergosterol peroxide, also demonstrate an anticancer effect, in this case based on the cytotoxic effect with respect to cancer cell lines and their growth inhibition. Ergosterol peroxide, or other sterols demonstrating this effect, was isolated, inter alia, from *Paecilomyces tenuipes* and *Cordyceps sinensis* species. *P. tenuipes* species also exhibited an anticancer activity in vivo (Hong et al. 2007; Lindequist et al. 2005). Ergosterol peroxide was moreover isolated for the first time from *H. erinaceus* species, demonstrating an anticancer activity. It was also possible to obtain it from other species of edible mushrooms found to be therapeutic, such as, inter alia, *V. volvacea, Boletus badius*, *B. edulis, Suillus bovinus*, *Morchella esculenta*, and *A. mellea* (Krzyczkowski et al. 2009). Ergosterol peroxide isolated from the edible species *Sarcodon aspratus* (*S. aspratus*) also demonstrated an activity inhibiting the growth of promyelocytic leukemia cells (HL60). This effect was observed for the dose above 10 mM ergosterol peroxide. In turn, the administration of 25 mM (10.7 μg mL^{-1}) of ergosterol peroxide isolated from this species reduced viable cells number to 5% compared to the control group. This concentration induced an activation of apoptotic cells and nucleosomal DNA fragmentation in most HL60 cells after 24 h of incubation. These results clearly show that ergosterol peroxide induced apoptosis, and thus inhibited the growth of HL60 cells. However, the mechanism of this action is not fully understood (Takei et al. 2005). *Hypsizigus marmoreus* species also proved to be a source of ergosterol and ergosterol peroxide. It turned out that due to compounds from the group of sterols, the fruiting bodies of this species can inhibit induced by TPA (13-ethyl-12-*O*-tetradecanoylphorbol) the inflammatory swelling of the ear and cancer development in the two-stage carcinogenesis induced by TPA and DMBA (7,12-dimethylbenz[a]anthracene) in mice. Thus, this confirmed the anticancer potential of sterols, which may be protective factors against cancers development (Yaoita et al. 2002).

It was possible to isolate agaritine from the species related to *A. bisporus*—*A. blazei*, present in aqueous extracts obtained hot (Fig. 10.5).

Agaritine exhibited an activity with respect to human leukemic monocyte lymphoma (U937) cells, inducing their apoptosis in vitro. The activity also concerned other leukemic cell lines including HL60, MOLT4, and K562. Its activity involved the damage of cancer cells DNA by their fragmentation and cytochrome c release from them. This species is used as an adjuvant during chemotherapy. Currently, it is believed that the consumption of mushrooms of *Agaricales* taxon containing agaritine is not only completely safe, but beneficial to human health. *A. bisporus* species was also effective in leukemia inhibition, e.g., HL60 cells inducing their apoptosis. Except anti-leukemia activity, *A. blazei* also demonstrated an anticancer effect in relation to stomach cancer KATO III cells and lung cancer LU99 cells. This activity was possible due to the content of steroid—blazein—that induces morphotic changes in the mentioned cell as a result of their apoptosis. An activity of six other steroids with anticancer activity was also examined in *A. blazei* species. They were isolated from this species in the form of acetone extracts. After steroids examination, it was also found they exhibit antimutagenic potential (Endo et al. 2010; Lindequist et al. 2005; Patel and Goyal 2012; Roupas et al. 2012). Bioactive steroid—ergosta-4,6,8 (14), 22-tetraen-3-one (ergone)—of antiproliferative and cytotoxic effect against HepG2 cells was isolated from edible *Russula cyanoxantha* species. Affected by ergone, HepG2 cells were subjected to apoptosis, inter alia, due to an inhibition of the cell cycle (phase G2/M), chromatin condensation and nucleus fragmentation which was mainly possible due to caspase activation (Endo et al. 2010; Patel and Goyal 2012; Roupas et al. 2012).

Fig. 10.5 Agaritine

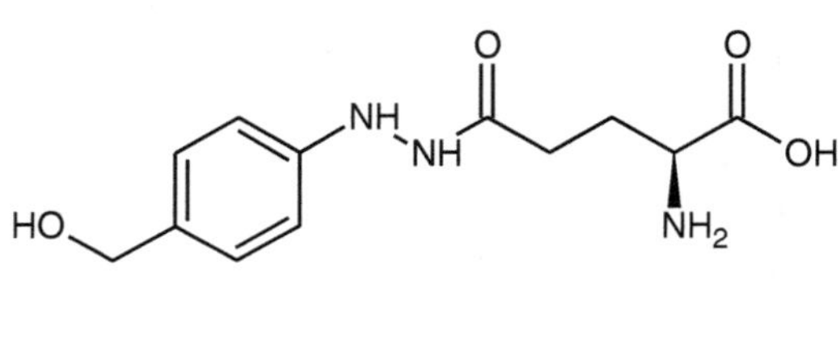

Fig. 10.6 Theanine

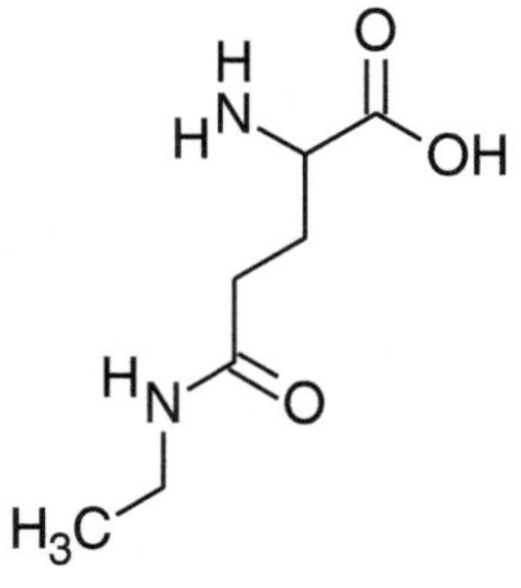

The results of controlled studies are also interesting, in which the edible mushrooms exhibit synergistic beneficial effect reducing the incidence of cancers, along with the consumption of green tea. Theanine located primarily in green tea, next to the above mentioned active substances, is increasingly often the substance considered in cancer therapy (Fig. 10.6).

B. badius is a popular species, which becomes a source of theanine as a result of fermentation. Theanine demonstrates an activity similar to the synthetic anticancer substances, such as, e.g., irinotecan, doxorubicin, or cisplatin; thus its efficacy during the treatment is assumed. Numerous studies both on animal models, and in the later stages on human cell lines, consistently confirm the efficacy of edible mushrooms use in breast cancer treatment. Most importantly, the examined edible mushrooms do not damage the healthy cells, but only the cancer ones inhibiting their proliferation (Patel and Goyal 2012; Roupas et al. 2012). Also *Calvatia caelata* species contains ubiquitin-like peptide with a molecular weight of 8 kDa. In the studies, it exhibits an antiproliferative activity with respect to human breast cancer cells and also affects murine splenocytes exhibiting an antimitogenic effect (Ng 2004).

Laccase present in *Agrocybe cylindracea* species proved to be an important enzyme with anticancer activity. Laccases are also found in the species like *H. erinaceus*, *A. blazei*, *L. edodes*, *P. ostreatus*, *C. cibarius*, or *Pleurotus eryngii*. However, their anticancer activity was not analyzed. Laccase is used for the purification of polluted water, as biosensor, or fabrics dye. It is a lignin degrading enzyme; in mushrooms it additionally participates in morphogenesis. This compound also catalyses the reactions of organic compounds oxidation, including phenolic compounds. The studies demonstrated that laccase present in *A. calvacea* demonstrates an antiproliferative activity with respect to breast cancer MCF-7 (IC_{50} = 6.5 mM) and hepatoma cell HepG2 (IC_{50} = 5.6 μM). The activity of laccase isolated from this species was the highest maintaining the pH in the range 3–4; at pH of about 9 its activity disappeared completely. Also the temperature increase to 50 °C profitably affected its activity, but further temperature increase was unprofitable, since a slow decomposition of the enzyme is noted above this temperature (Hu et al. 2011; Zhang et al. 2010). An activity with respect to cancer cells MCF-7 and HepG2 was also demonstrated for laccase isolated from the edible species *Clitocybe maxima*. IC_{50} values for these cancer cells are 3.0 mM and 12.3 mM, respectively (Zhang et al. 2010).

Another example of an enzyme demonstrating an anticancer activity is tyrosinase extracted from *A. bisporus*, characterized by a high degree of similarity to that found in human organism. Thus, this species is a perfect and concurrently inexpensive source of tyrosinase (Kampmann et al. 2015; Labus et al. 2011; Zaidi et al. 2014). The studies conducted on enzyme isolated from *A. bisporus* demonstrated that it has a protective effect on human lymphoma cell lines, preventing them from the negative effect of damaging factors such as dihydrogen peroxide (Shi et al. 2002). Moreover, gene-protecting activity of this species resulting from tyrosinase presence was examined. A putative protective effect with respect to DNA is associated with the pathway of tyrosine transformation into *l*-DOPA, and then this metabolite conversion to dopaquinone (Jani et al. 2016; Shi et al. 2002).

The presence of peptide significant from the therapeutic point of view, i.e., prosomatostatin, demonstrating an anticancer activity especially in pancreatic cancer treatment, was found in the fruiting bodies of *A. mellea* (Muszyńska et al. 2011).

Fruiting bodies and mycelia of *Antrodia camphorata* species are used as an anticancer agent in traditional Eastern medicine due to the abundance of bioactive substances. Anticancer activity is mainly due to ubiquinone derivative—antroquinonol. Importantly, antroquinonol exhibits selective activity acting only on cancer cells, leaving the healthy ones (Hu et al. 2016).

In Vitro Cultures

Taking into account the rapid development of biotechnology, and the need for research on anticancer substances, it seems to be necessary to examine the biomass from edible mushrooms in vitro cultures. This is extremely needed, since the fruiting bodies are characterized by seasonal occurrence, while the in vitro cultures can be prepared in reproducible conditions and subjected to experiments in a continuous manner. Typically, the same compounds as in the fruiting bodies are produced in in vitro cultures; however, appropriate modifications of the methods of applied media culturing can focus them on the production of the desired metabolites. Thus, it is essential to compare the composition of the fruiting bodies and in vitro cultures, and possibly to examine the media due to the possibility of active mushroom metabolites accumulation in them.

As in the case of fruiting bodies, the most popular are mushroom polysaccharides detected and isolated from mycelial cultures. One of the first and most active polysaccharides isolated from in vitro cultures of *L. edodes* was lentinan. It exhibits an anticancer activity corresponding with the potency of immunostimulating activity to that isolated from the fruiting bodies (Chihara et al. 1970). It was also possible to isolate another polysaccharide—KS-2—from in vitro cultures of the same species. It is α-mannan containing peptide fragment in its molecule. It demonstrated inhibitory activity with respect to Ehrlich cancer and sarcoma 180 developments in the mouse, both after oral or intraperitoneal administration. It also induced interferon biosynthesis in animal studies (Fujii et al. 1978; Ooi and Liu 2000). It was found during the therapy with the use of this polysaccharide that the best results are obtained in case of its combination with cytostatic agents, i.e., in so-called immunochemotherapy (Wheat et al. 1983). Another glucan exhibiting an anticancer activity was isolated, in turn, from in vitro cultures of *G. frondosa*. It acts by inhibiting an activity and secretion of metalloproteinases in the case of skin cancers (Noriko et al. 2003). Also a new protein-polysaccharide complex—GFPPS1b—was isolated from the in vitro cultures of this species. Its activity involves an induction of apoptosis in cancer cells by an inhibition of cell cycle phase in G2/M. The inhibitory effect mainly concerned proliferation of SGC-7901 (human gastric adenocarcinoma), as in the case of this species fruiting bodies. As a result of GFPS1b application, tumor cells are subject to apoptosis, which was determined analyzing the structure of the damaged cells (Patel and Goyal 2012; Zong et al. 2012).

Attention should also be paid to in vitro culture of *A. blazei* species. It was possible to isolate polysaccharides, which demonstrate an immunomodulatory and anticancer activity from the aqueous extract of mycelial cultures. This activity in this case was related to macrophages stimulation to TNF-α secretion, which is a pro-inflammatory cytokine leading the cancer cells to apoptosis. It may be concluded from the studies that the therapeutic effect is especially due to glucomannans with the main chain of β-1,2-*d*-mannopyranoside. Their structure is different from those of polysaccharides obtained from the fruiting bodies of *A. blazei*, which are β-1,6-glucans. Glucomannans isolated from mycelial cultures, when combined with a protein, form the Antitumor Organic Substance Mie complex. This complex exhibits a strong cytotoxic potential compared to murine sarcoma 180 cells, Ehrlich cancer (Ito et al. 1997; Lindequist et al. 2005).

Also *P. ostreatus* species is a source of water-insoluble polysaccharide—β--glucan—and soluble heteroglycan with anticancer activity. Water-soluble heteroglycan (PS) of a high molecular weight was isolated from the mycelium of this species. It consists of *d*-mannose, *l*-fucose, and *d*-glucose. An anticancer activity is mainly preventive in the nature, due to the strong antioxidant potential of heteroglycan of this species (Patra et al. 2013). Attention should also be paid to strobilurin present in in vitro cultures of *Strobilurus tenacellus*. It is mainly used as a plant protection agent, but also demonstrated an anticancer potential in the studies (Balba 2007). Due to the strong health-promoting potential and limited availability from the natural environment, mycelial cultures were produced from *C. sinensis* which are the source of sterols and other health-promoting substances, including those with anticancer properties, are successfully isolated from the culturing mycelium of this species (Lindequist et al. 2005). The above-mentioned vitamin D_2, which is an agent in cancer prevention, was isolated from the mycelia from in vitro cultures of edible *P. ostreatus*, *P. citrinopileatus*, and *Pleurotus salmoneostramineus* species. The studies involved the irradiation of fresh cultures in vitro with UV-B, then the analysis of the qualitative composition of mycelia in terms of vitamin D and polysaccharides content. The study also involved an analysis of antioxidant properties of irradiated mycelia—their ethanol extracts, which is also reflected in an anticancer effect. The composition of in vitro cultures was also evaluated for the presence of potentially strong antioxidant compounds, including total content of phenols, ergothioneine, or flavonoids. The studies confirmed the beneficial effect of 2 h irradiation of mycelia of all three species on an increase in vitamin D_2 content, maintaining a satisfactory antioxidant potential (Huang et al. 2015).

Conclusions

Edible mushrooms and their in vitro cultures are increasingly willingly studied and used as an element of health-oriented prevention. In addition to the typical prophylactic use, a plurality of bioactive compounds provides their therapeutic effect which is supported by a number of reliable studies. They are used in cancer therapy due to a multidirectional activity inhibiting the formation, growth, and development of cancers at different stages. This therapy is very popular due to its good tolerance and reduction of the side effects of radiotherapy and chemotherapy for the human organism. The most promising are the attempts of breast cancer treatment with the use of edible mushrooms, which confirms the abundance of reports on this subject. The number of examined mushroom species and their anticancer effect allows looking again at this group of organisms, and appreciating their not only therapeutic but also preventive importance, protecting against the disease. However, further studies on the effect of edible mushrooms and their in vitro cultures on particular cancers are needed. Anticancer compounds derived from mushrooms are usually responsible for immune system stimulation, cytotoxic effect with respect to cancer cells, or are used as a therapy combined with clinical one. The advantage is on the side of the extracts obtained from this mushroom material, but it does not change the fact that just the whole fruiting bodies or mycelia are the source of bioactive compounds. Continuous analysis of the structure of the compounds and their anticancer activity mechanisms in mushrooms, as well as the stimulation of in vitro cultures towards the production of these substances, may contribute to the development of commercial production of mushroom anticancer medicaments.

References

Ahmad MS, Ahmad S, Gautam B, Afzal M (2013) Antigenotoxic and anticlastogenic potential of *Agaricus bisporus* against MMS induced toxicity in human lymphocyte cultures and in bone marrow cells of mice. Egypt J Med Hum Genet 14:395–402

Balba H (2007) Review of strobilurin fungicide chemicals. J Environ Sci Health B 42:441–451

Byerrum R, Clarke D, Lucas E, Ringler R, Stevens J, Stock CC (1957) Tumor inhibitors in boletus edulis and other Holobasidiomycetes. Antibiot Chemother 7:1–4

Carrizo ME, Capaldi S, Perduca M, Irazoqui FJ, Nores GA, Monaco HL (2005) The antineoplastic lectin of the common edible mushroom (*Agaricus bisporus*) has two binding sites, each specific for a different configuration at a single epimeric hydroxyl. J Biol Chem 280:10614–10623

Chen S, Oh SR, Phun S, Hur G et al (2006) Anti-aromatase activity of phytochemicals in white button mushrooms (*Agaricus bisporus*). Cancer Res 66:12026–12034

Chen SY, Ho KJ, Hsieh YJ, Wang LT, Mau JL (2012) Contents of lovastatin, γ-aminobutyric acid and ergothioneine in mushroom fruiting bodies and mycelia. LWT Food Sci Technol 47:274–278

Cheung PCK (2013) Mini-review on edible mushrooms as source of dietary fiber: preparation and health benefits. Food Sci Human Wellness 2:162–166

Chihara G, Hamuro J, Maeda Y, Arai Y, Fukuoka F (1970) Fractionation and purification of the polysaccharides with marked antitumor activity, especially lentinan, from *Lentinus edodes* (Berk.) sing. (an edible mushroom). Cancer Res 30:2776–2781

Elmastas M, Isildak O, Turkekul I, Temur N (2007) Determination of antioxidant activity and antioxidant compounds in wild edible mushrooms. J Food Compos Anal 20:337–345

Endo M, Beppu H, Akiyama H, Wakamatsu K, Ito S, Kawamoto Y, Shimpo K, Sumiya T, Koike T, Matsui T (2010) Agaritine purified from *Agaricus blazei Murrill* exerts anti-tumor activity against leukemic cells. Biochim Biophys Acta 1800:669–673

Fujii T, Maeda H, Suzuki F, Ishida N (1978) Isolation and characterization of a new antitumor polysaccharide, KS-2, extracted from culture mycelia of Lentinus edodes. J Antibiot (Tokyo) 31:1079–1090

Hong IP, Nam SH, Sung GB, Chung IM, Hur H, Lee MW, Kim MK, Guo SX (2007) Chemical components of *Paecilomyces tenuipes* (peck) Samson. Mycobiology 35:215–218

Hu DD, Zhang RY, Zhang GQ, Wang HX, Ng TB (2011) A laccase with antiproliferative activity against tumor cells from an edible mushroom, white common *Agrocybe cylindracea*. Phytomedicine 18:374–379

Hu YH, Chen CM, Xu L et al (2015) Postharvest application of 4-methoxy cinnamic acid for extending the shelf life of mushroom (*Agaricus bisporus*). Postharvest Biol Technol 104:33–41

Hu YD, Zhang BB, Xu GR, Liao XR, Cheung PCK (2016) A mechanistic study on the biosynthetic regulation of bioactive metabolite antroquinonol from edible and medicinal mushroom *Antrodia camphorata*. J Funct Foods 25:70–79

Huang SJ, Lin CP, Tsai SY (2015) Vitamin D_2 content and antioxidant properties of fruit body and mycelia of edible mushrooms by UV-B irradiation. J Food Compos Anal 42:38–45

Ito H, Shimura K, Itoh H, Kawade M (1997) Antitumor effects of a new polysaccharide-protein complex (ATOM) prepared from *Agaricus blazei* (Iwade strain 101) "Himematsutake" and its mechanisms in tumor-bearing mice. Anticancer Res 17:277–284

Jani AR, Nag S, Setty SR (2016) Visualization of intracellular tyrosinase activity in vitro. Bio Protoc 6:e1794

Kampmann M, Hoffrichter AC, Stalinski D, Wichmann R (2015) Kinetic characterization of tyrosinase containing mushroom (*Agaricus bisporus*) cells immobilized in silica alginate. J Mol Catal B Enzym 116:124–133

Kanaya N, Kubo M, Liu Z, Chu P, Wang C, Yuan YC, Chen S (2011) Protective effects of white button mushroom (*Agaricus bisporus*) against hepatic steatosis in ovariectomized mice as a model of postmenopausal women. PLoS One 6:e26654

Kosanić M, Ranković B, Rančić A, Stanojković T (2016) Evaluation of metal concentration and antioxidant, antimicrobial, and anticancer potentials of two edible mushrooms *Lactarius deliciosus* and *Macrolepiota procera*. J Food Drug Anal 24:477–484

Krzyczkowski W, Malinowska E, Suchocki P, Kleps J, Olejnik M, Herold F (2009) Isolation and quantitative determination of ergosterol peroxide in various edible mushroom species. Food Chem 113:351–355

Labus K, Turek J, Bryjak J (2011) Efficient *Agaricus bisporus* tyrosinase immobilization on cellulose-based carriers. Biochem Eng J 56:232–240

Li YR, Liu QH, Wang HX, Ng TB (2008) A novel lectin with potent antitumor, mitogenic and HIV-1 reverse transcriptase inhibitory activities from the edible mushroom *Pleurotus citrinopileatus*. Biochim Biophys Acta 1780:51–57

Lindequist U, Niedermeyer THJ, Jülich WD (2005) The pharmacological potential of mushrooms. Evid Based Complement Alternat Med 2:285–299

Mantovani MS, Bellini MF, Angeli JPF, Oliveira RJ, Silva AF, Ribeiro LR (2008) β-glucans in promoting health: prevention against mutation and cancer. Mutat Res 658:154–161

Maseko T, Howell K, Dunshea FR, Ng K (2014) Selenium-enriched *Agaricus bisporus* increases expression and activity of glutathione peroxidase-1 and expression of glutathione peroxidase-2 in rat colon. Food Chem 146:327–333

Meng X, Liang H, Luo L (2016) Antitumor polysaccharides from mushrooms: a review on the structural characteristics, antitumor mechanisms and immunomodulating activities. Carbohydr Res 424:30–41

Moro C, Palacios I, Lozano M, D'Arrigo M, Guillamón E, Villares A, Martínez JA, García-Lafuente A (2012) Anti-inflammatory activity of methanolic extracts from edible mushrooms in LPS activated RAW 264.7 macrophages. Food Chem 130:350–355

Muszyńska B, Sułkowska-Ziaja K, Wołkowska M, Ekiert H (2011) Chemical, pharmacological, and biological characterization of the culinary-medicinal honey mushroom, *Armillaria mellea* (Vahl) P. Kumm. (Agaricomycetideae). A review. Int J Med Mushrooms 13:167–175

Muszyńska B, Sułkowska-Ziaja K, Łojewski M, Opoka W, Zając M, Rojowski J (2013) Edible mushrooms in prophylaxis and treatment of human diseases. Med Inter Rev 101:170–183

Ng TB (2004) Peptides and protein from fungi. Peptides 25:1055–1073

Noriko K, Kiyoshi K, Hiroaki N (2003) Effect of Maitake (*Grifola frondosa*) D-fraction on the activation of NK cells in cancer patients. J Med Food 6:371–377

Novaes MR, Valadares F, Reis MC, Gonçalves DR, Menezes MC (2011) The effects of dietary supplementation with *Agaricales* mushrooms and other medicinal fungi on breast cancer: evidence-based medicine. Clinics 66:2133–2139

Ooi VE, Liu F (2000) Immunomodulation and anti-cancer activity of polysaccharide-protein complexes. Curr Med Chem 7:715–729

Öztürk M, Duru ME, Kivrak S, Mercan-Doğan N, Türkoglu A, Özler MA (2011) In vitro antioxidant, anticholinesterase and antimicrobial activity studies on three *Agaricus* species with fatty acid compositions and iron contents: a comparative study on the three most edible mushrooms. Food Chem Toxicol 49:1353–1360

Patel S, Goyal A (2012) Recent developments in mushrooms as anti-cancer therapeutics: a review. 3 Biotech 2:1–15

Patra S, Patra P, Maity KK, Mandal S, Bhunia SK, Dey B, Devi KSP, Khatua S, Acharya K, Maiti TK, Islam SS (2013) A heteroglycan from the mycelia of *Pleurotus ostreatus*: structure determination and study of antioxidant properties. Carbohydr Res 368:16–21

Phillips KM, Ruggio DM, Horst RL et al (2011) Vitamin D and sterol composition of 10 types of mushrooms from retail suppliers in the United States. J Agric Food Chem 59:7841–7853

Roupas P, Keogh J, Noakes M, Margetts C, Taylor P (2012) The role of edible mushrooms in health: evaluation of the evidence. J Funct Foods 4:687–709

Ruthes AC, Smiderle FR, Iacomini M (2015) D-glucans from edible mushrooms: a review on the extraction, purification and chemical characterization approaches. Carbohydr Polym 117:753–761

Shao S, Hernandez M, Kramer JKG, Rinker DL, Tsao RTJ (2010) Ergosterol profiles, fatty acid composition, and antioxidant activities of button mushrooms as affected by tissue part and developmental stage. J Agric Food Chem 58:11616–11625

Shi YL, Benzie IF, Buswell JA (2002) Role of tyrosinase in the genoprotective effect of the edible mushroom. *Agaricus bisporus*. Life Sci 70:1595–1608

Singdevsachan SK, Auroshree P, Mishra J, Baliyarsingh B, Tayung K, Thatoi H (2016) Mushroom polysaccharides as potential prebiotics with their antitumor and immunomodulating properties: a review. Bioact Carbohydr Dietary Fibre 7:1–14

Singh SS, Wang H, Chan YS et al (2015) Lectins from edible mushrooms. Molecules 20:446–469

Słomiński A, Semak I, Zjawiony J et al (2005) Enzymatic metabolism of ergosterol by cytochrome P450scc (CYP11A1) to biologically active 17α,24-dihydroxyergosterol. Chem Biol 12:931–939

Takei T, Yoshida M, Ohnishi-Kameyama M, Kobori M (2005) Ergosterol peroxide, an apoptosis-inducing component isolated from *Sarcodon aspratus* (Berk.) S. Ito. Biosci Biotechnol Biochem 69:212–215

Teichmann A, Dutta PC, Staffas A, Jägerstad M (2007) Sterol and vitamin D2 concentrations in cultivated and wild grown mushrooms: effects of UV irradiation. LWT Food Sci Technol 40:815–822

Tian Y, Zhao Y, Huang J, Zeng H, Zheng B (2016) Effects of different drying methods on the product quality and volatile compounds of whole shiitake mushrooms. Food Chem 197:714–722

Tsai SY, Wu TP, Huang SJ, Mau JL (2008) Antioxidant properties of ethanolic extracts from culinary-medicinal button mushroom *Agaricus bisporus* (J. Lange) Imbach (Agaricomycetideae) harvested at different stages of maturity. Int J Med Mushrooms 10:127–137

Varrot A, Basheer SM, Imberty A (2013) Fungal lectins: structure, function and potential applications. Curr Opin Struct Biol 23:678–685

Wang HX, Liu WK, Ng TB, Ooi VEC, Chang ST (1996) The immunomodulatory and antitumor activities of lectins from the mushroom *Tricholoma mongolicum*. Immunopharmacology 31:205–211

Wang H, Ng TB, Ooi VEC (1998) Lectins from mushrooms. Mycol Res 102:897–906

Wheat RW, Woodruff WW 3rd, Haltiwanger RS (1983) Occurrence of antigenic (species-specific?) partially 3-O-methylated heteromannans in cell wall and soluble cellular (nonwall) components of *Coccidioides immitis* mycelia. Infect Immun 41:728–734

Xu X, Yan H, Chen J, Zhang X (2011) Bioactive proteins from mushrooms. Biotechnol Adv 29:667–674

Yaoita Y, Yoshihara Y, Kakuda R, Machida K, Kikuchi M (2002) New sterols from two edible mushrooms, *Pleurotus eryngii* and *Panellus serotinus*. Chem Pharm Bull 50:551–553

Yilmaz N, Solmaz M, Türkekul İ, Elmastaş M (2006) Fatty acids composition in some wild edible mushrooms growing in the middle Black Sea region of Turkey. Food Chem 99:168–174

Yuan JP, Kuang HC, Wang JH, Liu X (2008) Evaluation of ergosterol and its esters in the pileus, gill, and stipe tissues of agaric fungi and their relative changes in the comminuted fungal tissues. Appl Microbiol Biotechnol 80:459–465

Zaidi KU, Ali AS, Ali SA (2014) Purification and characterization of melanogenic enzyme tyrosinase from button mushroom. Enzym Res 2014:120739

Zhang M, Cui SW, Cheung PCK, Wang Q (2007) Antitumor polysaccharides from mushrooms: a review on their isolation process, structural characteristics and antitumor activity. Trends Food Sci Technol 18:4–19

Zhang GQ, Wang YF, Zhang XQ, Ng TB, Wang HX (2010) Purification and characterization of a novel laccase from the edible mushroom *Clitocybe maxima*. Process Biochem 45:627–633

Zhu F, Du B, Bian Z, Xu B (2015) Beta-glucans from edible and medicinal mushrooms: characteristics, physicochemical and biological activities. J Food Compos Anal 41:165–173

Zong A, Cao H, Wang F (2012) Anticancer polysaccharides from natural resources: a review of recent research. Carbohydr Polym 90:1395–1410

Chapter 11
Genomics and Artificial Intelligence Working Together in Drug Discovery and Repositioning: The Advent of Adaptive Pharmacogenomics in Glioblastoma and Chronic Arterial Inflammation Therapies

Glaucia C. Pereira

Abstract The field of pharmacogenomics investigates how genomics may modulate pathological trends using information on both genotype and phenotype, with the aim of designing personalised healthcare. Homoeostasis is partially regulated through the expression of core protein groups whose functionality is determined at gene level and modulated by environmental factors. Harmful changes in physiology may promote several dis-functionalities. In prior work gene expression was used as a biomarker to assess both pathological propensity and disease progression. A growing body of pharmacogenomics research has developed new compounds, on one hand, and on the other, it has proposed novel therapeutic applications for the existing ones. Over the past decades, collective efforts have significantly increased the number of omics information available. However, efficient and deterministic in silico mechanisms that efficiently analyse and detect trends on the basis of often unknown and limited physiological information responding to challenging clinical questions are still lacking. In this context, computational automation via artificial intelligence methodologies has proven to be accurate, robust to noise, cost efficient, and dynamic dealing with massive databases and forecasting on the basis of the available information. Moreover, this set of computational techniques, based on well-established mathematical models, provide efficient ways of determining trends based on both a priori knowledge and dynamically acquired information, working successfully on incomplete datasets. Therefore, in this chapter we assess developmental similarities between two major causes of worldwide death: glioblastoma and chronic arterial inflammation; and discuss the potential applicability of two artificial intelligence approaches for drug discovery and repositioning. According to the

G.C. Pereira (✉)
Biotechnology, Icelandic Institute for Intelligent Machines, Reykjavik 101, Iceland

Department of Computer Sciences, Polytechnic Institute, University Autonoma of Madrid, Madrid 28049, Spain

Department of Bioengineering, Imperial College London, London SW7 2AZ, UK
e-mail: glaucia.daconceicaopereira12@alumni.imperial.ac.uk

S. Malik (ed.), *Biotechnology and Production of Anti-Cancer Compounds*,
DOI 10.1007/978-3-319-53880-8_11

World Health Organization (WHO) a glioblastoma multiform is the most malignant glial-type tumour (graded level IV in the WHO scale); and inflammatory diseases affecting the cardiovascular network are the cause of high mortality. As suggested, these two pathologies have several developmental similarities and share common genetic variants. Therefore, we additionally seek to discuss the main promoters presented in the current literature, aiming at benefiting from their similarities in drug discovery and repositioning, via automatic artificial intelligence pattern recognition, forecasting, and computational design.

Keywords Genomics • Artificial intelligence • Drug discovery • Drug repositioning • Adaptive pharmacogenomics • Glioblastoma • Chronic arterial inflammation • Genetic fingerprints • Deep neural networks • Reinforcement learning • Inflammatory signalling cascades • Cytokines • Transcription factors

Abbreviation

ADME	Absorption, distribution, metabolism, and excretion
ApoE	Apolipoprotein E
ATP	Adenosine triphosphate
ATRX	Alpha-thalassaemia X
CSNK2A1	Casein kinase 2 alpha 1
DCN	Deep convolutional networks
DLNN	Deep learning neural networks
DNA	Deoxyribonucleic acid
DNN	Deep neural networks
EGFR	Epidermal growth factor receptor
GRR	Glycine-rich regions
IDH	Isocitrate dehydrogenase
IL	Interleukin
LDL	Low-density lipoprotein
LPA	Apolipoprotein A
LZ	Leucine zipper
NfkB	Nuclear factor-kappa B
oxi-LDL	Oxidized low-density lipoprotein
PTEN	Phosphatase and tensin homolog
RHD	Rel homology domain
SMC	Smooth muscle cells
SMCs	Vascular smooth muscle cells
SPP	Salesperson problem
STAT3	Signal transducer and activator of transcription 3
TAD	Transactivation domain
TERT	Telomerase reverse transcriptase
TNP-470	Trinitrophenol 470

TP53	Tumour protein p53
VCAM1	Vascular cell adhesion molecule 1
WHO	World Health Organization

Introduction

According to statistics from the National Cancer Institute (*Cancer of the Brain and Other Nervous System—SEER Stat Fact Sheets* 2013) a glioblastoma multiform has an incidence of c. 3 out of 100,000 adults annually, accounting for about 52% of all reported primary brain cancer, with a critical survival time rate varying from a few weeks to c. 31 months following diagnosis, in at least two well-defined groups of patients (Louis et al. 2016). On the other hand, cardiovascular diseases are the main cause of global death. According to the WHO (2016), around 17,500,000 people died from cardiovascular diseases and related pathologies globally in 2012, accounting for about 31% of the global mortality. Chronic inflammation, more specifically chronic arterial inflammation, is a major cause of severe cardiovascular pathologies (Mozaffarian et al. 2016). In December 2015, the mentioned figures were still increasing (Cardiovascular Disease Statistics 2015—BHF 2015), besides the collective efforts to reduce the impact of chronic cardiovascular pathologies. This motivates discussion on the identification of biomarkers related to both chronic (arterial) inflammation and glioblastoma multiform. The aim is to debate the applicability of the reported similarities in drug discovery and repositioning, supported by artificial intelligence methodologies. Here, we present newest trends in pharmacogenomics and offer an overview of two major artificial intelligence methodologies largely applied to pattern recognition and forecasting—deep neural networks and reinforcement learning—with the purpose of reporting on and stimulating novel research in the field, while clarifying some of the most important trends currently investigated by the research community. Moreover, we seek to unveil the main benefits of correlating information related to genetic fingerprints supporting drug discovery and genotype-phenotype correlations and similarities that lead to the identification of innovative usage of well-established chemical compounds—drug repositioning. In section "Two Major Causes of Worldwide Mortality" glioblastoma formation and chronic inflammation are characterised, while biomarkers widely reported in the literature are presented. Genetic common variants associated with both glioblastoma multiform and chronic arterial inflammation are presented. In section "Computational Pattern Recognition and Forecasting" the main computational trends on which this work centers are discussed. Primary concepts underlying the most commonly used deep neural network methodology are presented and exemplified, and reinforcement learning is described under the perspective of a logistic model. In section "Drug Discovery and Repositioning: Proposing New Adaptive Mechanisms in Pharmacogenomics" literature findings on pharmacogenomics and artificial intelligence techniques are correlated, aiming at highlighting the potential of adaptive pharmacogenomics in the context of in silico drug discovery and repositioning. Finally, in section "Conclusions" a summary of this review is presented and future research directions indicated.

Two Major Causes of Worldwide Mortality

Drawbacks in Current Glioblastoma Therapies: Why Is It Difficult to Treat? What Characterises the Disease?

A glioblastoma multiform or simply glioblastoma is according to the World Health Organization (WHO) the most malignant glial-type tumour (Louis et al. 2016), graded level IV. These lesions are characterised by significant cells' nuclear pleomorphism, disform cytoplasms, and increasing cells' proliferation. This malignant formation shows significant reduction in cell morphological differentiation and remarkable necrotic expansion. As a result, these high-grade astrocytomas commonly form lipid-reach plaques surrounded by vasogenic oedemas, which result from local angiogenesis promoted by abnormal increase in tissue ATP demand (Crouch et al. 1993).

Proposing therapies in grade IV tumours is always a challenge, because of chemotherapy and radiotherapy resistance, along with the commonly challenging tumour locations. Therefore, deciding which combination of treatments may increase the survival rate among patients with glioblastoma multiform depends on several factors, including early diagnosis and assessment of the Karnofsky Score (Hulshof et al. 2001; Caloglu et al. 2009; Metellus et al. 2011; Stark et al. 2012; Kumar et al. 2013). Data extracted from Louis et al. (2016) is presented in Table 11.1. This illustrates the mean survival time for two age-based groups of patients diagnosed with glioblastoma, with and without isocitrate dehydrogenase (*IDH*) mutations. Moreover, the figures show genotypic alterations found in *IDH*-wild-type and *IDH*-mutant patients (*TERT, TP53, ATRX, EGFR*, and *PTEN*).

Numerous works have reported on *IDH* mutations found in low-grade gliomas augmenting the production of *2-hydroxyglutarate*. However, in grade IV glial-type tumours the role of *IDH* mutations is related to increasing gene methylation (Cohen et al. 2013; Killela et al. 2014). The current literature has increasingly suggested that epigenetics is a key factor in the development of a variety of human disorders, including cancer and chronic inflammation. DNA-methylation is a family of epigenetics-based modifications in gene expression, which while not implying underlying changes in DNA sequences results in phenotypic disorders. In glioblastomas, DNA-methylation is thought to be closely related to lowering the cells' differentiation potential (Thon et al. 2013; Lai et al. 2014). Moreover, it has been found that the overexpression of *epidermal growth factor (EGF) receptor/HER1* common variants in glioblastoma correlates with increasing cell proliferation, tissue invasion, neo-angiogenesis, tumour cell chemoresistance, and abnormal apoptosis (Gadji et al. 2009).

Table 11.1 Key characteristics of isocitrate dehydrogenase (*IDH*) wild-type and isocitrate dehydrogenase (*IDH*) mutant glioblastomas

	IDH-wild type	IDH-mutant
Synonym	Primary glioblastoma	Secondary glioblastoma
Precursor lesion	~90%	~10%
Mean age when diagnosed	~62 years old	~44 years old
Male-to-female ratio	1.42:1	1.05:1
Mean length of clinical history	4 months	15 months
Mean overall survival		
Surgery + radiotherapy	9.9 months	24 months
Surgery + radiotherapy + chemotherapy	15 months	31 months
Location	*Supratentorial*	*Preferentially frontal*
Necrosis	Extensive	Limited
TERT promoter mutations	72%	26%
TP53 mutations	27%	81%
ATRX mutations	Exceptional	71%
EGFR amplification	35%	Exceptional
PTEN mutations	24%	Exceptional

Extracted from Louis et al. (2016)—copyright permission obtained

In addition to epigenetics and gene-mutation-related disease progression, this fatal malignant lesion tends to invade white matter tracts, and proposing case-based therapies is particularly challenging whether the glioma infiltrates the corpus callosum, proliferating along both cerebral hemispheres (Naik et al. 2015).

Established Biomarkers Predicting the Progression of Chronic Arterial Inflammation

Chronic inflammatory processes have several similarities, regardless of the affected tissue. To exemplify, in both epithelial and endothelial tissues subjected to inflammatory signalling networks the chronic deposition of metabolites implicated in immune responses is observed (Hansson and Hermansson 2011; Foley 2013; Jovanovic et al. 2014). In general, inflammatory dysfunction leads to tissue remodelling, which is a physiological adaptation aiming at maintaining homeostasis. In arterial inflammation, following endothelium invasion, when plaque deposition at the intimal layer reaches c. 40% of the normal arterial circumference and the arterial wall expands for keeping the physiological blood flow rate, the growing plaque named atheroma migrates into the lumen, and the disease progresses towards severe stages, which is classified by the WHO based on the level of arterial constriction and plaque composition, resulting in different levels of propensity to rupture (Solberg et al. 1968; Lusis 2000; Kronzon 2006). Figure 11.1a illustrates a plaque-free arterial segment and Fig. 11.1b shows an arterial segment where plaque has invaded the arterial lumen.

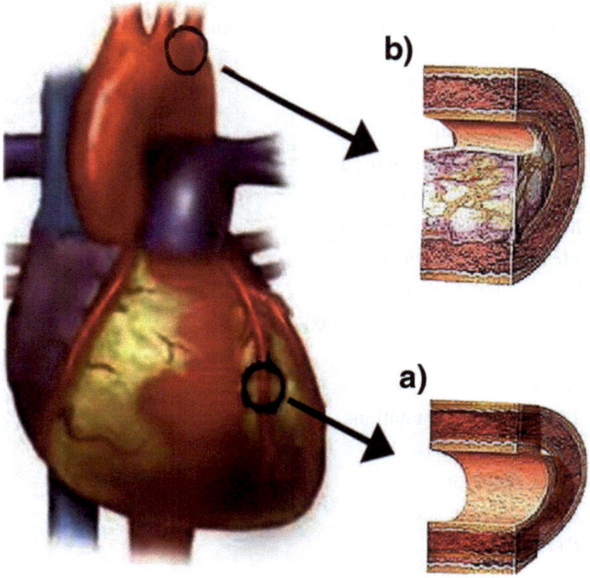

Fig. 11.1 Progression of coronary artery blockage by fat-rich material: (**a**) plaque-free region; and (**b**) arterial segment where atheroma formed

Among the main plaque compounds are macrophages resulting from *oxidized low-density lipoprotein* (*oxi-LDL*) rich monocyte differentiation and proliferating smooth muscle cells (SMC) that migrate from the tunica media into the intimal layer.

Over the last decades, numerous biomarkers (Atkinson et al. 2001) were identified and used to differentiate functional and dis-functional signalling leading to atheroma formation. To mention a few, hyperlipidaemia corroborates in raising the concentration of a major vulnerable plaques' compound (Ross and Harker 1976)—lipid; unbalance in the expression of anti- and pro-inflammatory cytokines (Van der Poll et al. 2001; Tedgui and Mallat 2006) as the ones found in the *interleukins*' family promotes chronic inflammatory conditions; and abnormal apoptosis (Kaiser et al. 1997; Rennier and Ji 2013) in regions of low wall shear stress (Slager et al. 2005; Reneman et al. 2006; Li et al. 2014) is thought to be a promoter in *low-density lipoprotein (LDL)* migration, by raising endothelial permeability. Ultimately, biomarker identification conducts to a better understanding of the dynamics underlying chronic inflammatory diseases, and the observed tendencies can be used to improve personalised therapies over the different phases of disease progression.

Similarities Between Glioblastoma and Chronic Arterial Inflammation Signalling Cascades

Chronic inflammation has been here described in the context of cardiovascular disorders. Unremarkably, patients suffering from myocardial infarct, stroke, and atherogenesis present a history of cholesterol-rich oxidative plaque deposition, in their arteries. However, that is not the only commonly found direction taken by inflammatory pathogenesis. Numerous forms of cancer are closely associated to an oxidative environment that promotes acute inflammatory responses (Tafani et al. 2011). In glioma formation and differentiation, the recent literature has reported on how the interactions between the immune system and the glioma may result in an immunosuppressed microenvironment, leading to tumour growth (Galvão and Zong 2013).

In diverse tissue, during inflammation, pro- and anti-inflammatory *cytokines* modulate signalling cascades via which the immune system eliminates the threat caused by infectious agents and local tissue impairment—this is an important aspect of the human physiology. However, resident inflammation leads to a chronic state that may cause diseases as atherosclerosis and tumour cells differentiation (Allavena et al. 2008; Mantovani et al. 2008; Grivennikov et al. 2010; Qian and Pollard 2010; He et al. 2012). In the current literature, it has been broadly reported the role of continuous exposure to oxidative stress and the collective effects of genotypic alterations in atherogenesis. However, tumorigenesis is thought to be highly correlated with similar genetic networks and cellular abnormal functioning. In chronic arterial inflammation, *oxidative-LDL* encapsulated in macrophages forms plaques that are targeted by the immune system raising homeostatic impairment. However, analysing the role of key biomarkers under different physiological environments is a challenge. Indeed, key promoters in cardiovascular threats enhanced by inflammation are suppressors in tumorigenesis, by triggering both *nuclear factor-kappa B (NfkB)* and *signal transducer and activator of transcription 3 (STAT3)* activation (Galvão and Zong 2013), which stimulate the expression of anti-apoptotic genes.

In atherogenesis therapies, which are mostly based on chronic arterial inflammation antagonists, *NfkB* and *STAT3* signalling cascades can modulate abnormal apoptosis and, consequently, reduce intercellular gaps responsible for enhancing the arterial wall migration of inflammatory metabolites. In tumorigenesis, this can induce cell death resistance, compromising the effectiveness of existing therapies aiming at tumour cell elimination. Furthermore, *NFkB* and *STAT3* pathways are thought to be related to an increase in cell proliferation. The underlying argument in favour of inflammation as a promoter in glioma progression is that insights were found that the *NFkB* inflammatory pathway causes proneural cells to differentiate towards mesenchymal glioblastoma (Bhat et al. 2013; Kim et al. 2013), which is the most aggressive and untreatable form of glioma. Indeed, as suggested above, the activation of *NfkB* and the underlying gene expression result in death-resistant cells that are less sensitive to radiotherapy. A schematic view of the *NfkB* family is shown in Fig. 11.2. The *n*-terminal defines the *nuclear factor-kappa B* family.

Mesenchymal glioblastoma differentiation is also associated with tumour angiogenesis, a metastasis promoter in diseased state. Moreover, the findings presented by Bhat et al. (2013) and their subsequent research at the University of Texas MD Anderson Cancer Center lend support to the claim that immune cells are key elements in glioblastoma progression towards its most aggressive form. These lesions are highly infiltrated by microglia, macrophages, and myeloid-derived suppressor cells. Therefore, understanding the tumour cell phenotype within glioblastoma subtypes

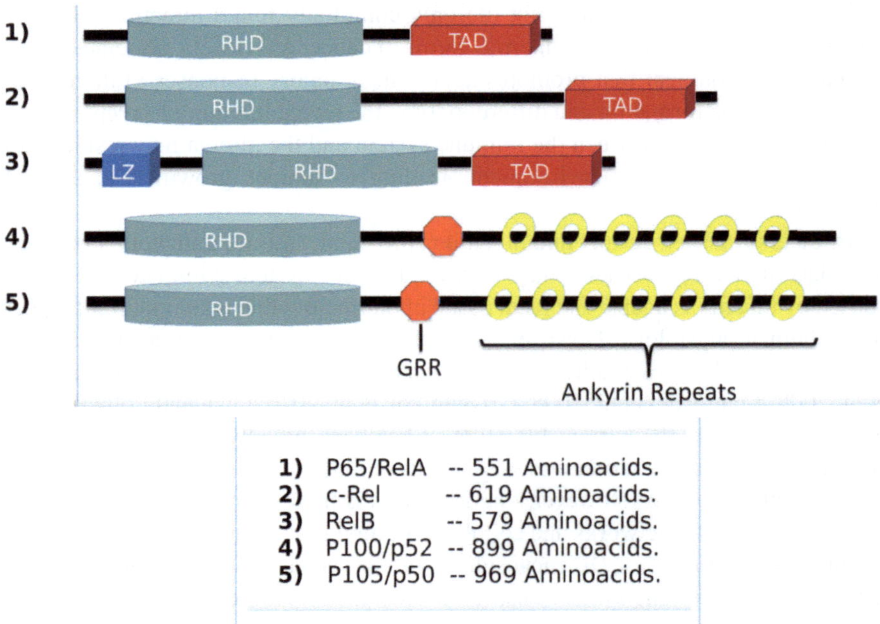

Fig. 11.2 Illustration of the *NfkB* family. The *NfkB* family members control the transcription of *cytokines* that regulate inflammatory signalling cascades, playing an important role in chronic inflammation, and regulate cellular differentiation, survival, and proliferation, thereby influencing various aspects of proneural cell differentiation into mesenchymal glioblastoma. The abbreviations *RHD*, *TAD*, *GRR*, and *LZ* stand for *Rel homology domain*, *transactivation domain*, *glycine-rich regions*, and *leucine zipper*, respectively

and associating the disease with different physiological conditions in which the immune system plays a fundamental role may lead to both discovering new targets leading to novel therapies and usage of existing compounds commonly applied to different disease treatments—drug discovery and repositioning.

Harmful Genotype Variants Associated with Glioblastoma and Chronic Arterial Inflammation

Genotype variants are among the most common mechanisms via which different species move forward along evolutionary stages. Genetic variants can occur between individuals from the same specie, and more locally among individuals in the same geographical location or community. Along these lines, the combination of alleles located on chromosomes determining a specific characteristic can change due to gene flow, genetic mutation, and meiotic recombination (Garcia-Ramos and Kirkpatrick 1997; Waples 1998; Clancy 2008; Wilkins and Holliday 2009). Via gene flow, genetic material is interchanged between populations through migratory processes, which contributes to the species evolutionary process by mixing together a set of beneficial genetic traits. Meiotic recombination makes it possible to raise the extent of directly inherited genetic variability during offspring built up, while mutations may occur either spontaneously or result from both micro- and macroscopic exposure to environmental factors as radiation.

As above suggested, genetic variability is fundamental for human evolution. However, regardless of the fact that natural selection reduces propensity for harmful genetic traits prevailing among consecutive generations (Barton and Partridge 2000; Kingsolver et al. 2001; Nielsen 2005; Barreiro et al. 2008; Beatty and Desjardins 2009; Ingvarsson 2010), making beneficial traits more likely to be expressed in a population, mutagens and naturally occurring *DNA* variation can lead to the propagation of existing harmful phenotypes and the onset of new harmful traits. In this context, over the past decades, numerous genes have been identified as key contributors in the initiation and progression of human pathologies as cancer, immune diseases, and neurological disorders.

Hypothesis underlying the concept of common variant is based on a few common allelic variants that may contribute to disease susceptibility, while rare variant stands for *DNA* sequence variations at any gene that can be found even once in the whole population. Genetic mapping based on disease common-common variant analysis may contribute with determining propensity to suffer from specific disorders, while disease common-rare variant may help understanding individual conditions towards designing personalised healthcare (Pritchard and Cox 2002; Peng and Kimmel 2007; Schork et al. 2009; Cirulli and Goldstein 2010; Robinson 2010). Indeed, according to the literature, commonly found genetic modifications related to coding and regulatory sequences are associated to complex polygenic diseases (Doris 2002; Miyawaki et al. 2012). In this case, individual gene variations will have cumulative or additive effects on the disease phenotype (Dominiczak and McBride 2003; Kathiresan et al. 2008; Purcell et al. 2009; Bullard et al. 2010; Lvovs et al. 2012; Dudbridge 2013; Escott-Price et al. 2015). Targeting genotype-phenotype relations in polygenic conditions is not as straightforward as this can be in single-gene inherited disorders (Lvovs et al. 2012; Dudbridge 2013). However, the literature has reported on several biomarkers supporting research in the field. *Apolipoprotein A (LPA)* variants have been associated to atherosclerosis (Aikawa and Schoen 2014), *apolipoprotein E (ApoE)* was found to be related to the development of Alzheimer (Corder et al. 1994; Bu 2009; Liu et al. 2013; Yu et al. 2014), and *interleukin families (IL)* were associated with a variety of chronic inflammatory diseases and cancer (Baggiolini and Clark-Lewis 1992; Sandborn 2000; Pizarro et al. 2006; Dinarello 2009, 2011; Dusatkova et al. 2009; Scheller et al. 2011; Tang et al. 2012; Interleukin-6 Receptor Mendelian Randomisation Analysis (IL6R MR) Consortium 2012; Egawa et al. 2013; Hölttä et al. 2013; Yao et al. 2014; Di Paolo and Shayakhmetov 2016).

Sahin et al. (2014) have reported on increasing levels of *IL17* serum mononuclear cell expression in active and inactive Crohn's disease patients to that of healthy individuals. This work shows results suggesting a correlation between overexpression of *IL17* in blood mononuclear cells and raising inflammation. Overall, the current literature supports the view that *IL17* expressed in T helper lymphocytes contributes to inflammatory signalling networks in several chronic immune disorders, such as multiple sclerosis, rheumatoid arthritis, systemic sclerosis, and psoriasis. Some biomarkers are well identified. However, the underlying mechanisms via which inflammation progresses into chronic disorders are still under investigation (Tesmer et al. 2008).

Another major study reporting on polymorphic-driven pathologies (Dudbridge 2013) suggests that ApoE ε4 allele carriers present high risk of developing cerebral amyloid angiopathy and age-associated cognitive impairment. The evidences presented thus far support the idea that common disease-common variant analysis plays an important role in forecasting propensity. One question that needs to be asked, however, is at each extent environmental factors and polygenicity play together, modulating risks.

Finally, throughout the former sections, we have presented numerous biomarkers involved in both chronic arterial inflammation and glioblastoma multiform formation. Along those lines, the interleukin family mentioned above is involved in important regulatory networks related to both the migration from acute to chronic arterial inflammation and proneural gene-regulatory cell differentiation, in mesenchymal glioblastoma formation. In a study on biomarkers associated to atherosclerosis progression, Zhang et al. (2013) found that *NfkB*-dependent *IL17* expression causes the expression of *vascular cell adhesion molecule 1 (VCAM1)* in aortic vascular smooth muscle cells (SMCs), which under disease condition stimulates the accumulation of inflammatory metabolites in the endothelium, while the arguments presented by Bhat et al. (2013) and Kim et al. (2013) suggest that similar interleukin family members are linked to the *NFkB* inflammatory pathways causing both mesenchymal glioblastoma formation and increasing cell resistance to radiotherapy undergoing apoptosis, as the result of damage caused by overexposure to mutagens like toxic chemicals and radiation. Together, these findings stimulate investigation on effective immune therapies repurposing for mitigating glioma progression.

Computational Pattern Recognition and Forecasting

What Computational Modelling Can Do in Biomedicine

Living beings are complex systems whose physiological balance depends on a series of interconnected processes (Marieb and Hoehn 2007; Sherwood 2010; Silverthorn 2010; Thiriet 2014), which are constantly monitored by regulatory mechanisms as the ones found in the immunological system. In computer sciences, this characterises a so-called mission-critical system (Nivel et al. 2014, 2015).

Models of living systems are based on complexity reduction, which depends on the establishment of a limited set of assumptions that makes it possible to mathematically describe the time-space behaviour of the underlying phenomena, via a well-defined closed set of equations.

At cellular level, cell division and differentiation undertake the coordinate accomplishment of several metabolic and biochemical processes, at molecular level (Lange 2005; Wang and Thampatty 2006). Therefore, in practice, determining the overall response to changes in the biological environment that regulates and promotes such processes would be impracticable. To overcome this drawback, variable selection and complexity reduction start in vivo, by identifying relevant biomarkers and variables of interest (Yang et al. 2005; Schwender et al. 2006; Li and Leder 2007), resulting in a better understanding of the phenomenon to be mathematically described and simulated in a computational environment.

Under the perspective of computational abstraction, deterministic strategies have proven to be capable of dealing with a wide variety of problems. Computational fluid mechanics has been widely applied in cardiovascular modelling, specially simulating blood flow characteristics that modulate wall shear stress, and consequently influencing signalling cascades associated with arterial inflammation (Yao et al. 2000; Steinman 2002; Kafi et al. 2017). However, the regulation of basic physiological phenomena depends on the capability of resisting and learning to adapt to fluctuations in both internal and external environments. This is among the strengths of major artificial intelligence techniques, as the artificial neural networks (Hopfield 1988; Gago et al. 2010; Amato et al. 2013; Schmidhuber 2015). In these circumstances determinism lacks in both flexibility and ability of dealing with either noisy or incomplete information. On the top of that, by the end of the twentieth and the beginning of this century, the world experienced a rapid increase in biotechnological innovation. Genomics contributed to significantly raise the number of available information on gene sequencing and their underlying functionality, via the encoded protein (Lange 2005; Hawkins et al. 2010; Graveley et al. 2011; Cerami et al. 2012; Xu et al. 2012; Killela et al. 2014; Weedon et al. 2014; Escott-Price et al. 2015; Selle and Barrangou 2015). Experimentally, gene-gene and protein-protein correlations have been determined via high-throughput methods. This provided the scientific community with valuable information that can be used to elucidate the characteristics of signalling cascades and regulatory networks involved in pathology (Chen and Xu 2003; Ng et al. 2003; Titz et al. 2004; Liu et al. 2005; Shoemaker and Panchenko 2007; Yu et al. 2012). For computational modellers, this is a complex scenery that may involve several interconnected factors, even after reduction of variables via data mining techniques routinely employed by data scientists. Additionally, as suggested above, even in very specific contexts, as in cancer cell's differentiation, incomplete information on signalling pathways underpinning pathogenicity requires the definition of computational models that are robust to noise and capable of handling the available data adaptively. Therefore, the relevance and applicability of computational tools depend on the characteristics of the biophysical phenomena that justify the model.

Deep Learning Neural Networks: What Is It All About?

Deep learning is a subfield of artificial intelligence (Lecun et al. 2015). In the context of artificial neural networks (Hsieh and Chen 2009; Cao et al. 2010a, b), the concept of deep learning is employed as a way of increasing the level of abstraction or the order of the interpolation the artificial neural network represents, which is used to approximate a mathematical function (Hornik et al. 1989; 1990; Hornik 1991; Park and Sandberg 1991; Debao 1993; Mhaskar 1996; Yingwei et al. 1997; Poggio and Girosi 2002; Ferrari and Stengel 2005; Yang et al. 2013) implicitly derived from correlations between dependent and independent variables. In general classification, whether realised via conventional or deep learning methods, the mentioned correlations are exemplified in the work undertaken by Lasko et al. (2013), in which longitudinal serum uric acid measurements are associated with gout versus the leukaemia phenotype signatures, for training a learning algorithm. A trained algorithm was there applied to clinical data in order to determine whether the gout or the leukaemia phenotype signature manifested in registered uric acid concentrations, over time, and to assess the accuracy of the method. In this example, deep learning neural networks (DLNN) may increase accuracy in predictions, by better representing non-linearities found in the behaviour of the underlying function that correlates both phenotypic targets and uric acid measurements. Moreover, deep learning offers the advantage of raising the level of model specialisation, without increasing the complexity of the related mathematical description, by separating complex problems into simpler and small-scale questions, in a hierarchical manner (Dietterich 2000; Craig and Tinaikar 2006; Lampert 2010; Lecun et al. 2015). This technique is thought to be based on the brain functional structure, where information is stored via simpler specialised units—the neurons—which together determine highly complex cognitive processes via synaptic communication, as illustrated in Fig. 11.3.

It is a common mistake to challenge artificial intelligence methodologies, particularly deep artificial neural networks; on the grounds that the existing methodologies lack clarity. The most common argument is based on the need for a formal explanatory description on the manner these networks cluster and classify data in different ways, along the lines of lacking logical outputted causal chains leading to the achieved solution. These observations must be interpreted with caution, because as for any mathematical model, the structure of the underlying tool is established via a reproducible algorithm, and the inherent decision chain subject to recording. Both theory and experimentation indicate that judging the relevance of keeping records on the decision processes highly depends on computational space complexity against explanatory needs (Moller and Smolka 1965; Mackworth and Freuder 1985; Cooper 1990; Nebel 1996; Blondel and Tsitsiklis 2000; Hutter 2000; Arora and Barak 2009; Bennett and Hauser 2013). On rigorous basis, theorem proofs are commonly based on direct deductions and reductio ad absurdum. In this case, a complete causal chain of logical arguments is presented. Every single step towards demonstrating a hypothesis to be true or false is supported by priorly built

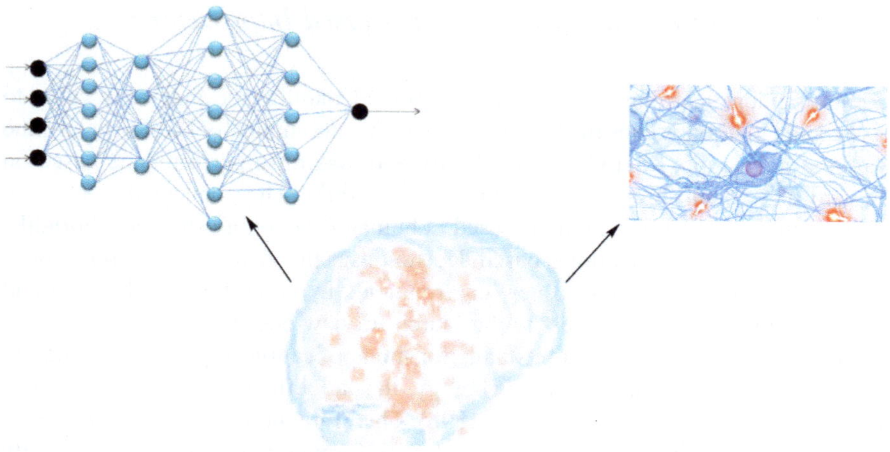

Fig. 11.3 The Artificial Neural Network paradigm: from brain to computational model nodes

knowledge and axioms—accepted as self-evident mathematical assertions. In deep learning theory, regardless of the application and underlying methodologies, the figures are not different. Convolutional neural networks for instance inherit the properties of the integrals, as for instance the commutability, associativeness, and distributiveness. These networks are subject to the characteristics of the convolved functions (f and g, as in $f \times g$), and depend on the continuity of the space on which f and g are defined (Liu et al. 1996; Mallat 2016). This technique became popular in pattern recognition and image classification for being space invariant, robust to both noisy data and incomplete information. Furthermore, raising efficiency may depend on conjugating the capabilities of different mathematical strategies, in an appropriate manner, giving rise to ensembled-like methods. The resulting technique may suffer from similar weaknesses that characterise the mathematical theory they derive from or form a more robust approach, because some drawbacks might have been resolved via the additional methodological strengths achieved.

In the current literature, there is a variety of works dedicated to the theoretical assessment of the mathematical tools from which deep neural networks derive (Lecun et al. 2015). Mallat (2016) presents a clear overview on the mode of action of deep convolutional networks (DCN), illustrating the most commonly found mathematical drawbacks that directly translate into DCN efficiency limitations. In Candes and Donoho (1999), the properties of multilayer neural networks with one single hidden layer are described via ridge function decomposition theory. According to Mallat (2016), this mathematical rigour does not extend to multi-hidden layer structures. Instead, deep neural networks generalise the former through linear convolutions followed by non-linearities (Le Cun et al. 1990; Ciresan et al. 2010, 2012; Mallat 2016). In general lines, caution has to be exercised, because often, the most serious disadvantages in using deep neural network methodologies are closely related to bounded rationality and the black-box approach, as this can lead to the inability of recognising inconsistencies in the achieved solution.

Reinforcement Learning: Experience-Based Improvement

Reinforcement learning is a branch of artificial intelligence that adds value to associated methodologies by defining the ways to automatically learn to determine optimal behaviour (Kaelbling et al. 1996), and maximise performance based on reward and penalisation (Dayan and Balleine 2002). The concept derives from behaviourism, which is commonly used in human psychology studies (Thorndike 1898). Reinforcement learning-based decision uses the idea that individuals optimise their way of adapting to new environments by combining exploration and exploitation. Exploration refers to the search for new possibilities within a given environment, while exploitation stands for the application of prior knowledge to perform the required tasks. In psychology, the law effect states that the consequences shape behaviour (Thorndike 1898). Therefore, human beings try different actions, assess their effects, and respond by reinforcing behaviour associated with positive outcomes—Fig. 11.4. Individuals adapt their course of action based on how their prior knowledge applies to new environments, in order to maximise benefits (Gallistel 2005).

In computational models, simple reward-feedback allows an intelligent algorithm to bring about instructions on how to behave adaptively. This reward is named reinforcement signal. The mentioned algorithm is required to contain steps responsible for deciding on the best course of action, based on the current state of the intelligent agent that this is associated to, and improve decision based on a weighting element representing either reward or penalty (Kaelbling et al. 1996; Hutter 2000; Sutton and Barto 2012). The main idea applies to different optimisation strategies, having long argued that the cost function is the reinforcement mechanism there applied (Byrne and Bogle 1999; Panagopoulos et al. 2002; Heckelei 2003;

Fig. 11.4 Reinforcement learning: illustration of how individuals respond to both positive and negative stimulus

Moeini and Afshar 2013; Helbig and Engelbrecht 2014; Brownlee and Wright 2015). However, while in the theory of optimisation reward and penalty are used to drive the search for an optimal solution, reinforcement learning applies this concept to support a mathematical entity to improve the search itself. Indeed, while traditional supervised learning receives pairs of input/output, associating this to new inputs in order to make conclusions, commonly found reinforcement learning strategies have a priori few or no information about the system of interest (Katehakis and Veinott 1987).

The well-known Boltzmann machine (Markov random fields) can be explained under the perspective of reinforcement learning, considering infinite horizon algorithms (Keerthi and Gilbert 1988; Hansen et al. 1996; Yasuda and Tanaka 2009; Fischer and Igel 2014). Let's assume that we are searching for the optimal solution of a problem that comprises in finding a route passing by n cities, starting and ending at the same location, whose cost in kilometres should be minimum. In optimisation, this problem is named "the salesperson problem" (SPP), and its model highly applied in logistics, operational research, and theoretical computer science (Tang and Miller-Hooks 2005). To date, at each realisation, by increasing the number n of cities, the computational time for finding a solution for the SPP may grow exponentially in n, which characterises the algorithms solving this problem as nondeterministic polynomial (Garey and Johnson 1979). Giving an initial candidate for a solution, conventional gradient-descent algorithms search for a candidate that improves the former guess, at each resolution (search) step (Hopfield 2007). However, finding the optimal solution highly depends on the characteristics of the search space. That is, this approach works perfectly well in concave-surface-like spaces resulting in one single global extrema. However, if the search space is characterised by a set of local maxima and minima, the gradient-descent-based approaches may stagnate at a local extrema and never find the optimal result. To overcome this drawback, a Boltzmann machine uses an adaptive search called simulated annealing, which uses distribution of probabilities to decide on whether a guess may be accepted as a candidate for solution (Tsallis 1988; Fischer and Igel 2014). In this case, suboptimal guesses are occasionally accepted which may allow the algorithm to escape from local extrema or stagnation points. Therefore, an apparently bad decision will actually maximise the reward (reduce the cost), in the long term, while the classical gradient descent is based on immediate cost-reward relation (Lima et al. 2005a, b). Reinforcement learning offers the capability of using this response to adjust the search. This adaptive technique comprises a graph and a search strategy. The graph contains the geometric representation of the problem—each node corresponds to a city and the weighted edges contain paths associated to given distances. The search strategy uses the topological information given by the graph and the Boltzmann-Gibbs distribution of probabilities, to learn how to estimate the value of a particular node state while adjusting to the neighbour nodes and influencing its neighbours themselves. This fragmented influence is propagated over time, maximising rewards—or equivalently minimising costs. Now, let's assume that either the whole paths in the graph are unknown a priori or the configuration may change dynamically, over time. As a result, new information is available dynamically, as in the classical problem of the gambler in a row of slot machines. In such case, the probabilistic decision-making process described above has to adjust to the environmental changes here defined.

Drug Discovery and Repositioning: Proposing New Adaptive Mechanisms in Pharmacogenomics

The Role of Genotype-Phenotype Variants in Pharmacology and Drug Effectiveness

Following the advent of the human genome, omics has added great value to clinics. Different genetic inherited diseases have distinct fingerprints in their copy number profiles, and, to date, variations in the chromosomal *DNA* sequence have been associated to different types of cancer (Wang et al. 2005; Andersson et al. 2010; Sanson et al. 2011; Leary et al. 2012; Rajaraman et al. 2012; Vivanco et al. 2012; Bhat et al. 2013). In glioblastoma multiforme for instance, copy number variation occurs in a few chromosomes, while other malignant tumour formations are characterised by copy number variation across the whole genome. More recently, the scientific community have concentrated efforts in understanding the role of de novo mutations in genetically heterogeneous conditions. Okur et al. (2016) performed whole-exome sequencing for identifying neurodevelopmental disorders related to *casein kinase 2 alpha 1 (CSNK2A1)* variants. Their results indicated that de novo missense and canonical splice-site mutations in the *CSNK2A1* are associated with a variety of structural disorders including microcephaly. This is outstanding, as in the current literature, *CSNK2A1* variants have also been associated to different malignant tumours (Wang et al. 2010; Rabjerg et al. 2016; Ruiz-Narváez et al. 2016).

Thanks to collective efforts to unveil the genetic profile of diseases, a considerable amount of information is publicly available for scientific usage. The challenge now is on both the development and improvement of efficient correlations based on the available information, aiming at translating genotype-phenotype relations into clinics, either predicting if genetic fingerprints and mutations might influence how a patient responds to novel drug treatment or identifying potential combined therapies resulting from genetic similarities. According to Zanger et al. (2001, 2004, 2008; Anzenbacher and Zanger 2012), regardless of the fact genotype-phenotype relationships have been under investigation for decades, information about genotype-based drug absorption, distribution, metabolism, and excretion (ADME) has started being translated into clinics in the last 15 years. In his works, Zanger refers to the two main directions that research in pharmacogenomics has taken: basic science and clinics. The first lends to unveiling genotype-phenotype correlations. The second correlates genetic factors with drug response phenotypes, to develop novel tools in clinical care.

Combining and Proposing New Therapies Based on Genetic Traits and Diseases' Similarities: The Benefits of Using Adaptive Methods

Thus far, we have referenced established research and described recent findings in biomedical studies, with concentration in the role of genomics in glioblastoma multiform and chronic (arterial) inflammation. Moreover, we have argued that recent trends on biomedical research can be approached via computational methods, which may benefit from diseases' genetic fingerprint similarities, in order to forecast drug response phenotypes subject to different physiological environments. What follows is an account of recent studies concentrating in combining the referenced trends and techniques, in order to advance research in drug discovery and repositioning.

Several studies have recently presented promising results on the application of artificial intelligence to a variety of problems ranging from determining patient response to pharmaceutical compounds to the identification of novel usage for established therapies (Zhao et al. 2009; Napolitano et al. 2013; Williams et al. 2015; Aliper et al. 2016; Gawehn et al. 2016; Liang et al. 2016). Recently, Aliper et al. (2016) presented results on the application of deep neural networks (DNN) classifying drugs' therapeutic categories based on their transcriptional profiles. In this work, both gene-level transcriptome data and transcriptome data obtained via a scoring algorithm were used to identify pharmacological properties of pharmaceutical compounds across varied physiological conditions. These results contributed advancing recent developments on the application of deep learning, identifying key cellular responses associated with both chemical and genetic fingerprints. This may result in finding novel usage for established drugs, by combining multidisciplinary knowledge and information available thanks to worldwide efforts for providing relevant databases and new technologies—Fig. 11.5.

Along these lines, recent communication by Williams et al. (2015) reported on the benefits of using computational automation via intelligent adaptive algorithms, aiming at reducing both the time and the costs associated with the establishment of new therapies while complying with well-defined efficiency and safety policies. According to the authors, it urges providing solutions for alleviating the lack of treatments for diseases that are currently neglected due to costs, such as tropical and orphan diseases. In this work, Williams et al. present results on the application of econometric models demonstrating that artificial intelligence methodologies excel standard drug screening selecting compounds, in terms of both time and resources. The proposed methodologies that are designed and validated computationally give rise to standardised assays quickly and cheaply engineered via synthetic biology. This potentially results in more targets engineered within a given budget. Indeed, the authors report on repositioning a variety of drugs against specific biomarkers for tropical diseases. To exemplify, an anticancer compound named *TNP-470*—which is a synthetic analogue of the antibiotic fumagillin—was identified as an inhibitor of the dihydrofolate reductase enzyme

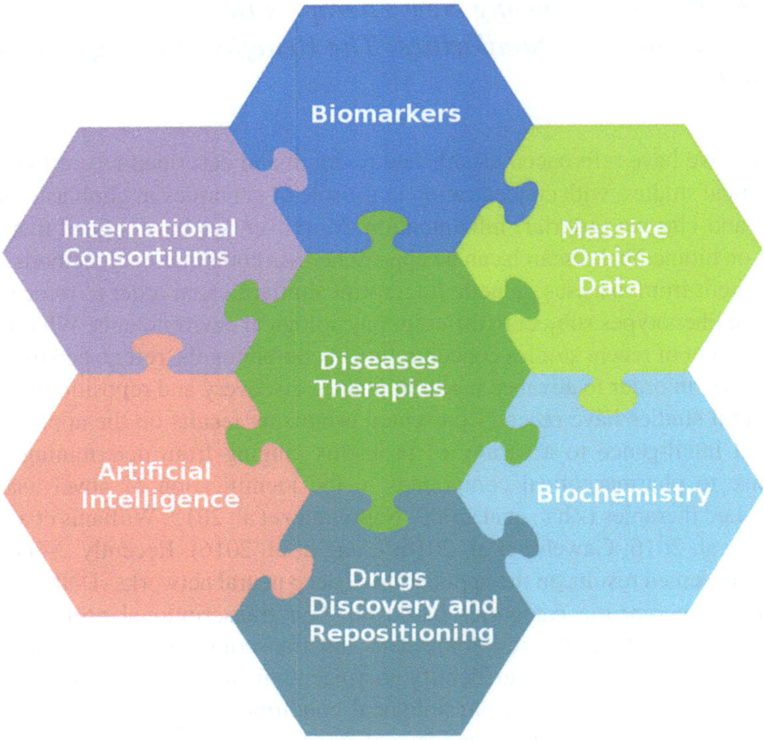

Fig. 11.5 Schematic view of key elements involved in the advent of novel therapies

synthase mediated by the malaria-causing parasite-specific additional sequences (inserts) in the dihydrofolate reductase-thymidylate domain.

By drawing on the concept of data mining and machine learning algorithms as variable selection and pattern recognition tools, the literature (Fox and Kriegl 2006; Wishart et al. 2006) has been able to show that computation adds value to drug discovery and repurposing, by automatically searching for molecules boosting response rates in cancer immunotherapy while dynamically combining information unknown a priori. Along these lines, throughout this chapter, we have discussed on genotypic factors that may contribute to identify efficient therapies for glioblastoma multiform, based on gene variants associated with inflammatory diseases, which are enhanced by dis-functional immune responses. General artificial intelligence methods are promising as tools for both correlating the mentioned biomarkers and assessing individual responses to chemical compounds. In our current research, a further question we particularly aim to investigate is the relevance of combining both the here mentioned well-established methodologies and their referred applications with the intelligent design of recombinant *DNA*, towards the development of novel compounds that benefit from treats derived from different organisms, including chemicals produced via secondary metabolism.

Conclusions

The aim of this chapter was to examine and discuss the biomarkers that characterise two major causes of the worldwide mortality—glioblastoma multiforme and chronic (arterial) inflammation. In this investigation, we assessed the benefits of correlating both genetic fingerprints that may lead to drug discovery and genotypic similarities resulting in potential drug repositioning. As a result, this study discussed state-of-the-art applications of artificial intelligence methodologies for in silico developments translated into advances in personalised health care.

We described the main characteristics of both glioma tumours and chronic (arterial) inflammation, and reported on results from the literature that present biomarkers for both glioblastoma and chronic inflammatory lesions. To mention a few, lipidemia-gene-promoter carriers are thought to be in risk of developing atherosclerosis, as a direct consequence of immune disorders; lipid-rich plaque progression correlates with the development of high-grade astrocytomas surrounded by vasogenic oedemas causing increase in local *ATP* demand; *isocitrate dehydrogenase* mutations may influence the average life expectancy in two age-based groups of patients diagnosed with glioblastoma; and both *cytokines* as the *IL* family and the *NfkB* signalling pathways have been correlated with progression from acute to chronic inflammation and proneural cell differentiation into mesenchymal glioblastoma. Moreover, we have pointed out that under varied disease condition, a promoter in chronic arterial inflammation may function as a suppressor in tumorigenesis, which requires caution during biomarker identification in drug repositioning.

One of the more significant findings illustrated in this study is that information about harmful genotype variants associated with glioblastoma and chronic arterial inflammation can be exploited via computational pattern recognition and forecasting tools in order to dynamically identify which chemical compounds have a potential for repurposing, determine in which case this can be applied, and forecast individual response to novel treatments.

Our final comment is on the fact that the physiological trends and methodologies described here may also have a fundamental role in the design of personalised healthcare combining the referenced methodologies and applications with the intelligent design of recombinant *DNA*, aiming at in silico development of novel compounds that can be engineered via laboratory assays benefiting from synthetic biology and organisms' secondary metabolism drug production.

Acknowledgments The author is grateful for all those who directly and indirectly contributed to her research. The author is particularly grateful for the support given by the sponsors and collaborators: Prof. Malik, UFMA (http://www.huufma.br/site/english/); Mr. James, GridPro (http://www.gridpro.com/); Landspitali (http://www.landspitali.is/um-landspitala/languages/english/); and the Icelandic Institute for Intelligent Machines (http://www.iiim.is).

References

Aikawa E, Schoen FJ (2014) Calcific and degenerative heart valve disease. In: Willis M, Homeister J, Stone J (eds) Cellular and molecular pathobiology of cardiovascular disease. Elsevier, Amsterdam. doi:10.1016/B978-0-12-405206-2.00009-0

Aliper A, Plis S, Artemov A, Ulloa A, Mamoshina P, Zhavoronkov A (2016) Deep learning applications for predicting pharmacological properties of drugs and drug repurposing using transcriptomic data. Mol Pharm 13(7):2524–2530. doi:10.1021/acs.molpharmaceut.6b00248

Allavena P, Sica A, Garlanda C, Mantovani A (2008) The Yin-Yang of tumor-associated macrophages in neoplastic progression and immune surveillance. Immunol Rev 222:155–161. doi:10.1111/j.1600-065X.2008.00607.x

Amato F, López A, Peña-Méndez EM, Vaňhara P, Hampl A, Havel J (2013) Artificial neural networks in medical diagnosis. J Appl Biomed 11:47–58. doi:10.2478/v10136-012-0031-x

Andersson U, Schwartzbaum J, Wiklund F, Sjostrom S, Liu Y, Tsavachidis S, Ahlbom A, Auvinen A, Collatz-Laier H, Feychting M, Johansen C, Kiuru A, Lonn S, Schoemaker MJ, Swerdlow AJ, Henriksson R, Bondy M, Melin B (2010) A comprehensive study of the association between the EGFR and ERBB2 genes and glioma risk. Acta Oncol 49(6):767–775. doi:10.3109/0284186X.2010.480980

Anzenbacher P, Zanger UM (2012) Metabolism of drugs and other xenobiotics. Wiley-VCH, Germany. doi:10.1002/9783527630905

Arora S, Barak B (2009) Computational complexity: a modern approach. Complexity 6(January):594

Atkinson AJ, Colburn WA, DeGruttola VG, DeMets DL, Downing GJ, Hoth DF, Oates JA, Peck CC, Schooley RT, Spilker BA, Woodcock J, Zeger SL (2001) Biomarkers and surrogate endpoints: preferred definitions and conceptual framework. Clin Pharmacol Ther 69(3):89–95. doi:10.1067/mcp.2001.113989

Baggiolini M, Clark-Lewis I (1992) Interleukin-8, a chemotactic and inflammatory cytokine. FEBS Lett:97–101. doi:10.1016/0014-5793(92)80909-Z

Barreiro LB, Laval G, Quach H, Patin E, Quintana-Murci L (2008) Natural selection has driven population differentiation in modern humans. Nat Genet 40(3):340–345. doi:10.1038/ng.78

Barton N, Partridge L (2000) Limits to natural selection. Bioessays 22(12):1075–1084. doi:10.1002/1521-1878(200012)22:12<1075::AID-BIES5>3.0.CO;2-M

Beatty J, Desjardins EC (2009) Natural selection and history. Biol Philos 24(2):231–246. doi:10.1007/s10539-008-9149-3

Bennett CC, Hauser K (2013) Artificial intelligence framework for simulating clinical decision-making: a Markov decision process approach. Artif Intell Med 57(1):9–19. doi:10.1016/j.artmed.2012.12.003

Bhat KPL, Balasubramaniyan V, Vaillant B, Ezhilarasan R, Hummelink K, Hollingsworth F, Wani K, Heathcock L, James JD, Goodman LD, Conroy S, Long L, Lelic N, Wang S, Gumin J, Raj D, Kodama Y, Raghunathan A, Olar A, Joshi K, Pelloski CE, Heimberger A, Kim SH, Cahill DP, Rao G, DenDunnen WFA, Boddeke HWGM, Phillips HS, Nakano I, Lang FF, Colman H, Sulman EP, Aldape K (2013) Mesenchymal differentiation mediated by NF-κB promotes radiation resistance in Glioblastoma. Cancer Cell 24(3):331–346. doi:10.1016/j.ccr.2013.08.001

Blondel VD, Tsitsiklis JN (2000) A survey of computational complexity results in systems and control. Automatica 36(9):1249–1274. doi:10.1016/S0005-1098(00)00050-9

Brownlee AEI, Wright JA (2015) Constrained, mixed-integer and multi-objective optimisation of building designs by NSGA-II with fitness approximation. Appl Soft Comput J 33:114–126. doi:10.1016/j.asoc.2015.04.010

Bu G (2009) Apolipoprotein E and its receptors in Alzheimer's disease: pathways, pathogenesis and therapy. Nat Rev Neurosci 10(5):333–344. doi:10.1038/nrn2620

Bullard JH, Mostovoy Y, Dudoit S, Brem RB (2010) Polygenic and directional regulatory evolution across pathways in Saccharomyces. Proc Natl Acad Sci U S A 107(11):5058–5063. doi:10.1073/pnas.0912959107

Byrne RP, Bogle IDL (1999) Global optimisation of constrained non-convex programs using reformulation and interval analysis. Comput Chem Eng 23(9):1341–1350. doi:10.1016/S0098-1354(99)00295-1

Caloglu M, Yurut-Caloglu V, Karagol H, Bayir-Angin G, Turan FN, Uzal C (2009) Prognostic factors other than the performance status and age for glioblastoma multiforme: a single-institution experience. J BUON 14(2):211–218

Cancer of the Brain and Other Nervous System—SEER Stat Fact Sheets (2013). https://seer.cancer.gov/statfacts/html/brain.html#risk. Accessed 10 Dec 2016

Candes EJ, Donoho DL (1999) Ridgelets: a key to higher-dimensional intermittency? Philos Trans R Soc A 357(1760):2495–2509. doi:10.1098/rsta.1999.0444

Cao F, Lin S, Xu Z (2010a) Approximation capability of interpolation neural networks. Neurocomputing 74(1–3):457–460. doi:10.1016/j.neucom.2010.08.018

Cao F, Yuan Y, Ding C (2010b) Exact and approximate interpolation for neural networks with single hidden layer. In: 2010 Sixth International Conference on Natural Computation, pp 273–277. doi: 10.1109/ICNC.2010.5583824

Cardiovascular Disease Statistics 2015 BHF (2015). https://www.bhf.org.uk/publications/statistics/cvd-stats-2015. Accessed 10 Dec 2016

Cerami E, Gao J, Dogrusoz U, Gross BE, Sumer SO, Aksoy BA, Jacobsen A, Byrne CJ, Heuer ML, Larsson E, Antipin Y, Reva B, Goldberg AP, Sander C, Schultz N (2012) The cBio cancer genomics portal: an open platform for exploring multidimensional cancer genomics data. Cancer Discov 2(5):401–404. doi:10.1158/2159-8290.CD-12-0095

Chen Y, Xu D (2003) Computational analyses of high-throughput protein-protein interaction data. Curr Protein Pept Sci 4:159–181. doi:10.2174/1389203033487225

Ciresan DC, Meier U, Gambardella LM, Schmidhuber J (2010) Deep big simple neural nets excel on handwritten digit recognition. Neural Comput 22(12):1–14. doi:10.1162/NECO_a_00052

Cireşan DC, Meier U, Gambardella LM, Schmidhuber J (2012) Deep big multilayer perceptrons for digit recognition. In: Lecture Notes in Computer Science (including subseries Lecture Notes in Artificial Intelligence and Lecture Notes in Bioinformatics). 7700 LECTU, pp 581–598. doi: 10.1007/978-3-642-35289-8-31

Cirulli ET, Goldstein DB (2010) Uncovering the roles of rare variants in common disease through whole-genome sequencing. Nat Rev Genet 11(6):415–425. doi:10.1038/nrg2779

Clancy S (2008) RNA Functions. Nat Educ 1(1):102

Cohen AL, Holmen SL, Colman H (2013) IDH1 and IDH2 mutations in gliomas. Curr Neurol Neurosci Rep 13(5):345. doi:10.1007/s11910-013-0345-4

Cooper GF (1990) The computational complexity of probabilistic inference using bayesian belief networks. Artif Intell 42(2–3):393–405. doi:10.1016/0004-3702(90)90060-D

Corder EH, Saunders AM, Risch NJ, Strittmatter WJ, Schmechel DE, Gaskell PC, Rimmler JB, Locke PA, Conneally PM, Schmader KE (1994) Protective effect of apolipoprotein E type 2 allele for late onset Alzheimer disease. Nat Genet 7(2):180–184. doi:10.1038/ng0694-180

Craig D, Tinaikar R (2006) Divide and conquer: rethinking IT strategy. McKinsey on IT:1–7

Crouch SPM, Kozlowski R, Slater KJ, Fletcher J (1993) The use of ATP bioluminescence as a measure of cell proliferation and cytotoxicity. J Immunol Methods 160(1):81–88. doi:10.1016/0022-1759(93)90011-U

Dayan P, Balleine BW (2002) Reward, motivation, and reinforcement learning. Neuron 36(2):285–298. doi:10.1016/S0896-6273(02)00963-7

Debao C (1993) Degree of approximation by superpositions of a sigmoidal function. Approx Theory Its Appl 9(3):17–28. doi:10.1007/BF02836480

Di Paolo NC, Shayakhmetov DM (2016) Interleukin 1α and the inflammatory process. Nat Immunol 17(8):906–913. doi:10.1038/ni.3503

Dietterich T (2000) The divide-and-conquer manifesto. In: Arimura H, Jain S, Sharma A (eds) Algorithmic learning theory. Springer, Berlin, pp 13–26. doi:10.1002/ieam.196

Dinarello CA (2009) Immunological and inflammatory functions of the interleukin-1 family. Annu Rev Immunol 27:519–550. doi:10.1146/annurev.immunol.021908.132612

Dinarello CA (2011) Interleukin-1 in the pathogenesis and treatment of inflammatory diseases. Blood 117(14):3720–3732. doi:10.1182/blood-2010-07-273417

Dominiczak AF, McBride MW (2003) Genetics of common polygenic stroke. Nat Genet 35(2):116–117. doi:10.1038/ng1003-116

Doris PA (2002) Hypertension genetics, single nucleotide polymorphisms, and the common disease: common variant hypothesis. Hypertension 39(2 Pt 2):323–331. doi:10.1161/hy0202.104087

Dudbridge F (2013) Power and predictive accuracy of polygenic risk scores. PLoS Genet 9(3):e1003348. doi:10.1371/journal.pgen.1003348

Dusatkova P, Hradsky O, Lenicek M, Bronsky J, Nevoral J, Kotalova R, Bajerova K, Vitek L, Lukas M, Cinek O (2009) Association of IL23R p.381Gln and ATG16L1 p.197Ala with Crohn disease in the Czech population. J Pediatr Gastroenterol Nutr 49(4):405–410. doi:10.1097/MPG.0b013e31819344ee

Egawa M, Mukai K, Yoshikawa S, Iki M, Mukaida N, Kawano Y, Minegishi Y, Karasuyama H (2013) Inflammatory monocytes recruited to allergic skin acquire an anti-inflammatory M2 phenotype via basophil-derived interleukin-4. Immunity 38(3):570–580. doi:10.1016/j.immuni.2012.11.014

Escott-Price V, Sims R, Bannister C, Harold D, Vronskaya M, Majounie E, Badarinarayan N, Morgan K, Passmore P, Holmes C, Powell J, Brayne C, Gill M, Mead S, Goate A, Cruchaga C, Lambert JC, Van Duijn C, Maier W, Ramirez A, Holmans P, Jones L, Hardy J, Seshadri S, Schellenberg GD, Amouyel P, Williams J (2015) Common polygenic variation enhances risk prediction for Alzheimer's disease. Brain 138(12):3673–3684. doi:10.1093/brain/awv268

Ferrari S, Stengel RF (2005) Smooth function approximation using neural networks. IEEE Trans Neural Netw 16(1):24–38. doi:10.1109/TNN.2004.836233

Fischer A, Igel C (2014) Training restricted Boltzmann machines: an introduction. Pattern Recogn 47(1):25–39. doi:10.1016/j.patcog.2013.05.025

Foley JF (2013) Focus issue: understanding mechanisms of inflammation. Sci Signal 6(258):eg2. doi:10.1126/scisignal.2003938

Fox T, Kriegl JM (2006) Machine learning techniques for in silico modeling of drug metabolism. Curr Top Med Chem 6(15):1579–1591. doi:10.2174/156802606778108915

Gadji M, Crous AMT, Fortin D, Krcek J, Torchia M, Mai S, Drouin R, Klonisch T (2009) EGF receptor inhibitors in the treatment of glioblastoma multiform: old clinical allies and newly emerging therapeutic concepts. Eur J Pharmacol 625(1–3):23–30. doi:10.1016/j.ejphar.2009.10.010

Gago J, Landín M, Gallego PP (2010) Strengths of artificial neural networks in modeling complex plant processes. Plant Signal Behav 5(6):743–745. doi:10.1016/j.jplph.2009.07.007

Gallistel CR (2005) Deconstructing the law of effect. Games Econ Behav 52(2):410–423. doi:10.1016/j.geb.2004.06.012

Galvão RP, Zong H (2013) Inflammation and gliomagenesis: bi-directional communication at early and late stages of tumor progression. Curr Pathobiol Rep 1(1):19–28. doi:10.1007/s40139-012-0006-3

Garcia-Ramos G, Kirkpatrick M (1997) Genetic models of adaptation and gene flow in peripheral populations. Evolution 51(1):21–28. doi:10.2307/2410956

Garey MR, Johnson DS (1979) Computers and intractability: a guide to the theory of np-completeness (Series of Books in the Mathematical Sciences). W. H. Freeman & Co, New York, p 340. doi:10.1137/1024022

Gawehn E, Hiss JA, Schneider G (2016) Deep learning in drug discovery. Mol Inform 35(1):3–14. doi:10.1002/minf.201501008

Graveley BR, Brooks AN, Carlson JW, Duff MO, Landolin JM, Yang L, Artieri CG, van Baren MJ, Boley N, Booth BW, Brown JB, Cherbas L, Davis CA, Dobin A, Li R, Lin W, Malone JH, Mattiuzzo NR, Miller D, Sturgill D, Tuch BB, Zaleski C, Zhang D, Blanchette M, Dudoit S, Eads B, Green RE, Hammonds A, Jiang L, Kapranov P, Langton L, Perrimon N, Sandler JE, Wan KH, Willingham A, Zhang Y, Zou Y, Andrews J, Bickel PJ, Brenner SE, Brent MR, Cherbas P, Gingeras TR, Hoskins RA, Kaufman TC, Oliver B, Celniker SE (2011) The developmental transcriptome of *Drosophila melanogaster*. Nature 471(7339):473–479. doi:10.1038/nature09715

Grivennikov SI, Greten FR, Karin M (2010) Immunity, inflammation, and cancer. Cell 140(6):883–899. doi:10.1016/j.cell.2010.01.025

Hansen LK, Andersen LN, Kjems U, Larsen J (1996) Revisiting Boltzmann learning: parameter estimation in Markov random fields. In: Proceedings of international conference on acoustics speech and signal processing.

Hansson GK, Hermansson A (2011) The immune system in atherosclerosis. Nat Immunol 12(3):204–212. doi:10.1038/ni.2001

Hawkins RD, Hon GC, Ren B (2010) Next-generation genomics: an integrative approach. Nat Rev Genet 11(7):476–486. doi:10.1038/nrg2795

He D, Li H, Yusuf N, Elmets CA, Athar M, Katiyar SK, Xu H (2012) IL-17 mediated inflammation promotes tumor growth and progression in the skin. PLoS One 7(2):e32126. doi:10.1371/journal.pone.0032126

Heckelei T (2003) Estimation of constrained optimisation models for agricultural supply analysis based on generalised maximum entropy. Eur Rev Agric Econ 30(1):27–50. doi:10.1093/erae/30.1.27

Helbig M, Engelbrecht AP (2014) Population-based metaheuristics for continuous boundary-constrained dynamic multi-objective optimisation problems. Swarm Evol Comput 14:31–47. doi:10.1016/j.swevo.2013.08.004

Hölttä V, Klemetti P, Salo HM, Koivusalo A, Pakarinen M, Westerholm-Ormio M, Kolho K-L, Vaarala O (2013) Interleukin-17 immunity in pediatric Crohn disease and ulcerative colitis. J Pediatr Gastroenterol Nutr 57(3):287–292. doi:10.1097/MPG.0b013e3182979252

Hopfield JJ (1988) Artificial neural networks. IEEE Circuits Dev Mag 4(5):3–10. doi:10.1109/101.8118

Hopfield J (2007) Hopfield network. Scholarpedia 2(5):1977. doi:10.4249/scholarpedia.1977

Hornik K (1991) Approximation capabilities of multilayer feedforward networks. Neural Netw 4(2):251–257. doi:10.1016/0893-6080(91)90009-T

Hornik K, Stinchcombe M, White H (1989) Multilayer feedforward networks are universal approximators. Neural Netw 2(5):359–366. doi:10.1016/0893-6080(89)90020-8

Hornik K, Stinchcombe M, White H (1990) Universal approximation of an unknown mapping and its derivatives using multilayer feedforward networks. Neural Netw 3(5):551–560. doi:10.1016/0893-6080(90)90005-6

Hsieh S-H, Chen C-H (2009) Adaptive image interpolation using probabilistic neural network. Expert Syst Appl 36(3):6025–6029. doi:10.1016/j.eswa.2008.06.124

Hulshof MC, Koot RW, Schimmel EC, Dekker F, Bosch DA, Gonzalez Gonzalez D (2001) Prognostic factors in glioblastoma multiforme. 10 years experience of a single institution. Strahlenther Onkol 177(6):283–290

Hutter M (2000) A theory of universal artificial intelligence based on algorithmic complexity. arXiv preprint cs/0004001:1–8. doi: 10.1007/b138233

Ingvarsson PK (2010) Natural selection on synonymous and nonsynonymous mutations shapes patterns of polymorphism in *Populus tremula*. Mol Biol Evol 27(3):650–660. doi:10.1093/molbev/msp255

Interleukin-6 Receptor Mendelian Randomisation Analysis (IL6R MR) Consortium (2012) The interleukin-6 receptor as a target for prevention of coronary heart disease: a mendelian randomisation analysis. Lancet 379(9822):1214–1224. doi:10.1016/S0140-6736(12)60110-X

Jovanovic K, Siebeck M, Gropp R (2014) The route to pathologies in chronic inflammatory diseases characterized by T helper type 2 immune cells. Clin Exp Immunol 178(2):201–211. doi:10.1111/cei.12409

Kaelbling LP, Littman ML, Moore AW (1996) Reinforcement learning: a survey. J Artif Intell Res 4:237–285. doi:10.1613/jair.301

Kafi O, Khatib NE, Tiago J, Sequeira A (2017) Numerical simulations of a 3D fluid-structure interaction model for blood flow in an atherosclerotic artery. Math Biosci Eng 14(1):179–193. doi:10.3934/mbe.2017012

Kaiser D, Freyberg MA, Friedl P (1997) Lack of hemodynamic forces triggers apoptosis in vascular endothelial cells. Biochem Biophys Res Commun 231(3):586–590. doi:10.1006/bbrc.1997.6146

Katehakis MN, Veinott AF Jr (1987) The multi-armed bandit problem: decomposition and computation. Math Oper Res 12(2):262–268. doi:10.1287/moor.12.2.262

Kathiresan S, Willer C, Peloso G, Demissie S (2008) Common variants at 30 loci contribute to polygenic dyslipidemia. Nat Genet 41(1):56–65. doi:10.1038/ng.291.Common

Keerthi SS, Gilbert EG (1988) Optimal infinite-horizon feedback laws for a general class of constrained discrete-time systems: stability and moving-horizon approximations. J Optim Theory Appl 57(2):265–293. doi:10.1007/BF00938540

Killela PJ, Pirozzi CJ, Healy P, Reitman, Z. J., Lipp, E., Rasheed, B. A., Yang, R., Diplas, B. H., Wang, Z., Greer, P. K., Zhu, H., Wang, C. Y., Carpenter, A. B., Friedman, H., Friedman, A. H., Keir, S. T., He, J., He, Y., McLendon, R. E., Herndon 2nd, J. E., Yan, H. and Bigner, D. D. (2014) Mutations in IDH1, IDH2, and in the TERT promoter define clinically distinct subgroups of adult malignant gliomas. Oncotarget 5(6):1515–1525. doi: 10.18632/oncotarget.1765

Kim YW, Koul D, Kim SH, Lucio-Eterovic AK, Freire PR, Yao J, Wang J, Almeida JS, Aldape K, Yung WKA (2013) Identification of prognostic gene signatures of glioblastoma: a study based on TCGA data analysis. Neuro-Oncology 15(7):829–839. doi:10.1093/neuonc/not024

Kingsolver JG, Hoekstra HE, Hoekstra JM, Berrigan D, Vignieri SN, Hill CE, Hoang A, Gibert P, Beerli P (2001) The strength of phenotypic selection in natural populations. Am Nat 157(3):245–261. doi:10.1086/319193

Kronzon I (2006) Aortic atherosclerotic disease and stroke. Circulation 114(1):63–75. doi:10.1161/CIRCULATIONAHA.105.593418

Kumar A, Kumar S, Vikram A, Hoffman TA, Naqvi A, Lewarchik CM, Kim YR, Irani K (2013) Histone and DNA methylation-mediated epigenetic downregulation of endothelial kruppel-like factor 2 by low-density lipoprotein cholesterol. Arterioscler Thromb Vasc Biol 33(8):1936–1942. doi:10.1161/ATVBAHA.113.301765

Lai RK, Chen Y, Guan X, Nousome D, Sharma C, Canoll P, Bruce J, Sloan AE, Cortes E, Vonsattel JP, Su T, Delgado-Cruzata L, Gurvich I, Santella RM, Ostrom Q, Lee A, Gregersen P, Barnholtz-Sloan J (2014) Genome-wide methylation analyses in glioblastoma multiforme. PLoS One 9(2):e89376. doi:10.1371/journal.pone.0089376

Lampert CH (2010) An efficient divide-and-conquer cascade for nonlinear object detection. In: Proceedings of the IEEE Computer Society Conference on Computer Vision and Pattern Recognition, pp 1022–1029. doi: 10.1109/CVPR.2010.5540107

Lange BM (2005) Single-cell genomics. Curr Opin Plant Biol 8(3):236–241. doi:10.1016/j.pbi.2005.03.015

Lasko TA, Denny JC, Levy MA (2013) Computational phenotype discovery using unsupervised feature learning over noisy, sparse, and irregular clinical data. PLoS One 8(6):e66341. doi:10.1371/journal.pone.0066341

Le Cun J, Boser B, Denker JS, Henderson D, Howard RE, Hubbard W, Cun LD, Le B, Denker J, Henderson D (1990) Handwritten digit recognition with a back-propagation network, Advances in neural information processing systems. Morgan Kaufmann Publishers, San Francisco, pp 396–404. doi:10.1111/dsu.12130

Leary RJ, Sausen M, Kinde I, Papadopoulos N, Carpten JD, Craig D, O'Shaughnessy J, Kinzler KW, Parmigiani G, Vogelstein B, Diaz LA, Velculescu VE, Shaughnessy JO, Kinzler KW, Parmigiani G, Vogelstein B, Diaz LA Jr, Velculescu VE (2012) Detection of chromosomal alterations in the circulation of cancer patients with whole-genome sequencing. Sci Transl Med 4(162):162ra154. doi:10.1126/scitranslmed.3004742

Lecun Y, Bengio Y, Hinton G (2015) Deep learning. Nature 521(1):436–444. doi:10.1038/nature14539

Li X, Leder P (2007) Identifying genes preferentially expressed in undifferentiated embryonic stem cells. BMC Cell Biol 8:37. doi:10.1186/1471-2121-8-37

Li X, Yang Q, Wang Z, Wei D (2014) Shear stress in atherosclerotic plaque determination. DNA Cell Biol 33(12):830–838. doi:10.1089/dna.2014.2480

Liang Z, Huang JX, Zeng X, Zhang G (2016) DL-ADR: a novel deep learning model for classifying genomic variants into adverse drug reactions. In: From IEEE International Conference on Bioinformatics and Biomedicine, pp 9–12. doi: 10.1186/s12920-016-0207-4

Lima PMV, Morveli-Espinoza MM, Pereira GC, Franga FMG (2005a) SATyrus: a SAT-based neuro-symbolic architecture for constraint processing. In: Fifth International Conference on Hybrid Intelligent Systems (HIS'05), p 6. doi: 10.1109/ICHIS.2005.97

Lima PMV, Pereira GC, Morveli-Espinoza MMM, França FMG (2005b) Mapping and combining combinatorial problems into energy landscapes via Pseudo-boolean Constraints. In: Proceedings of the First International Conference on Brain, Vision, and Artificial Intelligence. Berlin, Springer-Verlag (BVAI'05), pp 308–317. doi: 10.1007/11565123_30

Liu WK, Chen Y, Jun S, Chen J-S, Belytschko T, Pan C, Uras RA, Chang CT (1996) Overview and applications of the reproducing Kernel Particle methods. Arch Comput Meth Eng 3(1):3–80. doi:10.1007/BF02736130

Liu Y, Liu N, Zhao H (2005) Inferring protein—Protein interactions through high-throughput interaction data from diverse organisms. Bioinformatics 21(15):3279–3285. doi:10.1093/bioinformatics/bti492

Liu C-C, Liu C-C, Kanekiyo T, Xu H, Bu G (2013) Apolipoprotein E and Alzheimer disease: risk, mechanisms and therapy. Nat Rev Neurol 9(2):106–118. doi:10.1038/nrneurol.2012.263

Louis DN, Perry A, Reifenberger G, von Deimling A, Figarella-Branger D, Cavenee WK, Ohgaki H, Wiestler OD, Kleihues P, Ellison DW (2016) The 2016 World Health Organization Classification of Tumors of the Central Nervous System: a summary. Acta Neuropathol 131(6):803–820. doi:10.1007/s00401-016-1545-1

Lusis AJ (2000) Atherosclerosis. Nature 407(6801):233–241. doi:10.1038/35025203

Lvovs D, Favorova OO, Favorov AV (2012) A polygenic approach to the study of polygenic diseases. Acta Nat 4(3):59–71

Mackworth AK, Freuder EC (1985) The complexity of some polynomial network consistency algorithms for constraint satisfaction problems. Artif Intell 25(1):65–74. doi:10.1016/0004-3702(85)90041-4

Mallat S (2016) Understanding deep convolutional networks. arXiv:1–17. doi: 10.1063/1.4939230

Mantovani A, Mantovani A, Allavena P, Allavena P, Sica A, Sica A, Balkwill F, Balkwill F (2008) Cancer-related inflammation. Nature 454(7203):436–444. doi:10.1038/nature07205

Marieb EN, Hoehn K (2007) Human anatomy and physiology. Physiology 7:664–668. doi:10.1038/nnano.2011.234

Metellus P, Nanni-Metellus I, Delfino C, Colin C, Tchogandjian A, Coulibaly B, Fina F, Loundou A, Barrie M, Chinot O, Ouafik L, Figarella-Branger D (2011) Prognostic impact of CD133 mRNA expression in 48 glioblastoma patients treated with concomitant radiochemotherapy: a prospective patient cohort at a single institution. Ann Surg Oncol 18(10):2937–2945. doi:10.1245/s10434-011-1703-6

Mhaskar HN (1996) Neural networks for optimal approximation of smooth and analytic functions. Neural Comput 8(1):164–177. doi:10.1162/neco.1996.8.1.164

Miyawaki S, Imai H, Takayanagi S, Mukasa A, Nakatomi H, Saito N (2012) Identification of a genetic variant common to moyamoya disease and intracranial major artery stenosis/occlusion. Stroke 43(12):3371–3374. doi:10.1161/STROKEAHA.112.663864

Moeini R, Afshar MH (2013) Constrained Ant Colony Optimisation Algorithm for the layout and size optimisation of sanitary sewer networks. Urban Water J 10(3):154–173. doi:10.1080/1573062X.2012.716445

Moller F, Smolka SA (1965) On the computational complexity of algorithms. Trans Am Math Soc 117:285. doi:10.2307/1994208

Mozaffarian D, Benjamin EJ, Go AS, Arnett DK, Blaha MJ, Cushman M, Das SR, de Ferranti S, Després J-P, Fullerton HJ, Howard VJ, Huffman MD, Isasi CR, Jiménez MC, Judd SE, Kissela BM, Lichtman JH, Lisabeth LD, Liu S, Mackey RH, Magid DJ, McGuire DK, Mohler ER, Moy CS, Muntner P, Mussolino ME, Nasir K, Neumar RW, Nichol G, Palaniappan L, Pandey DK, Reeves MJ, Rodriguez CJ, Rosamond W, Sorlie PD, Stein J, Towfighi A, Turan TN, Virani SS, Woo D, Yeh RW, Turner MB (2016) Heart disease and stroke statistics—2016 update. Circulation 133(4):e38–e360. doi:10.1161/CIR.0000000000000350

Naik PP, Somani RR, Wagulde S, Juvatkar P, Kale MK (2015) A review on—Glioblastoma multiform. Pharma Times 47(10):14–18

Napolitano F, Zhao Y, Moreira VM, Tagliaferri R, Kere J, D'Amato M, Greco D (2013) Drug repositioning: a machine-learning approach through data integration. J Cheminform 5(6):30. doi:10.1186/1758-2946-5-30

Nebel B (1996) Artificial intelligence: a computational perspective, Principles of knowledge representation, (AUGUST 1997). CSLI Publications, Stanford, pp 237–266

Ng SK, Zhang Z, Tan SH (2003) Integrative approach for computationally inferring protein domain interactions. Bioinformatics 19(8):923–929. doi:10.1093/bioinformatics/btg118

Nielsen R (2005) Molecular signatures of natural selection. Annu Rev Genet 39(1):197–218. doi:10.1146/annurev.genet.39.073003.112420

Nivel E, Thórisson KR, Steunebrink BR, Dindo H, Pezzulo G, Rodríguez M, Hernández C, Ognibene D, Schmidhuber J, Sanz R, Helgason HP, Chella A (2014) Bounded seed-AGI. In: Lecture Notes in Computer Science (including subseries Lecture Notes in Artificial Intelligence and Lecture Notes in Bioinformatics). Springer, Cham, pp 85–96. doi: 10.1007/978-3-319-09274-4_9

Nivel E, Thórisson KR, Steunebrink B, Schmidhuber J (2015) Anytime bounded rationality. In: Bieger J, Goertzel B, Potapov A (eds) Artificial general intelligence: 8th international conference, AGI 2015, AGI 2015, Berlin, Germany, July 22–25, 2015, Proceedings. Springer, Cham, pp. 121–130. doi: 10.1007/978-3-319-21365-1_13

Okur V, Cho MT, Henderson L, Retterer K, Schneider M, Sattler S, Niyazov D, Azage M, Smith S, Picker J, Lincoln S, Tarnopolsky M, Brady L, Bjornsson HT, Applegate C, Dameron A, Willaert R, Baskin B, Juusola J, Chung WK (2016) De novo mutations in CSNK2A1 are associated with neurodevelopmental abnormalities and dysmorphic features. Hum Genet 135(7):699–705. doi:10.1007/s00439-016-1661-y

Panagopoulos H, Åström KJ, Hägglund T (2002) Design of PID controllers based on constrained optimisation. IEE P-Contr Theor Ap 149(1):32. doi:10.1049/ip-cta:20020102

Park J, Sandberg IW (1991) Universal Approximation Using Radial-Basis-Function Networks. Neural Comput 3(2):246–257. doi:10.1162/neco.1991.3.2.246

Peng B, Kimmel M (2007) Simulations provide support for the common disease-common variant hypothesis. Genetics 175(2):763–776. doi:10.1534/genetics.106.058164

Pizarro TT, De La Rue SA, Cominelli F (2006) Role of interleukin 6 in a murine model of Crohn's ileitis: are cytokine/anticytokine strategies the future for IBD therapies? Gut 55(9):1226–1227. doi:10.1136/gut.2005.083121

Poggio T, Girosi F (2002) Networks for approximation and learning. Proc IEEE 78(9):1481–1497. doi:10.1109/5.58326

Pritchard JK, Cox NJ (2002) The allelic architecture of human disease genes: common disease-common variant...or not? Hum Mol Genet 11(20):2417–2423. doi:10.1093/hmg/11.20.2417

Purcell SM, Wray NR, Stone JL, Visscher PM, O'Donovan MC, Sullivan PF, Sklar P, Purcell Leader SM, Ruderfer DM, McQuillin A, Morris DW, O'Dushlaine CT, Corvin A, Holmans PA, Macgregor S, Gurling H, Blackwood DHR, Craddock NJ, Gill M, Hultman CM, Kirov GK, Lichtenstein P, Muir WJ, Owen MJ, Pato CN, Scolnick EM, St Clair D, Sklar Leader P, Williams NM, Georgieva L, Nikolov I, Norton N, Williams H, Toncheva D, Milanova V, Thelander EF, Sullivan P, Kenny E, Quinn EM, Choudhury K, Datta S, Pimm J, Thirumalai S, Puri V, Krasucki R, Lawrence J, Quested D, Bass N, Crombie C, Fraser G, Leh Kuan S, Walker N, McGhee KA, Pickard B, Malloy P, Maclean AW, Van Beck M, Pato MT, Medeiros H, Middleton F, Carvalho C, Morley C, Fanous A, Conti D, Knowles JA, Paz Ferreira C, Macedo A, Helena Azevedo M, Kirby AN, Ferreira MAR, Daly MJ, Chambert K, Kuruvilla F, Gabriel SB, Ardlie K, Moran JL (2009) Common polygenic variation contributes to risk of schizophrenia and bipolar disorder. Nature 460(7256):748–752. doi:10.1038/nature08185

Qian B-Z, Pollard JW (2010) Macrophage diversity enhances tumor progression and metastasis. Cell 141(1):39–51. doi:10.1016/j.cell.2010.03.014

Rabjerg M, Bjerregaard H, Halekoh U, Jensen BL, Walter S, Marcussen N (2016) Molecular characterization of clear cell renal cell carcinoma identifies CSNK2A1, SPP1 and DEFB1 as promising novel prognostic markers. APMIS 124(5):372–383. doi:10.1111/apm.12519

Rajaraman P, Melin BS, Wang Z, McKean-Cowdin R, Michaud DS, Wang SS, Chanock SJ (2012) Genome-wide association study of glioma and meta-analysis. Hum Genet 131(12):1877–1888. doi:10.1007/s00439-012-1212-0

Reneman RS, Arts T, Hoeks APG (2006) Wall shear stress—an important determinant of endothelial cell function and structure—in the arterial system in vivo: discrepancies with theory. J Vasc Res 43(3):251–269. doi:10.1159/000091648

Rennier K, Ji JY (2013) The role of death-associated protein kinase (DAPK) in endothelial apoptosis under fluid shear stress. Life Sci 93(5–6):194–200. doi:10.1016/j.lfs.2013.06.011

Robinson R (2010) Common disease, multiple rare (and distant) variants. PLoS Biol 8(1):e1000293. doi:10.1371/journal.pbio.1000293

Ross R, Harker L (1976) Hyperlipidemia and atherosclerosis. Science 193(4258):1094–1100. doi:10.1126/science.822515

Ruiz-Narváez EA, Lunetta KL, Hong C-C, Haddad S, Yao S, Cheng T-YD, Bensen JT, Bandera EV, Haiman CA, Troester MA, Ambrosone CB, Rosenberg L, Palmer JR (2016) Genetic variation in the insulin, insulin-like growth factor, growth hormone, and leptin pathways in relation to breast cancer in African-American women: the AMBER consortium. Npj Breast Cancer 2:16034

Sahin A, Calhan T, Cengiz M, Kahraman R, Aydin K, Ozdil K, Korachi M, Sokmen HM (2014) Serum interleukin 17 levels in patients with Crohn's disease: real life data. Dis Markers 2014. doi:10.1155/2014/690853

Sandborn WJ (2000) Therapy for Crohn disease. Curr Opin Gastroenterol 16(4):318–323. doi:10.1097/00001574-200007000-00005

Sanson M, Hosking FJ, Shete S, Zelenika D, Dobbins SE, Ma Y, Enciso-Mora V, Idbaih A, Delattre JY, Hoang-Xuan K, Marie Y, Boisselier B, Carpentier C, Wang XW, Di Stefano AL, Labussière M, Gousias K, Schramm J, Boland A, Lechner D, Gut I, Armstrong G, Liu Y, Yu R, Lau C, Di Bernardo MC, Robertson LB, Muir K, Hepworth S, Swerdlow A, Schoemaker MJ, Wichmann HE, Müller M, Schreiber S, Franke A, Moebus S, Eisele L, Försti A, Hemminki K, Lathrop M, Bondy M, Houlston RS, Simon M (2011) Chromosome 7p11.2 (EGFR) variation influences glioma risk. Hum Mol Genet 20(14):2897–2904. doi:10.1093/hmg/ddr192

Scheller J, Chalaris A, Schmidt-Arras D, Rose-John S (2011) The pro- and anti-inflammatory properties of the cytokine interleukin-6. Biochim Biophys Acta 1813(5):878–888. doi:10.1016/j.bbamcr.2011.01.034

Schmidhuber J (2015) Deep Learning in neural networks: an overview. Neural Netw 61:85–117. doi:10.1016/j.neunet.2014.09.003

Schork NJ, Murray SS, Frazer KA, Topol EJ (2009) Common vs. rare allele hypotheses for complex diseases. Curr Opin Genet Dev 19(3):212–219. doi:10.1016/j.gde.2009.04.010

Schwender H, Krause A, Ickstadt K (2006) Identifying interesting genes with siggenes. RNews 6:45–50

Selle K, Barrangou R (2015) Harnessing CRISPR-Cas systems for bacterial genome editing. Trends Microbiol 23(4):225–232. doi:10.1016/j.tim.2015.01.008

Sherwood L (2010) Human physiology: from cells to systems, Human physiology, 7th edn. Brooks/Cole, Pacific Grove, p 766

Shoemaker BA, Panchenko AR (2007) Deciphering protein-protein interactions. Part II. Computational methods to predict protein and domain interaction partners. PLoS Comput Biol 3(4):e43. doi:10.1371/journal.pcbi.0030043

Silverthorn D (2010) Human physiology. Pearson, London. doi:10.1016/j.brainres.2005.11.002

Slager CJ, Wentzel JJ, Gijsen FJH, Schuurbiers JCH, van der Wal AC, van der Steen AF, Serruys PW (2005) 'The role of shear stress in the generation of rupture-prone vulnerable plaques. Nat Clin Pract Cardiovasc Med 2(8):401–407. doi:10.1038/ncpcardio0274

Solberg LA, McGarry PA, Moossy J, Strong JP, Tejada C, Löken AC (1968) Severity of atherosclerosis in cerebral arteries, coronary arteries, and aortas. Ann N Y Acad Sci 149(2):956–973. doi:10.1111/j.1749-6632.1968.tb53849.x

Stark AM, Van De Bergh J, Hedderich J, Mehdorn HM, Nabavi A (2012) Glioblastoma: clinical characteristics, prognostic factors and survival in 492 patients. Clin Neurol Neurosurg 114(7):840–845. doi:10.1016/j.clineuro.2012.01.026

Steinman DA (2002) Image-based computational fluid dynamics modeling in realistic arterial geometries. Ann Biomed Eng 30(4):483–497. doi:10.1114/1.1467679

Sutton RS, Barto AG (2012) Reinforcement learning. Learning 3(9):322. doi:10.1109/MED.2013.6608833

Tafani M, Di Vito M, Frati A, Pellegrini L, De Santis E, Sette G, Eramo A, Sale P, Mari E, Santoro A, Raco A, Salvati M, De Maria R, Russo MA (2011) Pro-inflammatory gene expression in solid glioblastoma microenvironment and in hypoxic stem cells from human glioblastoma. J Neuroinflammation 8(1):32. doi:10.1186/1742-2094-8-32

Tang H, Miller-Hooks E (2005) Algorithms for a stochastic selective travelling salesperson problem. J Oper Res Soc 56(4):439–452. doi:10.1057/palgrave.jors.2601831

Tang C, Chen S, Qian H, Huang W (2012) Interleukin-23: as a drug target for autoimmune inflammatory diseases. Immunology 135(2):112–124. doi:10.1111/j.1365-2567.2011.03522.x

Tedgui A, Mallat Z (2006) Cytokines in atherosclerosis: pathogenic and regulatory pathways. Physiol Rev 86(2):515–581. doi:10.1152/physrev.00024.2005

Tesmer LA, Lundy SK, Sarkar S, Fox DA (2008) Th17 cells in human disease. Immunol Rev 223:87–113. doi:10.1111/j.1600-065X.2008.00628.x

Thiriet M (2014) Anatomy and physiology of the circulatory and ventilatory systems. In: Biomathematical and biomechanical modeling of the circulatory and ventilatory systems. Springer, New York, p 585. doi:10.1007/978-1-4614-9469-0

Thon N, Kreth S, Kreth FW (2013) Personalized treatment strategies in glioblastoma: MGMT promoter methylation status. OncoTargets Ther 6:1363–1372. doi:10.2147/OTT.S50208

Thorndike EL (1898) Animal intelligence: an experimental study of the associative processes in animals. Psychol Rev 2(4):1–107. doi:10.1097/00005053-190001000-00013

Titz B, Schlesner M, Uetz P (2004) What do we learn from high-throughput protein interaction data? Expert Rev Proteomics 1(1):111–121. doi:10.1586/14789450.1.1.111

Tsallis C (1988) Possible generalization of Boltzmann-Gibbs statistics. J Stat Phys 52(1–2):479–487. doi:10.1007/BF01016429

Van der Poll T, De Jonge E, Levi M (2001) Regulatory role of cytokines in disseminated intravascular coagulation. Semin Thromb Hemost 27(6):639–651. doi:10.1055/s-2001-18868

Vivanco I, Ian Robins H, Rohle D, Campos C, Grommes C, Nghiemphu PL, Kubek S, Oldrini B, Chheda MG, Yannuzzi N, Tao H, Zhu S, Iwanami A, Kuga D, Dang J, Pedraza A, Brennan CW, Heguy A, Liau LM, Lieberman F, Alfred Yung WK, Gilbert MR, Reardon DA, Drappatz J, Wen PY, Lamborn KR, Chang SM, Prados MD, Fine HA, Horvath S, Wu N, Lassman AB, DeAngelis LM, Yong WH, Kuhn JG, Mischel PS, Mehta MP, Cloughesy TF, Mellinghoff IK (2012) Differential sensitivity of glioma- versus lung cancer-specific EGFR mutations to EGFR kinase inhibitors. Cancer Discov 2(5):458–471. doi:10.1158/2159-8290.CD-11-0284

Wang JH-C, Thampatty BP (2006) An introductory review of cell mechanobiology. Biomech Model Mechanobiol 5(1):1–16. doi:10.1007/s10237-005-0012-z

Wang Y, Putnam CD, Kane MF, Zhang W, Edelmann L, Russell R, Carrión DV, Chin L, Kucherlapati R, Kolodner RD, Edelmann W (2005) Mutation in Rpa1 results in defective DNA double-strand break repair, chromosomal instability and cancer in mice. Nat Genet 37(7):750–755. doi:10.1038/ng1587

Wang Z, Liu H, Liu B, Ma W, Xue X, Chen J, Zhou Q (2010) Gene expression levels of CSNK1A1 and AAC-11, but not NME1, in tumor tissues as prognostic factors in NSCLC patients. Med Sci Monit 16(8):CR357–CR364

Waples RS (1998) Separating the wheat from the chaff: patterns of genetic differentiation in high gene flow species. J Heredity 89:438–450. doi:10.1093/jhered/89.5.438

Weedon MN, Cebola I, Patch A-M, Flanagan SE, De Franco E, Caswell R, Rodríguez-Seguí SA, Shaw-Smith C, Cho CH-H, Lango Allen H, Houghton JAL, Roth CL, Chen R, Hussain K, Marsh P, Vallier L, Murray A, International Pancreatic Agenesis Consortium, Ellard S, Ferrer J, Hattersley AT (2014) Recessive mutations in a distal PTF1A enhancer cause isolated pancreatic agenesis. Nat Genet 46(1):61–64. doi:10.1038/ng.2826

Wilkins AS, Holliday R (2009) The evolution of meiosis from mitosis. Genetics 181(1):3–12. doi:10.1534/genetics.108.099762

Williams K, Bilsland E, Sparkes A, Aubrey W, Young M, Soldatova LN, De Grave K, Ramon J, de Clare M, Sirawaraporn W, Oliver SG, King RD (2015) Cheaper faster drug development validated by the repositioning of drugs against neglected tropical diseases. J R Soc Interface 12(104):20141289. doi:10.1098/rsif.2014.1289

Wishart DS, Knox C, Guo AC, Shrivastava S, Hassanali M, Stothard P, Chang Z, Woolsey J (2006) DrugBank: a comprehensive resource for in silico drug discovery and exploration. Nucleic Acids Res 34(90001):D668–D672. doi:10.1093/nar/gkj067

World Health Organization (2016) WHO/Cardiovascular diseases (CVDs). WHO, Geneva

Xu X, Hou Y, Yin X, Bao L, Tang A, Song L, Li F, Tsang S, Wu K, Wu H, He W, Zeng L, Xing M, Wu R, Jiang H, Liu X, Cao D, Guo G, Hu X, Gui Y, Li Z, Xie W, Sun X, Shi M, Cai Z, Wang B, Zhong M, Li J, Lu Z, Gu N, Zhang X, Goodman L, Bolund L, Wang J, Yang H, Kristiansen K, Dean M, Li Y, Wang J (2012) Single-cell exome sequencing reveals single-nucleotide mutation characteristics of a kidney tumor. Cell 148(5):886–895. doi:10.1016/j.cell.2012.02.025

Yang YH, Xiao Y, Segal MR (2005) Identifying differentially expressed genes from microarray experiments via statistic synthesis. Bioinformatics 21(7):1084–1093. doi:10.1093/bioinformatics/bti108

Yang S, Ting TO, Man KL, Guan SU (2013) Investigation of neural networks for function approximation. Procedia Comput Sci 17:586–594. doi:10.1016/j.procs.2013.05.076

Yao H, Ang KC, Yeo JH, Sim EK (2000) Computational modelling of blood flow through curved stenosed arteries. J Med Eng Technol 24(4):163–168. doi:10.1080/03091900050163445

Yao X, Huang J, Zhong H, Shen N, Faggioni R, Fung M, Yao Y (2014) Targeting interleukin-6 in inflammatory autoimmune diseases and cancers. Pharmacol Ther 141(2):125–129. doi:10.1016/j.pharmthera.2013.09.004

Yasuda M, Tanaka K (2009) Approximate learning algorithm in Boltzmann machines. Neural Comput 21(11):3130–3178. doi:10.1162/neco.2009.08-08-844

Yingwei L, Sundararajan N, Saratchandran P (1997) A sequential learning scheme for function approximation using minimal radial basis function neural networks. Neural Comput 9(2):461–478. doi:10.1162/neco.1997.9.2.461

Yu X, Wallqvist A, Reifman J (2012) Inferring high-confidence human protein-protein interactions. BMC Bioinformatics 13(1):79. doi:10.1186/1471-2105-13-79

Yu J-T, Tan L, Hardy J (2014) Apolipoprotein E in Alzheimer's disease: an update. Annu Rev Neurosci 37:79–100. doi:10.1146/annurev-neuro-071013-014300

Zanger UM, Fischer J, Raimundo S, Stüven T, Evert BO, Schwab M, Eichelbaum M (2001) Comprehensive analysis of the genetic factors determining expression and function of hepatic CYP2D6. Pharmacogenetics 11(7):573–585. doi:10.1097/00008571-200110000-00004

Zanger UM, Raimundo S, Eichelbaum M (2004) Cytochrome P450 2D6: overview and update on pharmacology, genetics, biochemistry. Naunyn Schmiedebergs Arch Pharmacol 369(1):23–37. doi:10.1007/s00210-003-0832-2

Zanger UM, Turpeinen M, Klein K, Schwab M (2008) Functional pharmacogenetics/genomics of human cytochromes P450 involved in drug biotransformation. Anal Bioanal Chem 392(6):1093–1108. doi:10.1007/s00216-008-2291-6

Zhang H, Chen J, Liu X, Awar L, Mcmickle A, Bai F, Nagarajan S, Yu S (2013) IL-17 induces expression of vascular cell adhesion molecule through signalling pathway of NF-κB, but not Akt1 and TAK1 in vascular smooth muscle cells. Scand J Immunol 77(4):230–237. doi:10.1111/sji.12030

Zhao Y, Kosorok MR, Zeng D (2009) Reinforcement learning design for cancer clinical trials. Stat Med 28(26):3294–3315. doi:10.1002/sim.3720

Chapter 12
A Multiscale Haemorheological Computer-Based Model of Chronic Inflammation: An In-Depth Investigation of Erythrocytes-Driven Flow Characteristics in Atheroma Development

The Application of the Iterative Incompressible Immersed Boundary (ThreeIB) Method

Glaucia C. Pereira

Abstract The mortality caused by cardiovascular diseases is dramatically increasing. Atherosclerosis is amongst the main contributors to this extremely high cardiovascular disease mortality. Atherosclerosis is controlled by mechanical forces exerted by the flow of blood on the inner lining of arteries, the endothelium. In order to fight this lethal disease, a realistic computational model is required that offers an accurate understanding of the effect of blood flow on the arterial wall. In order to realistically describe complex blood flow patterns and their interaction with the arterial wall, we have developed an integrated computational technique that takes into account both the particulate cellular composition of the blood and the interactions between the particulate blood and the vessel wall, in macro-circulation. The cellular composition of the blood was modelled using a multiphase fluid dynamics method by computing both an Eulerian fluid domain for modelling blood plasma and a Lagrangian solid domain that represented the blood cells. Interactions of the blood with the vessel wall were realistically modelled using a novel iterative immersed boundary method. Both the multiphase technique and the immersed boundary method were validated by comparing obtained numerical results with the literature, and a high degree of similarity was found. Moreover, our multiscale integrated model was applied to characterise clinically relevant flow patterns from atherosclerotic arterial segments. Our model revealed that under realistic non-Newtonian multiphase

G.C. Pereira (✉)
Department of Bioengineering, Imperial College London, London SW7 2AZ, UK

Biotechnology, Icelandic Institute for Intelligent Machines (IIIM), Reykjavik 101, Iceland

Department of Computer Sciences, University Autonoma of Madrid, Madrid 28049, Spain
e-mail: glaucia.daconceicaopereira12@alumni.imperial.ac.uk

S. Malik (ed.), *Biotechnology and Production of Anti-Cancer Compounds*,
DOI 10.1007/978-3-319-53880-8_12

cell-containing blood flow conditions, the pressure at the stenosis site is c. 30% higher than predicted by single-phase Newtonian fluid dynamics model. This effect is most probably explained by a decrease of c. 28% in the peak velocity. The latest seems to be caused by the momentum interchange due to the collision of the particles (blood cells) with each other, with the vessel wall, and with the fluid phase (plasma). The most significant difference in velocities between the single-phase and the two-phase results was registered at the stenosis site. It was also found that high blood cell content (peak particle concentration) correlates with increasing flow laminarisation. This means a decrease in velocity, which is more evidenced at the site of the stenosis. Additionally, our findings indicated that wakes forming downstream of the stenosis might be much weaker in non-Newtonian blood flow simulations compared to the Newtonian models. Taken together, our mathematical model and the resulting computational results might offer a more accurate understanding of the effects of realistic blood flow patterns on atherosclerotic vessels. This is of crucial importance since this deadly disease is initiated and progressed by blood flow-related mechanical factors, such as the wall shear stress.

Keywords Haemorheology • Chronic inflammation • Non-Newtonian blood flow • Computational models of cardiovascular diseases • ThreeIB method • Computational forecast supporting clinics • Numerical methods • Immersed boundary • Iterative methods • Incompressibility • Multiphase flow

Abbreviations

Apoe	Apolipoprotein E
ApoE–/–	Homozygous apoE-deficient
ApoE+/–	Heterozygous apoE-deficient
ApoE+/+	Apolipoprotein E wild-type
CFL	Courant Friedrichs Lewy
DEM	Discrete element method
DNS	Direct numerical simulation
Hct	Haematocrit
IB	Immersed boundary
IBM	Immersed boundary method
LDL	Low-density lipoprotein
MRI	Magnetic resonance imaging
RANS	Reynolds Averaged Navier–Stokes
RBCs	Red blood cells
SCAD	Stable coronary arterial diseases
spd	Symmetric positive definite
TBL	Theory of the boundary layer
WSS	Wall shear stress

Introduction and Motivation

Atherosclerosis is a cardiovascular disease characterised by increasing arterial inflammation, which causes the blockage of the arterial lumen and the reduction of blood supply to important parts of the human body. This is amongst the main causes of global mortality (*Cardiovascular Disease Statistics 2015—BHF* 2015; British Heart Foundation 2016; WHO 2015; World Health Organization 2015, 2016). The initiation and development of atherosclerosis is driven by fluid-mechanical forces (Huang et al. 1995; Reese and Thompson 1998; Provenzano and Rutland 2002; Stone et al. 2003; Cunningham and Gotlieb 2005; Chatzizisis et al. 2007; Chiu et al. 2009; Lu and Kassab 2011). These forces are sensed by the endothelial cells and the resulting signals are either transduced into chemical responses or transmitted to the surroundings to regulate the cellular activity (Davies and Hollman 2002; Scarborough et al. 2010, 2011; Townsend et al. 2012; WHO 2012, 2013; World Health Organization 2012; JBS3 Board 2014; World Health Organization et al. 2014). Therefore, changes in flow patterns caused by the blood corpuscular compounds might reveal new relevant modifications in both the magnitude and the distribution of the mechanical forces sensed by the endothelium. Hence, understanding variations in blood flow characteristics under physiological conditions might yield new insights on atheroma development.

Jung et al. (2006) reported on changes in flow due to plasma–erythrocyte interaction, leading to important non-Newtonian effects, such as changes in the shear rate. However, in this study neither the arterial geometry nor the haematocrit is physiologically realistic. Prior research on geometrical features such as non-planarity and bifurcations indicates that the characteristics of the arterial network correlate with the flow patterns (Frazin et al. 1990; Moore et al. 1994; Sherwin et al. 2000; Giessibl 2003). To assess whether computational models based on physiological haematocrit and arterial shape perform better than Newtonian models that neglect these characteristics of the in vivo arterial network, we addressed the coupled effects of the realistic arterial geometry and the non-Newtonian assumption. Indeed, both idealized models of artery and 3D image reconstruction-based realistic structures are used in our study to represent the arterial segments. Moreover, the physiological haematocrit is also used. Furthermore, while Zhao et al. (2010) assess the effects of the blood particles in micro-circulation, we simulate the effect of erythrocytes on flow patterns in large arteries which are predilections sites for atheroma development. Therefore, we consider this study of importance, because it offers a realistic and clinically relevant understanding of atherosclerosis development.

In the literature, the reported changes in flow due to plasma–erythrocytes interactions (Jung et al. 2006; Ii et al. 2011; Zastawny et al. 2012; Chabi et al. 2015) lead to non-Newtonian effects that we address taking into account the effects of different levels of particle concentration—growing haematocrit. To the best of our knowledge, to date, there are few haemorheological studies that attempt to simulate the progression of atherosclerosis. Indeed, as suggested above, the great majority of general haemorheological flow analyses focuses on either experimentation or in computation at micro-circulation (Moore et al. 1994; Jung et al. 2006; Zhao et al. 2010). However, this does not provide significant advancement in the understanding of the mechanisms that drive atherosclerotic plaques development and its side effects, in silico; because atherosclerosis develops in large arteries.

In this work, we introduce a multiscale rheological model of blood flow in large arteries, in order to understand the relevance of haemorheology in computational models of atherosclerosis. Indeed, an in-depth assessment of flow characteristics modified by the blood particles is presented. Our model comprises an established multiphase technique (Tsuji et al. 1992; Crowe et al. 1998; Tsuji 2007), which is used to represent haemorheology, and a new immersed boundary (IB) method, to address the blood–artery interactions. This new fluid–structure interaction methodology for generalized incompressible flow derives from Lai and Peskin (2000) and Uhlmann (2005), was developed by the author Pereira in (2013), improved in Pereira et al. (2014), and is here advanced and coupled with the above-described multiphase approach. The multiphase modelling derives from Crowe et al. (1998), Tsuji (2007) and Zastawny et al. (2012). The coupling multiphase-IB is here applied to an in-depth investigation of erythrocytes-driven flow characteristics in atheroma development. As mentioned previously, our computational predictions are supported by both numerical and experimental findings from the literature (Roshko 1953; Berger and Wille 1972; Rogers and Kwak 1990; Belov et al. 1995; Huang et al. 1995; Liu et al. 1996; Reese and Thompson 1998; Lai and Peskin 2000; Provenzano and Rutland 2002; Williamson and Govardhan 2004; Linnick and Fasel 2005). The goal of this study is to address and resolve issues inherent to current computational models of blood flow, at predilection sites for atheroma development.

We have performed both single-phase (plasma) and two-phase (plasma–erythrocytes) simulations at predilection sites for atherosclerotic plaques development. We have found that the presence of erythrocytes can change the velocity distribution and its magnitude in a range between 5 and 45%, in function of the cross-sectional particle volume fraction. This correlates with the shear rate and, consequently, with the shear forces sensed by the endothelium. These shear forces control an array of cellular activities (Caro et al. 1971; SenBanerjee et al. 2004; Cheng et al. 2006; Tedgui and Mallat 2006; Davies 2009; Liu et al. 2012; Pereira 2016). Our results indicate that rising flow laminarization occurs as a result of increasing haematocrit, while the flow vorticity diminishes. On the contrary, both apparent viscosity and wall shear stress rise with rising particles volume fraction. This difference in the results indicates that blood rheology might be an important aspect of models of plaque progression leading to significant contributions in translational and clinical research.

The following sections are structured as follows. Firstly, we will summarise the literature on atherosclerosis development that is relevant for the models presented here. Secondly, we will introduce the main methodologies used in our study. Thirdly, we will present and discuss the results on both the validation and the application of our computational model of haemorheology. Fourthly, we will conclude by pointing out the issues that we have addressed, and mention what could be explored in future research. The goal in following this structure is to show where we stand for in the current literature and illustrate both our motivations and contributions to the field.

The State of the Art

Atherosclerosis

In the past decades, atherosclerosis and, more specifically, the mechanisms related to the development of atheroma have been the focus of numerous works. According to the literature, plaque develops preferentially near arterial branches and curvatures (Frazin et al. 1990; Moore et al. 1994; Sherwin et al. 2000; Jung et al. 2006; Alsheikh-Ali et al. 2010; Weber and Noels 2011) in humans. Animal models of atherosclerosis require the induction of plaque formation by genetic modifications—the physiological conditions necessary for the development of atherosclerosis are imposed through the use of transgenic species, the use of a fat rich diet, and the surgical reduction of the arterial diameter through the use of implanted cuffs, in regions that are not usually affected by plaque deposition. Moreover, as indicated above, animal models of atherosclerosis are commonly found in the literature, because the whole plaque's developmental cycle can be observed and studied within a reasonable timescale (Moore et al. 1994; Weber and Noels 2011). The goal in these animal model studies is to change flow patterns locally, in a way that the required biomechanical environment is artificially created. Researchers from diverse areas with different purposes—from finding bio-factors-related insights to uncover the mechanics of the circulation—have presented a variety of hypothesis on the mechanisms driving plaque growth, at different stages. Such hypotheses refer to, for instance, low-density lipoprotein (LDL) as a transporter of lipids from the liver to the circulation and how it interacts with the diseased endothelium, mainly in the initial stages of the disease.

In humans, according to the literature, atherosclerotic plaques can be broadly divided into two main categories: vulnerable (aka. rupture prone, unstable) and stable (aka. fibrous) plaques (Cheng et al. 2006; Alsheikh-Ali et al. 2010; Maurovich-Horvat et al. 2014). As the plaque develops it invades the lumen when it reaches c. 40% of the arterial circumference. At or after this stage, vulnerable plaques rupture and the dripping bio-material can block blood flow in downstream narrower vessel segments, leading to heart attack, stroke and gangrene. Alsheikh-Ali et al. in a 2010 review paper (Alsheikh-Ali et al. 2010) state that amongst the factors that primarily indicate a higher risk of rupture are a lipid core accounting to more than 40% of the total lesion area, a fibrous cap with a thickness less than 65 μm, a volume of macrophages of about 26%, and an intensive inflammatory process. On the other hand, stable plaques are characterised by a thick cap consisting of collagen fibres covering the lesion (Alsheikh-Ali et al. 2010). In general, the main components of plaques are: necrotic cells, foam cells (thought to be lipid-loaded macrophages), lipids including cholesterol crystals collagen, smooth muscle cells, calcium, elastin and other immune cells.

As mentioned, local flow disturbances have been associated with plaque development (Frazin et al. 1990; Moore et al. 1994; Huang et al. 1995; Reese and Thompson 1998; Malek et al. 1999; Sherwin et al. 2000; Provenzano and Rutland

2002; Cunningham and Gotlieb 2005; Jung et al. 2006; Chatzizisis et al. 2007; Brown et al. 2016). Flow regions that present physiologically high values of wall shear stress—which in large human arteries is between 1.5 and 2 Pa—induce endothelial cells alignment. This pattern is related to a healthy and functional endothelium, in which intercellular gaps are kept under normal levels. These regions are said to be athero-protective. Locations with enhanced migration of macromolecules and cells into the arterial wall are coincident with regions of low shear (Caro et al. 1971; Cheng et al. 2006). At those locations, endothelial cells alignment is no longer properly achieved and the process by which the cells are normally renewed (apoptosis) presents critical abnormalities, such as the acceleration of the cells removal, without their subsequent replacement. As a result, intercellular gaps are enlarged, which promotes the migration of macromolecules and cells that are related to plaque build-up.

Oscillatory shear regions are associated with plaque stabilisation (Caro et al. 1971; Cheng et al. 2006). At those locations, the plaques are characterised by a thicker fibrous cap, which makes them less prone to rupture (Alsheikh-Ali et al. 2010). At that stage in plaque development, the plaque growth might be already contained and the risk of plaque rupture reduced: plaques are said to be stabilised. The conclusion is that knowledge on blood–endothelium interaction is a fundamental part of the study of this disease. Amongst the biophysical processes that we are interested in, wall shear stress (WSS) is an important factor, because this is a key element in the evaluation of the influence of different flow patterns in arterial wall dynamics and, consequently, in plaque development. WSS correlates with velocity patterns and are sensed by endothelial cells. These mechanisms contribute to cellular activity. Additionally, the blood particles might change the flow characteristics (Reese and Thompson 1998; Provenzano and Rutland 2002; Cunningham and Gotlieb 2005; Li et al. 2005; Jung et al. 2006; Chatzizisis et al. 2007; Zastawny et al. 2012; Humphrey et al. 2014; Maurovich-Horvat et al. 2014; Pereira 2013). Therefore, at sites of increasing cellular activity due to inflammation (vulnerable plaques), the blood particles via changes in WSS might affect the outcomes of atherosclerosis.

In this work, we propose a model of haemorheology in which shear stress results from flow characteristics that are influenced by particles. The red blood cells (RBCs) represent about 45% of the total blood volume. Therefore, about 4006 RBCs of 6 μm can be found, per cross section, in a 6 mm long and 500 μm thick (diameter) mouse arterial segment, which results in a significant number of blood particles within such a small volume. Indeed, in the mentioned mouse artery, a volume fraction or Haematocrit (Hct in %) of about 0.45 corresponds to about 4.2 million RBCs (Pereira 2016; Pereira 2013; Pereira et al. 2014). Once more, in this case, particles might change both the magnitude and the distribution of shear forces applied on the endothelial cells. The resulting signals are either transduced into chemical responses or transmitted to the surroundings to regulate cellular activity (Humphrey 2001; Li et al. 2005).

In vivo, the signalling processes driven by endothelial sensing mechanisms account for the non-Newtonian nature of blood. Otherwise, physiological regulation would not be possible, because a natural lack in accounting for the effects of the blood compounds in homeostasis would indicate that this is the reason why humans are genetically susceptible to develop atherosclerosis. However, to date, it is not clear why humans are genetically suceptible to develop the disease. The former is already a reason why computational models of atherosclerosis might account for blood rheology, in order to be comparable with the in vivo phenomenon.

Moreover, although predilection sites for atheroma development have been identified, the reason why plaque develops around those regions is not fully understood. Therefore, while in large vessels the effects of particles are usually neglected due to the characteristic length of the domain, non-Newtonian models that account to local changes in flow characteristics due to blood compounds might contribute to research on atheroma initiation. This is because, at those sites (branches and curvatures), we might expect blood particles to influence both the magnitude and distribution of near wall forces, as they might behave as fluidized particles (Crowe et al. 1998; Tsuji 2007) following the fluid streamlines.

Our hypotheses about the relevance of representing haemorheology in macrocirculation is also based on the fact that when the endothelium becomes dysfunctional, and local homoeostasis is compromised, changes in mechanical forces due to particles may have an important effect on flow, locally, as the erythrocytes behave as fluidised particles, modifying the shear stress map. Indeed, while a significant increase in recirculation might occur in cases of severe stenosis, by adding the cumulative effects of particle, the net vorticity might be attenuated, if compared to single-phase patterns. Therefore, the behaviour of the blood particles and their interactions with the surroundings might change due to rising constriction. According to the literature (Alsheikh-Ali et al. 2010), in late stages of plaque deposition, a reduction higher than 80% in the arterial cross-sectional area can be observed. This corresponds to more than 50% of reduction in the arterial diameter. Therefore, we infer that under disease conditions, the blood particles should not be neglected. Hence, this work might contribute to the current literature seeking to understand the mechanisms related to both the built-up of vulnerable plaques and the overall plaque progression.

The Importance of the Influence of the Blood Particles in Atheroma Formation

Blood is a shear thinning fluid, which means that its dynamic viscosity (μ) decreases in function of increasing strain rate (DU/Dy). Strain rate-dependent viscosity can be a characteristic of non-Newtonian fluids. In Newtonian models, on the other hand, the blood viscosity is assumed constant (Classic Newtonian) or consider its time dependency (Generalised Newtonian). However, instantaneous strain rate approaches do not take into account the strain history. In these models, the restoring forces are neglected and the elastic component of shear is not taken into account. This is valid if either the observation time is smaller than the system response time or the length of reference is significantly bigger than the system's characteristic length. Therefore, this depends on the level at which the system's internal structure is observed. In those cases, forces arising from interactions between the system compounds can be neglected. In contrast, if the internal structure of the system influences time–space trends, a closer approximation to reality requires the combination of viscous (Newtonian) and elastic (Hooke) components, in order to mimic strain–stress relations.

A way of doing the above was proposed by Maxwell in 1867. The model is currently named the Maxwell formulation. In it, both the instantaneous and the history-dependent strain are combined. This considers dissipative effects and adds memory to the resulting stress equation. This sort of formulation represents non-Newtonian effects in a continuous manner. Overall, blood is represented as a single-phase fluid, and extra terms are added to the stress–strain relation.

We decided to numerically mimic haemorheology via particulate blood flow. This approach allows the system to be resolved separately, and coupled at the end of each integration time step, which reduces the computational time complexity (Lai and Peskin 2000; Uhlmann 2005; Pereira 2013). Moreover, disjointed resolutions have the advantage of providing a clear way of implementing the system particularities (Crowe et al. 1998; Lai and Peskin 2000; Uhlmann 2005; Tsuji 2007; Zastawny et al. 2012; Pereira 2013), using established computational fluid dynamics techniques. This reduces errors due to implementation and facilitates the identification of model inconsistencies. Moreover, good practices in programming lead to modular and, therefore, portable pieces of software, which can be easily adapted to different applications. However, during the coupling stage, imposing conservation constraints might be an issue. We have presented some alternatives for dealing with the most common drawbacks found in the literature, considering our subject of application. These alternative solutions will be presented in the section "Methodology".

The Representation of Non-Newtonian Effects via Multiphase Flow Techniques

The term Multiphase Flow is commonly used to refer to a flow of two or more non-miscible fluids or fluids (continuous phase) and particles (discrete phase). Following the work presented in Tsuji et al. (1992) and Crowe et al. (1998), these flows can be classified according to their relative concentrations (volume fractions) and the characteristic length of the system. Usually, the continuous phase is a liquid or a gas and the discrete phase is made up of solids, encapsulated liquid and gases. We follow a classification (Tsuji et al. 1992; Crowe et al. 1998) that divides the phases into three main categories: dispersed, dense and sparse. Throughout this work, the term dispersed will be used to refer to a flow regime in which the particles motion is predominantly governed by inertial, surface and body forces, which implies both that the particles size is significantly smaller than the size of the whole domain and that the particle volume fraction makes the inter-particle free distance (characteristic length) to be large enough, to minimise the particle–particle interaction effects. The term dense (volume fraction beyond about 10%) will be used to refer to a flow regime in which the particles motion is dominated by particle–particle interactions, and the term sparse (volume fraction up to about 0.1%) will indicate the negligible effect of particles on the surrounding fluid, because of their low concentration.

The computational time–space resolution of a multiphase flow relates the description of the domain via the Eulerian or the Lagrangian representation, the phase coupling, and the phase details (particle tracking or bulk description for instance), which strongly interconnects specific flow regimes with the appropriate numerical approach (Crowe et al. 1998; Tsuji 2007; Zastawny et al. 2012). Indeed, in sparse and dispersed regimes, the particles characteristics are usually represented through the Lagrangian approach, although the computational cost may occasionally become excessively high, whether the number of particles grows beyond some extent. In this case, the discrete phase can be statistically treated such that only sample particles are resolved over the whole domain (Crowe et al. 1998; Tsuji 2007).

Overall, in the Lagrangian representation, the particles characteristics are followed using the centre of mass of a particle or of a particles sampling, as reference (Crowe et al. 1998; Tsuji 2007). In contrast, the Eulerian approach consists in the representation of particle characteristics at the continuous phase grid nodes (Crowe et al. 1998; Tsuji 2007), which is more suitable for large-scale problems whether the whole phases can be represented as a continuum. A wide variety of methods can be used to simulate multiphase flows. To mention a few, the Direct Numerical Simulation (DNS) method is a Lagrangian-based approach that resolves the equation of motion of individual particles. Indeed, they are considered point-particles, leading to restrictions on their size, which has to be smaller than the characteristic grid spacing. Another usual technique is the Reynolds Averaged Navier–Stokes (RANS) equation method (Feng and Michaelides 2004). This approach is particularly suitable for dense regimes, where the characteristics of the discrete phase are averaged over a control volume. In the RANS method the discrete phase is treated as a continuum (Eulerian approach), which means that internal interactions might have minimum effects because of, for instance, a viscous-dominant flow regime. Moreover, resolution at the interphase region might be carefully addressed, to avoid the introduction of errors due to numerical diffusion. Variations of the method track a set of representative particles and statistically derives the discrete phase properties resulting in an Euler–Lagrange approach.

Central to the modelling of multiphase flows is the concept of phase coupling, which can be addressed by means of uncoupled or coupled models. The criteria for adopting an approach or another lies on the level of influence that the phases exert on each other and their internal behaviour. Indeed, if local particle–fluid interactions or their macroscopic effects are negligible, the continuous phase behaves like a single-phase flow and it is said to be uncoupled. Otherwise, it is addressed by means of different levels of coupling. The appropriateness of the strategy to be adopted depends on both the flow regime (Crowe et al. 1998; Jung et al. 2006; Tsuji 2007; Zastawny et al. 2012) and the particles tendency to follow the fluid streamlines (Fig. 12.1), which is addressed via the Stokes number (Chandran et al. 2011).

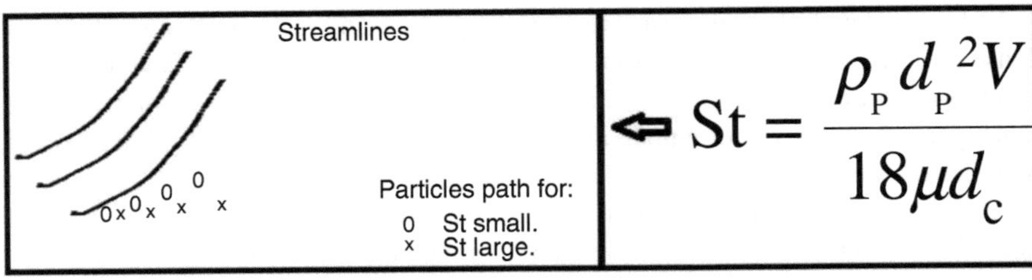

Fig. 12.1 The particles in flow behaviour as a function of the Stokes number St (Chandran et al. 2011). The equation derived from Chandran et al. (2011) relates the particle's density(ρ_p), diameter(d_p), and volume(V), with the fluid dynamic viscosity(μ) and the arterial diameter(d_c), via the empirical constant$(\frac{1}{18})$

Table 12.1 Coupling regimes as a function of both the flow regime and the Stokes number (Crowe et al. 1998; Chandran et al. 2011)

	Stokes number (St)	
Flow regime	St < 1	St > 1
Sparse	One-way	One-way
Dispersed	One-way	Two-way
Dense	Four-way	Four-way

Our approach assumes the continuous phase (the plasma) to be an incompressible Newtonian fluid. The discrete phase (the erythrocytes) is modelled in a dense regime. Therefore, our model uses a four-way coupling of two-phases interactions (fluid–particle and particle–particle), as previously described in the literature—Table 12.1.

The Representation of Fluid–Structure Interactions via Immersed Boundary Techniques

The so-called Immersed Boundary Method (IBM) aroused the interest of researchers working on numerical models of multiphase systems due to its capability of dealing with deformable bodies immersed in fluids. Peskin (1972, 1977) introduced the method to the solution of fluid domains discretized on uniform Eulerian meshes and solid domains represented by Lagrangian sets. The method was used to simulate intracoronary blood flow (Peskin 1972, 1977; Goldstein et al. 1993). The motivation to use this methodology lies in the fact that solid and fluid domains are resolved separately and the global solution is reached by applying the non-slip condition at boundaries. Therefore, the coupling between the Eulerian and the Lagrangian domain consists in applying convolutions that distribute the Lagrangian forces onto the nearest Eulerian points.

An alternative representation of the forcing scheme uses feedback mechanisms. This was proposed by Goldstein et al. (1993) and subsequently used in Saiki and Biringen (1996), Höfler and Schwarzer (2000) and Feng and Michaelides (2004).

This strategy is based on a spring-dashpot model connected to the Lagrangian points. This is used to correct the error in the Lagrangian coordinates. The challenge is to estimate the appropriate time discretisation (Saiki and Biringen 1996; Lai and Peskin 2000). A direct formulation for the force was presented by Fadlun et al. (2000) to deal with the former. Overall, the stiffness matrix for the implicit time discretisation is modified to include the desired velocity at each Lagrangian point (collocation method).

Uhlmann (2005) proposed a correction for the IB method. He combines both a projection-based approach for pressure correction as in Taira and Colonius (2007) and a collocation method to adjust the stiffness matrix as in Paige and Saunders (1975). Spurious oscillations of force were observed during the interpolation of the Eulerian forces into the Lagrangian domain. To fix the problem, Uhlmann used Dirac Delta-like functions determining hydrodynamic smoothed forces that might preserve the global order of convergence of the spatial scheme. The method facilitates the incorporation of deformable and moving structures. However, this lies on uniform Cartesian grids, which is not convenient for modelling systems whose representation requires the use of heterogeneous meshes, as the ones we present here. Therefore, we introduce a method that extends the finest methodology proposed by Uhlmann (2005), to the resolution of systems that are discretised on heterogeneous meshes. Additionally, we couple both the pressure projection (Taira and Colonius 2007) and the interpolation of forces iteratively. This is to control the numerical diffusion due to an implicit time-lag, which is introduced by the fractional time step approach (Uhlmann 2005) that comprises the overall integration scheme.

Methodology

The Mathematical Characterization of the Blood as a Multiphase Flow

Our methodology comprises the separation of the system into two major parts, the fluid phase (plasma) and the particles phase. As the majority of particles in blood consists of red blood cells (RBCs), we omit the other components. In our approach, particles are treated in a Lagrangian framework and the particle–particle interactions calculated using sets of particles, which are distributed along the whole blood volume. Moreover, the Eulerian domain is discretised via a staggered grid, in which the scalar quantities are placed at each cell centre, while vectors are represented at the edges (Uhlmann 2005; Pereira 2013).

The Fluid Phase

The fluid phase (plasma) is here modelled via the Newtonian incompressible Navier–Stokes equations. These equations read as follows.

Conservation of Momentum:

$$\alpha\left(\frac{\mathrm{d}u}{\mathrm{d}t}+u^{\circ}\nabla u\right)=\alpha\left(-\nabla p+\frac{1}{\mathrm{Re}}\Delta u+f\right) \tag{12.1}$$

Conservation of Mass:

$$\alpha(\nabla^{\circ}u)=0 \tag{12.2}$$

Throughout this work we will use the term *f* to represent different source terms. Here, this term represents the frictional contributions to the momentum (Eq. (12.1)). Moreover, the terms Re, u, α, p and t are the Reynolds number, the fluid velocity, the fluid volume fraction, pressure and time, respectively.

As indicated above, the source terms arising from forces exerted by the particles are included in the right-hand-side of Eq. (12.1). These forces are interpolated at each fluid cell, by considering the volume occupied by particles, and the net contribution added under the form of *f*. These source terms are calculated as in Eq. (12.3).

Source Terms (Uhlmann 2005; Pereira 2013):

$$f=\rho\sum_{n}^{k=1}(\alpha_k Drag_k) \tag{12.3}$$

The terms ρ, α_k and Drag_k are the fluid density, the particles volume fraction at a fluid mesh cell k and the drag force calculated at a fluid mesh cell k, respectively. The calculation of the forces arising from fluid–particles interactions are shown in section "The Particles Phase". This includes the drag force.

Physiological data from Huang et al. (1995) is used to characterise the plasma. The values of reference are indicated in Table 12.2.

The Particles Phase

Particles (the erythrocytes) are modelled as soft spheres, in a Lagrangian framework, and the Newtonian second law of motion applied in the characterisation of their behaviour (Eqs. 12.4–12.6) (Crowe et al. 1998; Tsuji 2007).

$$F=ma=m(\frac{\mathrm{d}u}{\mathrm{d}t}) \tag{12.4}$$

Table 12.2 Values of reference to characterise the plasma (Huang et al. 1995)

Model assumptions	
Parameters	Values of reference (Fluid)
Density (kg/m³)	1060
Dynamic viscosity $(\frac{kg}{\mathrm{m}\times\mathrm{s}})$	0.0037100
Kinematic viscosity (Pa × s)	0.0000035

Table 12.3 Values of reference to characterise the erythrocytes (Baskurt and Meiselman 2007)

Model assumptions	
Parameters	Values of reference (Particles)
Density (kg/m³)	1125
Diameter (m)	0.000006–0.000008
Friction coefficient	0.1
Coefficient of restitution	0.9
Youngs modulus (Pa)	4.4e + 3
Poisson ratio	0.49
Volume fraction	0.45

$$u = \int \frac{1}{m}(Drag + Lift + f)\mathrm{d}t \tag{12.5}$$

$$Drag + Lift = \alpha\alpha_k\rho_k \parallel u - u_k \parallel \circ(e_1 + e_2 + e_3) \tag{12.6}$$

The terms F, m, a, ρ_k, α_k, α, u and u_k are the net force acting on particles, the mass of the particle, the acceleration of the particle, the particle's density, the particles' volume fraction at a fluid mesh cell k, the fluid volume fraction, the fluid velocity and the particle's velocity, respectively. The terms e_1, e_2 and e_3 indicate the components of the fluid-particle contact forces (Tsuji 2007). Table 12.3 shows the relevant biophysical characteristics of the particles.

As indicated above, the forces acting on particles and their counterparts acting on both the fluid and the arterial wall are the drag and the lift (Drag and Lift), plus any relevant sinks and sources (f) that vary with the model assumptions. Here, the term f represents the contact forces arising from both particle–particle and particle–wall collisions, which also generate a torsion in the e_3 direction. This will be described in section "The Collisions" (Eq. (12.7)).

The Collisions

The particles deformation and the collisional forces result from both elastic and viscous components (Tsuji et al. 1992; Tsuji 2007). Similarly, forces due to particle–wall collisions are calculated using a spring-dashpot model (Tsuji et al. 1992; Tsuji 2007). These results in only part of the pre-collisional configuration being restored. A schematic view of two particles colliding is shown in Fig. 12.2.

The mathematical representation of the collisions are described in Eq. (12.7) (Tsuji 2007), which we use to model both the particles deformation and the particles interactions with themselves and the arterial wall. A classical spring-dashpot model of viscoelasticity is defined by setting $\alpha = 1$ in Eq. (12.7) (Tsuji 2007).

$$f = -(k\delta^{\alpha} + \eta u_k)\circ(e_1 + e_2 + e_3) \tag{12.7}$$

The terms f and k in Eq. (12.7) are, respectively, the collisional net force and the elasticity coefficient (Youngs modulus) throughout each material direction—which

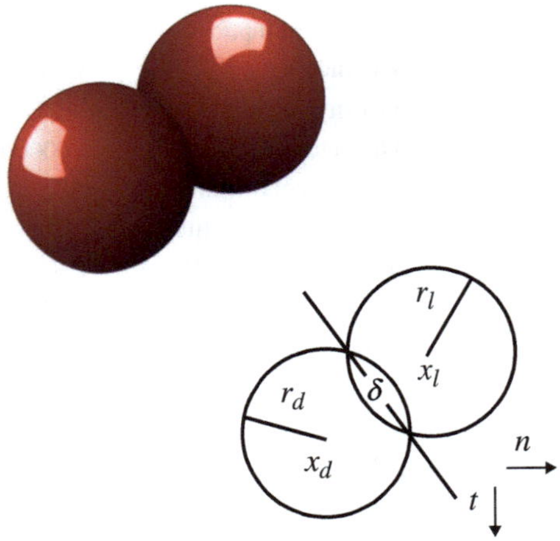

Fig. 12.2 A cross-sectional cut illustrating the particles collision plan (Tsuji 2007). The radius and centre of two particles (d and l) colliding are represented via the variables r_d, r_l, x_d and x_l. The normal and the tangential directions in reference to the canonical system of reference are represented by $n(c_2e_2)$ and $t(c_1e_1)$—e_1 and e_2 are the canonical vectors, and $c_1$and c_2 represent linear transformations in the plan of reference. The variable δ represents the resulting deformation due to collision (overlapping). This geometrical notion is used to formulate the equation that governs both the particle–particle and the particle–wall collisions

varies in heterogeneous systems. The particles' deformation due to collision is modelled via the overlapping δ. Overall, the term $k\delta^{\alpha}$ in Eq. (12.7) represents elastic effects.

The term ηu_kin Eq. (12.7) represents the viscous effects. The parameter u_k represents the components of the collisional system velocity (particle–particle or particle–wall), and η is the coefficient of restitution (Tsuji et al. 1992; Tsuji 2007), which represents the mechanical energy that is preserved (not dissipated) during collision. This term accounts for the fact that the particles reaction to the contact forces is not fully applied back on their counterparts—the viscous effect. Finally, the terms e_1, e_2 and e_3 indicate the components of the resulting contact force—directional term.

Overall, the coefficient of restitution, the Youngs modulus and the Poisson ratio of the wall—Table 12.4, together with the particles properties (Table 12.3), dictate the rate of dissipation and restitution during collisions.

The Mathematical Characterization of the Blood–Artery Interactions Using the Iterative Incompressible Immersed Boundary (ThreeIB) Approach

The coupled fluid–wall model is resolved iteratively. However, prior proceeding on the fluid–wall resolution, the Navier–Stokes equations shown in section "The Fluid Phase" (Eqs. (12.1) and (12.2)) are discretised. These equations are then decoupled, and at each integration time step k, the convection–diffusion terms are resolved

Table 12.4 Values of reference to characterise the arterial wall (Bozec et al. 2004)

Model assumptions	
Parameters	Values of reference (Wall)
Coefficient of restitution	0.9
Friction coefficient	0.1
Friction coefficient	0.1
Youngs modulus (Pa)	3.61e + 5
Poisson ratio	0.49
Wall thickness (as a function of the radius R)	0.04[a] R

[a]3.61e + 5 if genetically engineered mice and 2.66e + 5 if wild type

implicitly (Uhlmann 2005), in a fractional-step fashion, which results in the so-called intermediate velocity $\hat{u}$ (Uhlmann 2005)—Eq. (12.8). At this stage, the reference velocity consists of the fluid velocity u at the end of the integration time step $k-1$. The term Re is the Reynolds number.

$$\hat{u} = u^{k-1} + \Delta t(\frac{1}{\text{Re}}\Delta u^{k-1} - ((u\nabla)u)^{k-1}) \tag{12.8}$$

The next stage consists of calculating the source term f that represents the fluid–wall interactions—Eqs. (12.9)–(12.11).

$$U = \sum_{\Omega_x} \hat{u} I \Delta^n x \tag{12.9}$$

The term Uin Eq. (12.9) is the so-called Lagrangian velocity (Uhlmann 2005), because this is calculated at each Lagrangian point used to discretise the immersed solid (the wall). Equation (12.9) shows how this velocity is derived via numerical interpolation. The term I represents the interpolation basis function, which is in this work a Dirac delta function (Uhlmann 2005). Additionally, the velocity $\hat{u}$ is the interpoland, and $\Delta^n x$ represents the integration spacing in each of the whole n dimensions. The term Ω_x represents the set of Eulerian points (fluid cells) used as the base for this interpolation. These points are chosen to be within the neighbourhood of the Lagrangian points (solid points). Moreover, this neighbourhood is defined by a volume of influence defined by $\Delta^n x$, which is based on the Euclidian distance—the closer Eulerian points x are considered to influence calculations at a Lagangian point X and vice versa. Therefore, the interpolation here used is based on a colocation moving weighted least square method (Li and Liu 1996).

$$F = \frac{U_{\Omega_x} - U}{\Delta t} \tag{12.10}$$

The so-called Lagrangian or singular force F (Li and Liu 1996; Roma et al. 1999; Uhlmann 2005) derives from the application of the non-slip condition at the fluid–wall interface—Eq. (12.10). The term U_{Ω_x} in Eq. (12.10) is the desired velocity at the solid coordinates (non-slip velocity). The difference between the desired velocity U_{Ω_x} and the interpolated velocity U is called slip velocity. This is unit adjusted by dividing by Δt, to give rise to the Lagrangian force F at the solid points X.

The Lagrangian force F is then interpolated back onto the Eulerian cells x via interpolation—Eq. (12.11). The interpolation terms are as above explained in Eq. (12.9). However, the base of this interpolation derives from the Lagrangian quantities. This back-interpolation brings about a so-called Eulerian or volumetric force f (Roma et al. 1999)—Eq. (12.11). The force f is used to correct the fluid velocity field. As a result, the influence of the immersed solid in the global system is added to the resolution.

$$f = \sum_{\Omega_X} FI\Delta^n X \tag{12.11}$$

As stated, the above adjusts the intermediate velocity $\hat{u}$, which generates a new intermediate velocity $\hat{u}$ (Eq. (12.12))—fractional-step resolution.

$$\hat{u} = \hat{u} + f \tag{12.12}$$

Mass conservation is then prescribed via the constraint described in section "The Fluid Phase" (Eq. (12.2)). This is resolved through the Helmholtz projection scheme (Uhlmann 2005), resulting in the so-called Poisson equation (Eq. (12.13)).

$$u^k = \hat{u} - \Delta t \nabla \phi^k \tag{12.13}$$

The term u^k represents the final fluid velocity field at the end of the k-th integration time step. The term ϕ^k characterises an intermediate (or pseudo) pressure, which is calculated in Eq. (12.13) to correct both the fluid velocity field and the pressure (Eq. (12.14)).

$$p^k = p^{k-1} + \phi^k \tag{12.14}$$

Now, notice that adding to the velocity field $\hat{u}$ (Eq. (12.12)) the source term f from Eq. (12.11) and correcting the resulting velocity $\hat{u}$ via the pseudo pressure in Eq. (12.13) result in two sub-steps, in this fractional-step approach. This de-coupled calculation of both the source term and the pressure correction may result in numerical diffusion (Ii et al. 2011), because the latest step may significantly change the velocity field at the location where the non-slip condition is prescribed. This numerical error may become prominent with rising inertial effects—increasing Reynolds number; because the gradients of velocity at the fluid–solid interface may be more sensitive to numerical instabilities (Ii et al. 2011). Rising numerical diffusion could dramatically change the resulting velocity field, which may bring about significant changes in predicting the time–space trends of the physical system of interest. Moreover, in such circumstances, numerical convergence may not be attained (Ii et al. 2011).

Our proposed solution to the above-described numerical error is based on the iterative calculation of both the source terms that are added to Eq.(12.1) and the pseudo pressure in Eq. (12.13). This is summarised in Algorithm 12.1. The proposed scheme is based on two premises: it conserves mass integrally, and the error imposing the non-slip condition is kept below the tolerance *tol*. This preserves the numerical stability of the method, because the tolerance for the maximum error is determined by the desired global order of convergence of the time integration scheme. Indeed, because we use a classical second order in time and space Runge–Kutta strategy (Uhlmann 2005), the tolerance *tol* is set to dt^2. Moreover, the resolution is physically consistent by fully imposing the incompressibility constraint.

Algorithm 12.1 Illustration of the gradient method for the calculation of the coupling immersed boundary source terms and divergence-free velocity.

The next steps iteratively compute a conservative_IB_*force* as a function of both the *numerical_diffusion d* and an unknown constant σ, $\sigma_d = \sigma \frac{\partial x}{\partial n}$.
This IB force is conservative because it is adjusted by the mass conservation constraint (Eq. (12.2), section "The Fluid Phase"). This is finally added as a source term to the right-hand-side of Eq. (12.1) (section "The Fluid Phase"), in an immersed-boundary fashion. This avoids numerical diffusion and assures mass conservation by computing σ taking into account both $(\nabla \circ u) = 0$ and $d < tol = dt^2$.

Inputs: *temp_velocity_field:= latest_fluid_velocity_field*; initial guess $\sigma_{i=0}$; initial search directions; rg(*A*).

while $i \le \mathrm{rg}(A)$ **do**:

calculate the *Lagrangian_velocity U*; # Eq. (12.9) (section "The Mathematical Characterization of the Blood–Artery Interactions Using the Iterative Incompressible Immersed Boundary (ThreeIB) Approach").

calculate the residual $r_i := -A\sigma_i$ according to:

- the *poisson_equation*; # Eq. (12.13)—section "The Mathematical Characterization of the Blood–Artery Interactions Using the Iterative Incompressible Immersed Boundary (ThreeIB) Approach".
- a *numerical_displacement* $\frac{\partial X}{\partial n} := \left(\left(U_{\Omega_x} - U\right)^{\circ} n\right) dt$; # This is normal to the fluid–solid interface and # is a function of the *slip_velocity*.
- the *numerical_diffusion* $d := \frac{\partial X}{\partial n}$; # *d* is assumed to be proportional to $\frac{\partial X}{\partial n}$.

if ($r_1 \le tol$) *and* ($\nabla \circ u$): stop; # intermediate stop criteria test.

otherwise:
i++;

update the *temp_velocity_field*;
calculate the *search_directions*;
calculate σ_i as a function of the search_directions and the residual r_i;
endif.
endwhile.
Output: *final_fluid_velocity_field*;

Finding a constant σ — as described in Algorithm 12.1— that adjusts the source terms to assure both $(\nabla \circ u) = 0$ and $r_i < tol$ according to the order of convergence of the time resolution scheme consists in resolving a multi-restriction problem. Therefore, following a general gradient-based scheme (Freund et al. 1999; Nazareth 2009), a sequence of approximate solutions $\{\sigma_i\}$ is constructed. In this process, a residual vector $r_i = \; - A\sigma_i$ is progressively minimised within a tolerance level tol. The coefficient matrix A representing the linear transformation that result in both the solution of ($u^k = \hat{u} - \Delta t \nabla \phi^k$) and the calculation of the source terms is expected to be symmetric positive definite (spd) (Paige and Saunders 1975). Its range rg(A) defines the dimension of the search space and, consequently, the maximum number of iterations needed to find the desired solution (Nazareth 2009).

The global fractional-step scheme modified by an iterative sub-step-resolution is shown in section "The Coupling That Characterizes the Multiscale Approach".

The Coupling That Characterizes the Multiscale Approach

As illustrated in section "The Mathematical Characterization of the Blood–Artery Interactions Using the Iterative Incompressible Immersed Boundary (ThreeIB) Approach", in our scheme, the interactions between the plasma and the artery are simulated using a novel immersed boundary technique. The plasma is modelled as a Newtonian incompressible fluid using the Navier–Stokes equations, as explained in section "The Fluid Phase". The erythrocytes are represented as spherical particles, following the Newton second law of motion. The modelling of the particles behaviour was defined in section "The Particles Phase". Moreover, the forces resulting from both the plasma–erythrocytes and the erythrocytes–erythrocytes interactions are assumed to be elastic and based on the Hooke's law — DEM model of collisions described in section "The Collisions". Additionally, in this novel fractional-step multiphase flow methodology, which is coupled with an iterative immersed boundary method, the incompressibility constraint in Eq. (12.2) is imposed via a pressure projection technique. This derives from the Helmholtz theorem (Uhlmann 2005; Griffith et al. 2007; Flamini et al. 2016).

In summary, our global scheme comprises a fractional-step methodology in which, at each integration time step:

1. The fluid domain is resolved for the convection-diffusion terms (section "The Mathematical Characterization of the Blood–Artery Interactions Using the Iterative Incompressible Immersed Boundary (ThreeIB) Approach"— Eq. (12.8)).
2. The particle–particle and the particle–wall collisions treated (section "The Collisions"— Eq. (12.7)).
3. The pressure projection scheme applied iteratively and coupled with the resolution in sub-step number 4 (section "The Mathematical Characterization of the Blood–Artery Interactions Using the Iterative Incompressible Immersed Boundary (ThreeIB) Approach"— Eq. (12.13)).

4. The immersed or contact structures (the erythrocytes and the artery wall) are computed, along with the particles phase (sections "The Mathematical Characterization of the Blood–Artery Interactions Using the Iterative Incompressible Immersed Boundary (ThreeIB) Approach"— Eq. (12.11), "The Fluid Phase"— Eq. (12.3); and section "The Particles Phase"— Eq. (12.6), respectively).
5. The global system is fully updated and the numerical scheme advances towards the next integration time step.

The Characteristics of the Computational Domains and both Boundary and Initial Conditions

In this work, we simulated flow trends on four computational domains. They were named M1, M2, FC and MA.

Models M1 and M2 were used to both validate the multiphase methodology here presented and assess the effects of multiphase flow assumptions on blood flow in stenosed arteries — sections "Validating the Multiphase Technique (Shape-Based Assessment)" and "The Application of a New Multiscale Approach to Resolve Non-Newtonian Blood Flow in Diseased Arteries" respectively. We started by comparing our results with results from the literature that focus on the geometrical characteristics of the domain — section "Validating the Multiphase Technique (Shape-Based Assessment)". Moreover, a mesh sensitivity study was carried out in order to adjust mesh characteristics and other computational parameters. Model M1 was later used, in section "The Application of a New Multiscale Approach to Resolve Non-Newtonian Blood Flow in Diseased Arteries" to simulate particulate blood flow in constricted arteries using our multiscale method, and the results compared to single-phase flow.

Model FC was applied throughout the validation of a novel immersed boundary (IB) methodology — section "Validation: The Iterative Incompressible Immersed Boundary (ThreeIB) Method". Convergence analysis was carried out to assure that the integrity of the fractional time-step scheme was preserved in the coupling with the iterative immersed boundary approach, and the desired global order of convergence attained.

Model MA was used to assess the effects of haemorheology in mechanical forces that modulate arterial inflammation via our multiscale (the multiphase technique coupled with the ThreeIB) methodology — section "The Application of a New Multiscale Approach to Resolve Non-Newtonian Blood Flow in Diseased Arteries".

The above-mentioned computational domains and the related initial and boundary conditions are described below.

The Mild (M1) and the Severe (M2) Stenosis Structural Models

The computational domain consists of a venturi-like shape — Fig. 12.3. The radius of reference (R) at the inlet is 0.0035 m and the axial length (L_x) is 0.4 m. The entrance region was calculated under the assumption of laminar pipe flow (White

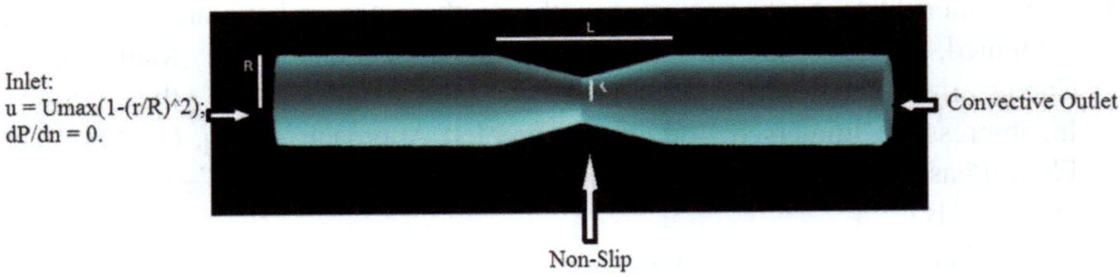

Fig. 12.3 Stenosis model. Schematic view of both the computational domain and the boundary conditions

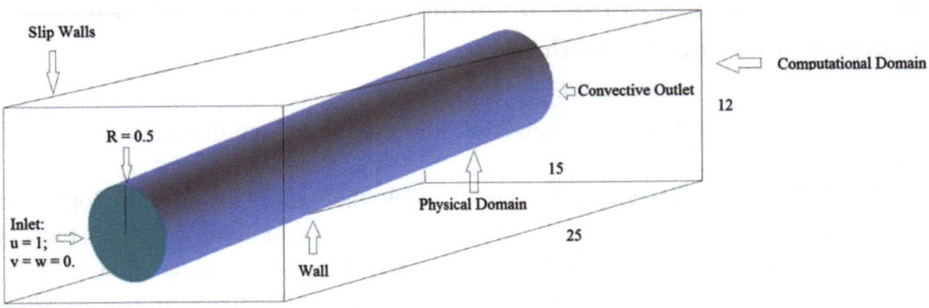

Fig. 12.4 Pipe model for the immersed boundary flow past a cylinder simulation. Schematic view of the computational domain, the physical domain (region of interest) and the boundary conditions

2010). The difference between M1 and M2 is on the peak level of constriction at the central location defined by the length L and the minimum radius r — Fig. 12.3. The mild stenosis (M1) peak cross-sectional area reduction is c. 43.75%, while the peak reduction in the severe (M2) model is c. 80% (Gilsbach and Hassler 1984). The x-component of velocity (**u**) at the inlet is a Poiseuille-like profile, and the y and z-components — **v** and **w,** respectively; along with the normal pressure gradient are set to zero. At the outlet, we defined a velocity Neumann condition, such that $du_i/dn = 0$ ($u_i \in \{\mathbf{u}, \mathbf{v}, \mathbf{w}\}$), and $P = 0$. This characterizes the so-called convective outlet. A non-slip condition is set at the fluid–artery wall interface. Additionally, considering the x-component of the velocity field, the maximum velocity ($U_{max} = \max(\mathbf{u},\mathbf{v},\mathbf{w})$) is prescribed over the whole fluid domain, at $t = 0$ (initial condition). Figure 12.3 shows the characteristics of the computational domain and the boundary condition.

Flow Past a Cylinder (FC)

The computational domain comprises a rectangular box with sides $L_x = 25$, $L_y = 12$ and $L_z = 15$. A pipe is introduced at the middle of this rectangular channel—Fig. 12.4 (out of scale). The pipe radius is $R = 0.5$, its length is 25 and its centre is at $C = (12.50, 6.00, 7.50)$. The dimensionless boundary condition is $\mathbf{u} = 1$, and $\mathbf{v} = \mathbf{w} = \mathbf{0}$,

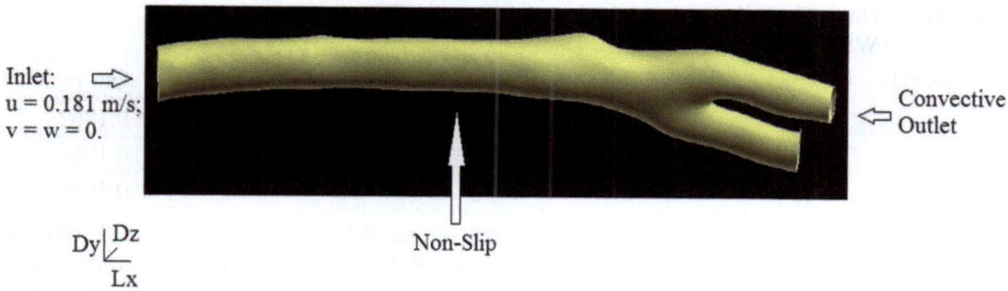

Fig. 12.5 Mice model of artery for the multiscale simulations. Schematic view of the computational domain and the boundary conditions

at the inlet. The outlet is convective, and slip wall is prescribed elsewhere. The Reynolds number is Re = 200. At $t = 0$, the velocity is set as $U_{max} = \max(\mathbf{u},\mathbf{v},\mathbf{w})$ everywhere. As above, **u, v** and **w** stand for the *x*, *y* and *z*-components of the Eulerian velocity field, respectively.

The Mouse Carotid Artery (MA)

The computational domain is defined on the basis of an Apoe+/− mouse arterial branch obtained from structural images (the structure was generated by the author on the basis of MRI-based coordinates obtained via Vikram Mehta PhD thesis—copyright permission obtained)—Fig. 12.5. The cross section at the entrance region is defined by the pair (D_y, D_z)–(0.00042 m, 0.00052 m). The maximum length in the axial direction (L_x) is c. 0.0061 m. Physiological values from the literature (Hochmuth et al. 1973; Fung and Cowin 1994; Dulińska et al. 2006) were used to derive boundary and initial conditions, which are defined as follows: at the inlet, the *x*-component of velocity is uniform and equal to $u = 0.181$ m/s, and the y and z-components of the velocity, along with the normal pressure gradient, are set to zero. We assume that the flux at the inlet is constant, convective outlet, and non-slip condition is prescribed at the artery wall. The initial condition ($t = 0$) is $U_{max} = 0.181$ m/s.

Results and Discussions

Validating the Multiphase Technique (Shape-Based Assessment)

Firstly, we presented results from computational simulations on idealized models of both a severe M2 and a mild M1 stenosis. These results are comparable with theoretical values from a control-volume approach (White 2010). However, the difference in results rises with increasing constriction (Tables 12.5 and 12.6), which indicates that the theory of the boundary layer (TBL) would be a most suitable approach for estimating the flow characteristics in complex geometries

Table 12.5 The peak wall shear stress (WSS) in a mild stenosis M1

Peak WSS (Pa)	
Axial refinement (ED) at the constriction	Our results[a]
16	0.1962
24	0.1961
62	0.1959

[a]Numerical WSS of reference: 0.2100 (Huang et al. 1995); Courant Friedrichs Lewy (CFL) number = 0.5; Radial Refinement (NFS) = 1.1; ED outside the constriction = 8

The peak WSS (Pa) at the throat is calculated using different mesh refinements. The refinement is defined by two parameters found in the commercial package used to construct the meshes (*Home|GridPro—Program Development Company—High Quality Grid Generation Solutions* 2016). These parameters are named ED (axial refinement) and NSF (radial refinement). The Reynolds number is 100

Table 12.6 Peak wall shear stress in a severe stenosis M2

Peak WSS (Pa)	
Axial refinement (ED) at the constriction	Our results[a]
16	0.4851
24	0.4989
62	0.4984

[a]Numerical WSS of reference: 0.8480 (Huang et al. 1995); Courant Friedrichs Lewy (CFL) number = 0.5; Radial Refinement (NFS) = 1.1; ED outside the constriction = 8

The peak WSS (Pa) at the throat is calculated using different mesh refinements defined by the parameters ED and NSF (*Home|GridPro—Program Development Company—High Quality Grid Generation Solutions* 2016). The Reynolds number is 100

(Thwaites 1949). This was confirmed through the numerical assessment of the application of the TBL (Table 12.4). A summary on how to calculate the Wall Shear Stress using the TBL can be found in (Thwaites 1949).

The figures have shown that in a mild stenosis M1—Table 12.5, the peak WSS varies c. 0.05% with varying axial refinement. This indicates that axial refinement does not critically affect the present set of simulations. Additionally, comparing our results with data from the literature (Huang et al. 1995; Reese and Thompson 1998), we found a difference of c. 6.5%. However, Reese and Thompson (1998) reported on 10–20% of discrepancies amongst numerical results found in the literature. Therefore, we infer that our methodology reproduces numerical data accurately.

Table 12.7 The peak wall shear stress (WSS) at the throat, compared to the literature values, on both a mild M1 and a severe M2 stenosis: Re = {100, 500}

Peak WSS (Pa)	Mild stenosis (M1)		Severe stenosis (M2)	
	Re = 100	Re = 500	Re = 100	Re = 500
Our results	0.85	6.50	5.00	40.66
Huang et al. (1995)	0.83	6.71	4.99	42.32
Provenzano and Rutland (2002)	0.84	6.88	4.05	37.88
Reese and Thompson (1998)	0.89	7.33	4.02	34.51

The above-reported discrepancies increase with increasing constriction—Table 12.6. Indeed, the numbers indicate that for a severe stenosis, with a peak reduction in cross-sectional area of about 80%, the differences in results are c. 41%. This increases further by increasing the Reynolds number. Therefore, the theory of the viscous boundary layer (Thwaites 1949) was applied in the calculation of the WSS—Table 12.7.

Computing the WSS on the basis of the near wall characteristics via the shape index (Reese and Thompson 1998) seems to be a more suitable approach. This has a prominent effect on results. Indeed, the formerly reported differences in data due to rising cross-sectional area reduction and Reynolds number are much diminished—Table 12.7. Certainly, our results mimic numerical results from the literature (Huang et al. 1995; Reese and Thompson 1998) in c. 97%, regardless of both Reynolds number and shape.

An experimental approach based on intracoronary ultrasound and angiographic images was applied by Stone et al. (2007) to correlate WSS patterns, progression of stable coronary arterial diseases (SCAD) and vascular remodelling, looking at the estimation of arterial regions where plaque accumulates. They found that constrictive remodelling was restricted to zones of low WSS, with and without plaque growth. Our simulations have correctly identified these regions, on the adopted geometrical models. This is important because we need to assure the correct identification of different flow characteristics, as they are associated with arterial wall remodelling. Indeed, according to Stone et al. (2007), expansive remodelling in low WSS locations correlates with active plaque progression. Therefore, we infer that our methodology correctly capture the flow characteristics that are thought to be relevant in atheroma (aka. atherosclerotic plaque) progression.

Validation: The Iterative Incompressible Immersed Boundary (ThreeIB) Method

Secondly, simulations of flow past an obstacle (Model FC) were used to validate our novel immersed boundary approach (Table 12.8). We report trends similar to those found in the literature (Roshko 1953; Berger and Wille 1972; Rogers and Kwak

Table 12.8 Validation of the bespoke immersed boundary approach

	Flow features		
	St	Cd	Cl
Our results	0.190	1.32	0.65
Belov et al. (1995)	0.193	1.19	0.64
Berger and Wille (1972)	0.18–0.19	–	–
Linnick and Fasel (2005)	0.197	1.34	0.69
Rogers and Kwak (1990)	0.185	1.23	0.65
Lai and Peskin (2000)	0.190	–	–
Liu et al. (1996)	0.192	1.31 ± 0.049	0.69
Roshko (1953)	0.19	–	–
Williamson and Govardhan (2004)	0.197	–	–

The following flow features were compared to the literature at Re = 200: the Strouhal number (St), the drag coefficient (Cd) and the lift coefficient (Cl)

1990; Belov et al. 1995; Liu et al. 1996; Lai and Peskin 2000; Williamson and Govardhan 2004; Linnick and Fasel 2005)—c. 94% of accuracy calculating the Strouhal number (St); c. 90% of accuracy computing the drag coefficient (Cd); and c. 97% of accuracy mimicking the lift coefficient (Cl). Overall, the values on Table 12.8 indicate similar levels of recirculation. Our Strouhal number indicates that the characteristics of wakes forming downstream of the immersed structure correspond to what is reported in the literature (Roshko 1953; Berger and Wille 1972; Rogers and Kwak 1990; Belov et al. 1995; Liu et al. 1996; Lai and Peskin 2000; Williamson and Govardhan 2004; Linnick and Fasel 2005). This correlates with the computed lift coefficient which reflects the frequency of oscillation of the periodic phenomenon resulting from the flow disturbances due to the immersed structure. Additionally, as expected, our predictions do not indicate that any resonance might occur. Resonance is a well-known physical phenomenon (Roshko 1953; Williamson and Govardhan 2004). In models of flow past obstacles, this may occur when the frequency of oscillation of the vortex forming downstream of the immersed body reaches the natural frequency of oscillation of the immersed body. This causes the immersed structure to oscillate with greater amplitude and may lead to the collapse of the entire system (Roshko 1953; Williamson and Govardhan 2004).

The former results are in agreement with our convergence study, which shows that both the time (Fig. 12.6a) and the space (Fig. 12.6b) order of convergence of the iterative IB method are preserved. Moreover, the figures show the effectiveness of the proposed scheme avoiding numerical diffusion. In Fig. 12.6a the solid-dots indicate the accuracy of the resolution with the iterative IB pressure projection coupling shown in section "The Mathematical Characterization of the Blood–Artery Interactions Using the Iterative Incompressible Immersed Boundary (ThreeIB) Approach" (15) and the stars represent the uncoupled solution, without the adjustment of the IB forcing for heterogeneous meshes. The latest does not preserve the global order of convergence, differing from the most accurate predictions in about two orders of magnitude (Fig. 12.6a)—from c. $O(\mathrm{d}t^3)$ to c. $O(\mathrm{d}t)$. As discussed

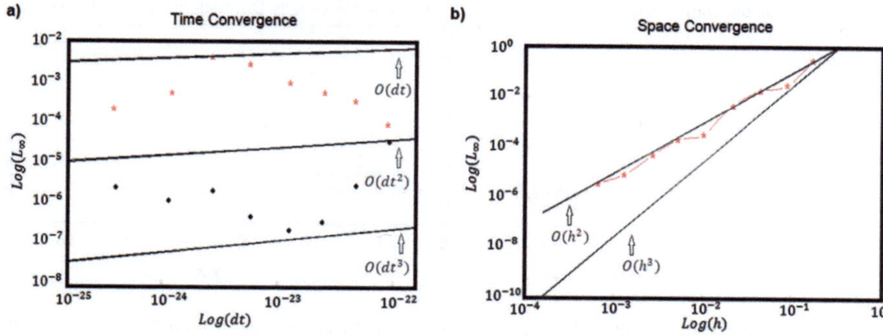

Fig. 12.6 Loglog plot of the infinity norm L_{∞} of the residual velocity normal to the fluid–solid boundary in function of time and space. (**a**) Loglog plot in function of time. The *solid-dots* represent the results achieved using the iterative IB and the *stars* represent the results from the application of the classical IB. (**b**) Loglog plot in function of mesh spacing h. The *stars* represent the iterative IB scheme. In both diagrams, support lines are used, because their slops help identifying the order of convergence of the achieved results

previously, this is a suitable approach if homogeneous finite difference meshes are used (Uhlmann 2005). However, if mesh heterogeneities are introduced, to better fit specific parts of the immersed structure, the conventional IB approach becomes inaccurate—Fig. 12.6a. Therefore, the characteristics of the numerical integration are not preserved. As a consequence, the resolution could diverge, or the achieved solution could not represent the behaviour of the physical system accurately. Again, this improvement in the IB model was achieved via a regularised (for meshes heterogenities) collocation Dirac delta-like function (Pereira 2013), which is based on the reproducing kernel particle fashion (Wang and Liu 2004). This assures the conservative interpolation of quantities as velocity, force, stresses, and torque between Eulerian and Lagrangian domains.

The Application of a New Multiscale Approach to Resolve Non-Newtonian Blood Flow in Diseased Arteries

Thirdly, the influence of the blood particles was assessed using our multiscale approach.

Results on single phase flow simulations are compared with data from (Huang et al. 1995) —Figs. 12.7 and 12.8. We found similar trends. Figure 12.7a shows the normalized pressure drop at the arterial stenosis M1. The minimum value is registered at the minimum cross-sectional area ($\frac{x}{X} = 0$). Downstream of the stenosis, the normalized pressure increases. However, flow does not recover the fully developed peak value observed upstream of the stenosis. We infer that this results from a shorter downstream arterial segment. To overcome this drawback, the computational domain

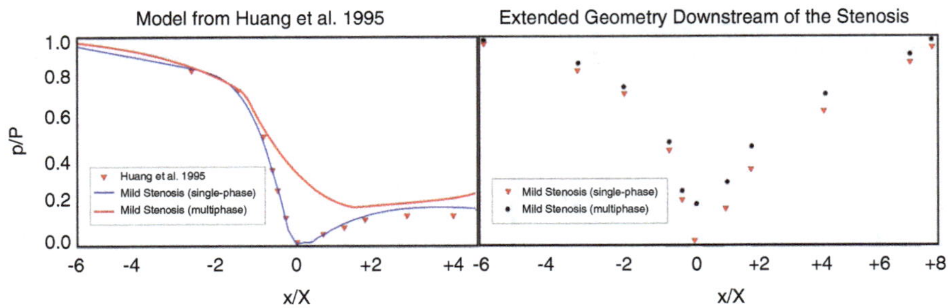

Fig. 12.7 Comparison of single-phase and multiphase pressure values around the site of a mild arterial stenosis. The site of stenosis is at $\frac{x}{X} = 0$. (**a**) Reference pressure values in function of location from Huang et al. (1995) (*red triangles*), compared to our single-phase flow (*blue line*), and to our plasma–erythrocytes (*red line*) pressure value results. (**b**) Pressure values in function of location from an extended geometry—*black dots* illustrates the plasma–erythrocytes pressure values and red triangles indicate the single-phase pressure results (plasma)

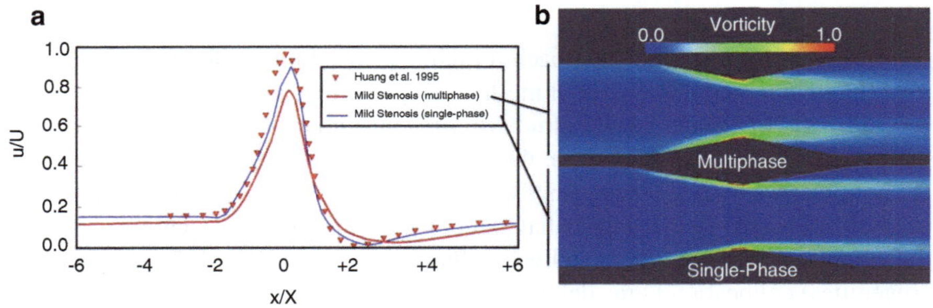

Fig. 12.8 Comparison of single-phase and multiphase flow velocities and flow patterns around the site of a mild arterial stenosis (model M1). (**a**) Normalized *centre-line* velocity at the arterial mild stenosis in function of location: reference values from Huang et al. (1995) (*red triangles*), our single-phase flow results (*blue line*), and our plasma–erythrocytes results (*red line*). (**b**) Normalised vorticity distribution map: multiphase (*top*) and single-phase (*bottom*) flow conditions

was extended to account for the flow-based re-entrance region. As a result, the flow progresses towards the expected fully developed pattern—Fig. 12.7b.

The normalised centreline velocity in Fig. 12.8a peaks at the maximum constrictions. The reported result corresponds to the value stated in the literature (Huang et al. 1995) in c. 90%. The location of peak velocity matches (Fig. 12.8) the minimum registered pressure (Fig. 12.7).

The above single phase (plasma) results are compared to data from two-phase (plasma–erythrocytes) flow simulations—Figs. 12.7 and 12.8. The addition of blood particles (erythrocytes) has a pronounced effect on the flow characteristics. First, the pressure drop is attenuated in c. 30%, at the maximum constriction, as seen in Fig. 12.7. We infer that this is caused by the flow deceleration due to particles. Moreover, the focal nature of the observed phenomenon suggests that the distribution

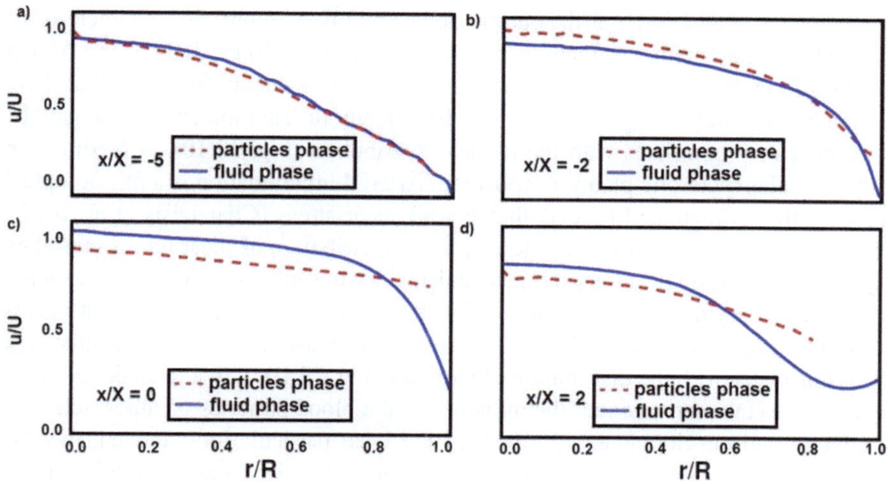

Fig. 12.9 Comparison of fluid phase (plasma) and particle phase (erythrocyte) normalized axial velocities in function of normalized radius r at different cross-sectional locations along the arterial segment M1—The arterial radius R is the normalization factor: (**a**) close to the inlet $x/X = -5$; (**b**) near the beginning of the arterial constriction $x/X = -2$; (**c**) at the minimum cross section (stenosis) $x/X = 0$; (**d**) downstream of the stenosis $x/X = 2$

of particles modulates flow laminarisation, as peak concentration might be registered around the maximum constriction. Second, the figures show that the two-phase peak velocity is c. 28% lower compared with the single phase peak value—Fig. 12.8a. This affects recirculation—Fig. 12.8b.

In presence of particles, recirculation develops much slower—Fig. 12.8b top, which compares to rising viscous effects. We expect this to result in differences in both the wall shear stress and the apparent viscosity. Single phase patterns are much more pronounced—Fig. 12.8b bottom. Figure 12.8 suggests that the Newtonian blood flow assumption leads to recirculation growing faster and the viscous phenomena being weakened.

Figure 12.9 illustrates differences in phase velocities, along the arterial segment. Near the inlet, both the fluid and the particles phase velocities are about the same, because the particle injection is set using the fluid inlet velocity profile—Fig. 12.9a. The more the flow progresses towards the stenosis, the more the fluid and the particles velocities differ. Regardless of the fact both velocity profiles become flatter—Fig. 12.9b, at the maximum constriction (Fig. 12.9c), the particles velocity almost completely plateau out. Finally, downstream of the constriction Fig. 12.9d, this gradually adjusts to the fluid velocity. The data gathered in this study suggests that arterial narrowing due to atherosclerotic plaque deposition causes a focal peak in particle concentration. This has fostered debate on whether the accumulation of erythrocytes could lead to significant differences in flow trends, at the location where plaque forms. Therefore, the key features that Figs. 12.7, 12.8 and 12.9 convey motivated further investigation.

In the literature, much of the current debate revolves around flow characteristics related to inflammatory signalling pathways (Lin et al. 2005; Tedgui and Mallat 2006; Liu et al. 2012). Erythrocytes might modulate mechanical forces that drive inflammation. Indeed, we observed that the velocity distribution and, consequently, the shear map correlates with the particle distribution (Fig. 12.10), at predilection sites for atherosclerotic plaques deposition. Arterial inflammation is a physiological process that is regulated by endothelial wall shear stress (Chatzizisis et al. 2007). Therefore, changes in either the intensity or the distribution of shear forces sensed by the endothelium might affect the mechanisms driving inflammation. We registered differences varying from c. 5% to c. 45% in the axial velocity, as shown in results from the single phase (Fig. 12.10a) versus the two-phase (Fig. 12.10b) simulations. This correlates with the distribution of particles (Fig. 12.10c). In a realistic model, it is crucial to take into account the influence of the blood particles on atherosclerosis initiation and development, regardless of the local particularities. Indeed, there is overwhelming evidence corroborating the notion that under disease condition, focal changes in arterial geometry and, consequently, in flow features might require the inclusion of the blood particles in the resulting computational model.

Changes in flow characteristics due to particle–particle and particle–wall dynamics was formerly reported by Sugiyama et al. (Ii et al. 2011). They correlate particle–particle and particle–wall interactions with the particle initial arrangement, flow rate and apparent viscosity in a Poiseuille flow. There, the initial arrangement of the particles resulted in different sequences of interactions, which affected the particle shape and motion. Therefore, the initial conditions might also modify the instantaneous flow rate. The findings reported by Sugiyama and his collaborators provided insights on the role of the erythrocytes in blood flow. Our results focus on growing haematocrit. As formerly reported, the figures might vary with changes in the particles' initial configuration (Ii et al. 2011). However, we have not assessed flow under different particle injection conditions.

The above-mentioned phenomena result from the momentum interchange between the particles and the flow, which is not considered in Newtonian approaches. Indeed, data from direct numerical simulations of flow past solids were analysed by Zastawny et al. (2012). There, shape-based correlations were designed to study fluid–structure interactions and the resulting drag, lift and torque. Their findings support our results by illustrating the variation of mechanical forces due to solids embedded in the fluid. However, their analysis does not focus on macro-circulation, which is an important element of flow studies related to the development of atherosclerosis.

As stated previously, haemorheological studies found in the literature usually focus on micro-circulation. Looking at the macro-circulation, we have found that rising haematocrit correlates with increased flow laminarization (Fig. 12.11b), as a consequence of both, decreasing velocity (Fig. 12.11b) and increasing shear stress (Fig. 12.11a). This was observed by comparing both the non-Newtonian (Fig. 12.11—Haematocrit (Hct) $\neq 0$) with the Newtonian blood flow simulations (Fig. 12.11—Haematocrit (Hct) $= 0$). Indeed, the apparent viscosity (Fig. 12.11a) rises in function of the haematocrit. These changes in flow features indicate an

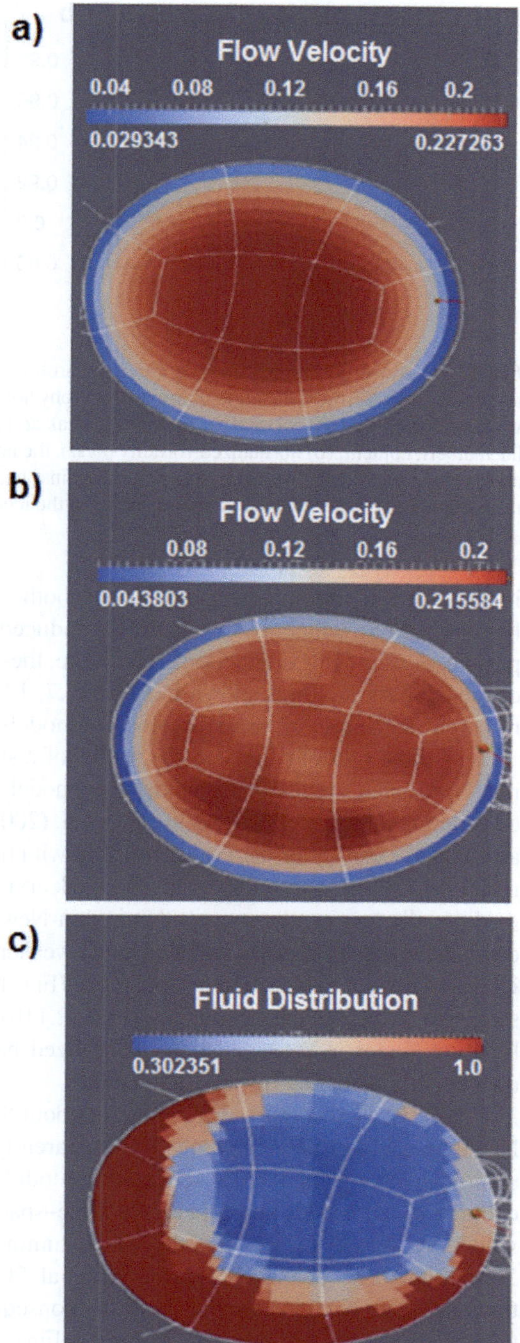

Fig. 12.10 Rheological flow in a mouse carotid artery (model MA), at Re = 58: Cross-sectional view of the stream-wise velocity. (**a**) Single-phase flow velocity in m/s; (**b**) multiphase flow velocity in m/s; (**c**) fluid distribution at an arterial cross section (0–100%) [Image published by the author in (Pereira 2016)—experimental model validation]

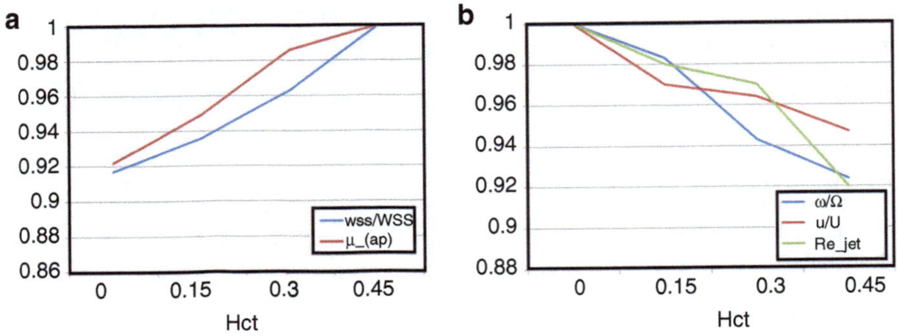

Fig. 12.11 Blood flow simulations on a mouse carotid artery (model MA), Re = 58. The haematocrit (Hct) varies from 0.0 (single-phase) to 0.45 (physiological): (**a**) evolution of normalized peak wall shear stress (wss/WSS) and the normalized peak apparent viscosity (μ_(ap)) in function of the haematocrit content; (**b**) normalized vorticity (ω/Ω), the normalized x-component of velocity (u/U) and the jet Reynolds number (Re_jet) plotted against haematocrit content. The y-axis represents the variation of the normalized variables shown in the legends

increase in the viscous effects because of both shear stress augmentation and flow laminarization, which is mechanically induced by the erythrocytes due to the plasma–erythrocytes interactions. Therefore, these results provide confirmatory evidences that the findings shown in Figs. 12.7, 12.8, 12.9 and 12.10 might motivate research on the role of haemorheology in models of atherosclerosis.

Jung et al. (2006) compared the results of a single and a two-phase transient 3D non-Newtonian simulation on an idealized model of human coronary artery curvature considering pulsatile effects. As in Jung et al. (2006), we also identified an increase in apparent viscosity (Fig. 12.11a) correlating with flow laminarization (Fig. 12.11b), as mentioned above. This corresponds to a peak erythrocytes concentration (Fig. 12.10). Additionally, our results suggest that in non-Newtonian simulations, wakes forming downstream of the stenosis might be much weaker (Fig. 12.11b)—(decreasing vorticity), if compared to Newtonian simulations (Fig. 12.11b)—(Hct = 0.0). This is a consequence of the decreasing velocity (Fig. 12.11b). Therefore, the erythrocytes might be a key factor in the evaluation of localized haemodynamical trends and, consequently, in modulating focal inflammation.

Figure 12.11 illustrates differences of about 8% comparing Newtonian and non-Newtonian blood flow features. This apparently low variation might represent a significant change in cellular activity that modulates arterial inflammation, because cellular functions highly depend on time–space changes in mechanical forces, which are sensed by the endothelium (Cunningham and Gotlieb 2005; Li et al. 2005; Chatzizisis et al. 2007; Humphrey et al. 2014). Therefore, the reported deviations in the flow characteristics, and the consequent changes in both the intensity and the distribution of mechanical forces (Figs. 12.10 and 12.11) along an arterial segment might affect plaque deposition. However, we are aware that on the basis of the evidences currently available, it seems fair to affirm that changes in flow due to the blood particles have a pronounced effect on inflammatory signalling pathways.

Further evidences supporting this hypothesis may lie in experimental research combining cell biology, microfluidics (Pereira 2016) and computation.

As suggested, another interesting finding here reported is on increasing apparent viscosity with increasing shear stress. It is well known that blood is a shear thinning fluid. However, in macro-circulation, haemorheology is not extensively investigated, because under the underlying conditions blood is thought to behave as a Newtonian fluid. Here, we have found insights that local conditions may change the figures even further and the reported apparent viscosity actually increases c. 8% with the increasing shear stress, which is a characteristic of shear thickening materials.

Conclusions, Final Remarks, and Future Research

In this work, we presented a multiscale haemorheological computer-based model of blood flow in arterial segments affected by the abnormal deposition of metabolites such as lipids in the context of atherosclerosis development. We have developed two modelling techniques: a multiphase flow approach for haemorheology and an immersed boundary methodology. We have validated both of these techniques (Roshko 1953; Berger and Wille 1972; Rogers and Kwak 1990; Belov et al. 1995; Huang et al. 1995; Liu et al. 1996; Reese and Thompson 1998; Lai and Peskin 2000; Provenzano and Rutland 2002; Williamson and Govardhan 2004; Linnick and Fasel 2005) and overcame fluid flow modelling related numerical drawbacks reported in the literature (Peskin 1977; Goldstein et al. 1993; Saiki and Biringen 1996; Fadlun et al. 2000; Höfler and Schwarzer 2000; Kim et al. 2001; Feng and Michaelides 2004; Uhlmann 2005; Taira and Colonius 2007; Ii et al. 2011). The two bespoke modelling techniques were coupled to obtain an integrated realistic multiscale computational fluid dynamics technique that describes physiological blood flow conditions relevant to atherosclerosis development.

Finally, our results indicate blood rheology varying with local mechanical and physiological conditions. Indeed, apparent viscosity may increase with particles local accumulation and decrease with particles redistribution. This requires further analysis and motivates research towards applying different numerical methods and experimental assays in order to assess the extents of the reported trend. Because the presented methodology has been extensively validated, showing trends comparable at a very high rate with established results from the literature, we infer that we have captured a novel haemorheological phenomenon occurring in large diseased arteries.

Validation: The Multiphase Technique

We presented results from DNS simulations in arterial segments, for both a mild and a severe stenosis. The results are comparable to numerical data from the literature (Huang et al. 1995; Reese and Thompson 1998; Provenzano and Rutland 2002) and pronounced similarities were observed. Moreover, we found that the theory of the boundary layer (Thwaites 1949) might be the most suitable approach for estimating the flow characteristics in complex geometries. Additionally, our computational model was improved via a mesh sensitivity study, which included the application of the viscous boundary layer theory (White 2010). This supported defining the characteristics of the computational domain and, consequently, the adjustment of the mesh characteristics.

Comparing the results of Newtonian and non-Newtonian simulations, we found that the velocity distribution and, consequently, the shear map correlates with the particle distribution. Indeed, changes in flow characteristics due to particle–particle and particle–wall dynamics correlated with both the flow rate and the apparent viscosity. Our results focused on growing haematocrit. However, we infer that the initial arrangement of the particles might result in different sequences of interactions, which might affect the particles shape and motion. Therefore, the initial conditions might also modify the instantaneous flow rate. Hence, the assessment of flow under different particle injection conditions might be the subject of future research.

Validation: The Iterative Incompressible Immersed Boundary (ThreeIB) Method

We validated a new methodology for simulating the interactions between fluids and solid structures (Roshko 1953; Berger and Wille 1972; Rogers and Kwak 1990; Belov et al. 1995; Liu et al. 1996; Lai and Peskin 2000; Williamson and Govardhan 2004; Linnick and Fasel 2005). The methodology comprised a fractional-step method with pressure projection to impose the mass conservation constraint, which was carried out iteratively, and coupled with the interpolation of forces, which arouse from the fluid–structure interactions. A regularised Dirac delta-like function was applied to distribute IB quantities between both the Eulerian and the Lagrangian domain.

In many applications and particularly in models of blood flow in diseased arteries, precision in determining the mechanical forces resulting from the flow characteristics is important. Therefore, we assured that mass was not spuriously created or consumed. Indeed, our methodology guaranteed that the force and torque were interpolated amongst the Lagrangian (solids) and the Eulerian (fluid) domains integrally.

Our results were comparable to numerical data from the literature where similar levels of recirculation were observed (Roshko 1953; Berger and Wille 1972; Rogers and Kwak 1990; Belov et al. 1995; Liu et al. 1996; Lai and Peskin 2000; Williamson and Govardhan 2004; Linnick and Fasel 2005). Moreover, the computed Strolhal number indicated that wakes forming past the obstacles and the characteristics of the resulting von Karman Vortex Street were analogous to what is found in the literature. Indeed, our results indicated similar levels of disturbances resulting from the balance between the lift and the drag forces. Moreover, the computed lift coefficient reflected the frequency of oscillation of the periodic phenomenon, resulting from flow variations. These disturbances were induced by the immersed structure.

Regarding future improvements, to speed up convergence, a pre-conditioner might be implemented using functional-based matrix treatment, to avoid breakdowns. Moreover, the inclusion of mathematical terms that account for mass transport and the concentration of chemical species might be of relevance, while evaluating the signalling pathways that contribute to inflammation. Moreover, continuous models of blood rheology may be tested.

The Application of a New Multiscale Approach Resolving Non-Newtonian Blood Flow in Diseased Arteries

Our simulations correctly identified regions of low Wall Shear Stress (WSS) that correlates with vascular remodelling, with and without plaque growth. This is important because remodelling in low WSS locations correlates with plaque progression (Stone et al. 2007). Therefore, we infer that our methodology correctly capture the flow characteristics that are associated with plaque progression.

The reported differences in flow rate and apparent viscosity resulted from the momentum interchange between the particles and the flow, which is not considered in Newtonian approaches. Here, we conducted an in-depth analysis of particulate blood flow and discussed its relevance in understanding atheroma formation. We found that rising haematocrit correlates with increasing flow laminarisation, resulting from decreasing velocity and increasing shear stress, if compared with Newtonian blood flow (Huang et al. 1995; Reese and Thompson 1998; Provenzano and Rutland 2002). This corresponded to a peak in the concentration of the erythrocytes. Additionally, our findings indicated that in non-Newtonian simulations, wakes forming downstream of the stenosis might be much weaker in non-Newtonian blood flow simulations. Therefore, the erythrocytes might be a key factor in modulating mechanical forces that drive focal inflammation.

As formerly stated, mechanical forces both affect homeostatic functions that maintain a functional endothelium and drive abnormal signalling pathways, under disease conditions. The wall shear stress correlates with velocity patterns and is sensed by endothelial cells (Humphrey 2001; Li et al. 2005; Chatzizisis et al. 2007; Maurovich-Horvat et al. 2014). This mechanism contributes to cellular activity. Therefore, at arterial segments experiencing increasing cellular agglomeration and activity due to inflammation, the blood particles alter flow characteristics and might affect the outcomes of atherosclerosis. This phenomenon is here simulated computationally. The results suggested that haemorheology might be considered in computational models of atherosclerosis.

Acknowledgements The author is grateful to all those who provided service and advice.
Dr. Kis (Imperial College London and the Institute for Integrated Economic Research) is thanked for revising the manuscript. Dr. Roma and Dr. Simeoni are respectively thanked for advising and kindly receiving the author during a research visit at the University of Sao Paulo (USP). Dr. Mehta (Imperial College London) is thanked for providing the MRI information that allowed the author to computationally build up the structural model MA, while receiving the author's advice and

contributions to his pulsatile flow simulations. Mr. Rochford (Imperial College London), Mr. James (GridPro—www.gridpro.com), Mr. Burbidge (Imperial College London), and Mr. Smet are thanked for the technical support.

This work is partially supported by the Santander Mobility scheme, the Ministerio de Ciencia y Innovacion de España (MICINN), the FAPESP (Brazil), and the EU/UK governance.

References

Alsheikh-Ali AA, Kitsios GD, Balk EM, Lau J, Ip S (2010) The vulnerable atherosclerotic plaque: scope of the literature. Ann Intern Med 153(6):387–395. doi:10.7326/0003-4819-153-6-201009210-00272

Baskurt OK, Meiselman HJ (2007) Hemodynamic effects of red blood cell aggregation. Indian J Exp Biol 45(1):25–31

Belov A, Martinelli L, Jameson A (1995) A new implicit algorithm with multigrid for unsteady incompressible flow calculations. In 33rd Aerospace Sciences Meeting and Exhibit. American Institute of Aeronautics and Astronautics (Aerospace Sciences Meetings) doi:10.2514/6.1995-49

Berger E, Wille R (1972) Periodic flow phenomena. Annu Rev Fluid Mech 4(1):313–340. doi:10.1146/annurev.fl.04.010172.001525

Bozec E, Lacolley P, Bergaya S, Boutouyrie P, Meneton P, Heriss-Legrand M, Boulanger CM, Alhenc-Gelas F, Kim H-S, Laurent S, Dabir H (2004) Arterial stiffness and angiotensinogen gene in hypertensive patients and mutant mice. J Hypertens 22(7):1299–1307

British Heart Foundation (2016) News from the BHF. https://www.bhf.org.uk/news-from-the-bhf. Accessed 1 July 2016

Brown AJ, Teng Z, Evans PC, Gillard JH, Samady H, Bennett MR (2016) Role of biomechanical forces in the natural history of coronary atherosclerosis. Nat Rev Cardiol 13(4):210–220. doi:10.1038/nrcardio.2015.203

Cardiovascular Disease Statistics 2015—BHF (2015). https://www.bhf.org.uk/publications/statistics/cvd-stats-2015. Accessed 10 Dec 2016

Caro CG, Fitz-Gerald JM, Schroter RC (1971) Atheroma and Arterial Wall shear observation, correlation and proposal of a shear dependent mass transfer mechanism for Atherogenesis. Proc R Soc Lond B Biol Sci 177(1046):109–133. doi:10.1098/rspb.1971.0020

Chabi F, Champmartin S, Sarraf C, Noguera R (2015) Critical evaluation of three hemodynamic models for the numerical simulation of intra-stent flows. J Biomech 48(10):1769–1776. doi:10.1016/j.jbiomech.2015.05.011

Chandran KB, Udaykumar HS, Reinhardt JM (2011) Image-based computational modeling of the human circulatory and pulmonary systems: methods and applications. Springer, New York. doi:10.1007/978-1-4419-7350-4

Chatzizisis YS, Coskun AU, Jonas M, Edelman ER, Feldman CL, Stone PH (2007) Role of endothelial shear stress in the natural history of coronary atherosclerosis and vascular remodeling: molecular, cellular, and vascular behavior. J Am Coll Cardiol 49(25):2379–2393. doi:10.1016/j.jacc.2007.02.059

Cheng C, Tempel D, Van Haperen R, Van Der Baan A, Grosveld F, Daemen MJAP, Krams R, De Crom R (2006) Atherosclerotic lesion size and vulnerability are determined by patterns of fluid shear stress. Circulation 113(23):2744–2753. doi:10.1161/CIRCULATIONAHA.105.590018

Chiu J-J, Usami S, Chien S (2009) Vascular endothelial responses to altered shear stress: pathologic implications for atherosclerosis. Ann Med 41(1):19–28. doi:10.1080/07853890802186921

Crowe CT, Sommerfeld M, Tsuji Y (1998) Multiphase flows with droplets and particles. CRC Press, New York. doi:10.1201/b11103-4

Cunningham KS, Gotlieb AI (2005) The role of shear stress in the pathogenesis of atherosclerosis. Lab Invest 85(1):9–23. doi:10.1038/labinvest.3700299

Davies PF (2009) Hemodynamic shear stress and the endothelium in cardiovascular pathophysiology. Nat Clin Pract Cardiovasc Med 6(1):16–26. doi:10.1038/ncpcardio1397
Davies MK, Hollman A (2002) British Heart Foundation. Heart (British Cardiac Society) 87(2):100. doi:10.1136/heart.87.2.100
Dulińska I, Targosz M, Strojny W, Lekka M, Czuba P, Balwierz W, Szymoński M (2006) Stiffness of normal and pathological erythrocytes studied by means of atomic force microscopy. J Biochem Biophys Methods 66(1–3):1–11. doi:10.1016/j.jbbm.2005.11.003
Fadlun EA, Verzicco R, Orlandi P, Mohd-Yusof J (2000) Combined immersed-boundary finite-difference methods for three-dimensional complex flow simulations. J Comput Phys 161(1):35–60. doi:10.1006/jcph.2000.6484
Feng ZG, Michaelides EE (2004) The immersed boundary-lattice Boltzmann method for solving fluid-particles interaction problems. J Comput Phys 195(2):602–628. doi:10.1016/j.jcp.2003.10.013
Flamini V, DeAnda A, Griffith BE (2016) Immersed boundary-finite element model of fluid–structure interaction in the aortic root. Theor Comput Fluid Dyn 30(1–2):139–164. doi:10.1007/s00162-015-0374-5
Frazin LJ, Lanza G, Vonesh M, Khasho F, Spitzzeri C, McGee S, Mehlman D, Chandran KB, Talano J, McPherson D (1990) Functional chiral asymmetry in descending thoracic aorta. Circulation 82(6):1985–1994. doi:10.1161/01.CIR.82.6.1985
Freund JE, Miller I, Miller M, Freund JE (1999) John E. Freund's mathematical statistics, 7th edn. Prentice Hall, Upper Saddle River
Fung YC, Cowin SC (1994) Biomechanics: mechanical properties of living tissues, 2nd ed. J Appl Mech 61(4):1007. doi:10.1115/1.2901550
Giessibl FJ (2003) Advances in atomic force microscopy. Rev Mod Phys 75(3):949–983. doi:10.1103/RevModPhys.75.949
Gilsbach JM, Hassler WE (1984) Intraoperative Doppler and real time sonography in neurosurgery. Neurosurg Rev 7(2–3):199–208. doi:10.1007/BF01780705
Goldstein D et al (1993) Modeling a non-slip flow boundary with an external force field. J Comput Phys 105:354–366
Griffith BE, Hornung RD, McQueen DM, Peskin CS (2007) An adaptive, formally second order accurate version of the immersed boundary method. J Comput Phys 223(1):10–49. doi:10.1016/j.jcp.2006.08.019
Hochmuth RM, Mohandas N, Blackshear PL (1973) Measurement of the elastic modulus for red cell membrane using a fluid mechanical technique. Biophys J 13(8):747–762. doi:10.1016/S0006-3495(73)86021-7
Höfler K, Schwarzer S (2000) Navier–Stokes simulation with constraint forces: finite-difference method for particle-laden flows and complex geometries. Phys Rev E Stat Phys Plasmas Fluids Relat Interdiscip Topics 61(6 pt B):7146–7160. doi:10.1103/PhysRevE.61.7146
Home|GridPro—Program Development Company—High Quality Grid Generation Solutions (2016). http://www.gridpro.com/. Accessed 10 Dec 2016
Huang H, Modi VJ, Seymour BR (1995) Fluid-mechanics of stenosed arteries. Int J Eng Sci 33(6):815–828
Humphrey JD (2001) Stress, strain, and mechanotransduction in cells. J Biomech Eng 123(6):638–641. doi:10.1115/1.1406131
Humphrey JD, Dufresne ER, Schwartz M a (2014) Mechanotransduction and extracellular matrix homeostasis. Nat Rev Mol Cell Biol 15(12):802–812. doi:10.1038/nrm3896
Ii S, Sugiyama K, Takeuchi S, Takagi S, Matsumoto Y (2011) An implicit full Eulerian method for the fluid-structure interaction problem. Int J Numer Methods Fluids 65(1–3):150–165. doi:10.1002/fld.2460
JBS3 Board (2014) Joint British Societies' consensus recommendations for the prevention of cardiovascular disease (JBS3). Heart 100:ii1–ii67. doi:10.1136/heartjnl-2014-305693
Jung J, Lyczkowski RW, Panchal CB, Hassanein A (2006) Multiphase hemodynamic simulation of pulsatile flow in a coronary artery. J Biomech 39(11):2064–2073. doi:10.1016/j.jbiomech.2005.06.023

Kim J, Kim D, Choi H (2001) An immersed-boundary finite-volume method for simulations of flow in complex geometries. J Comput Phys 171(1):132–150. doi:10.1006/jcph.2001.6778
Lai M-C, Peskin CS (2000) An immersed boundary method with formal second-order accuracy and reduced numerical viscosity. J Comput Phys 160(2):705–719. doi:10.1006/jcph.2000.6483
Li S, Liu WK (1996) Moving least-square reproducing kernel method part II: Fourier analysis. Comput Methods Appl Mech Eng 139(1–4):159–193. doi:10.1016/S0045-7825(96)01082-1
Li S, Huang NF, Hsu S (2005) Mechanotransduction in endothelial cell migration. J Cell Biochem 96(6):1110–1126. doi:10.1002/jcb.20614
Lin Z, Kumar A, SenBanerjee S, Staniszewski K, Parmar K, Vaughan DE, Gimbrone MA, Balasubramanian V, García-Cardeña G, Jain MK (2005) Kruppel-like factor 2 (KLF2) regulates endothelial thrombotic function. Circ Res 96(5):e48–e57. doi:10.1161/01.RES.0000159707.05637.a1
Linnick MN, Fasel HF (2005) A high-order immersed interface method for simulating unsteady incompressible flows on irregular domains. J Comput Phys 204(1):157–192. doi:10.1016/j.jcp.2004.09.017
Liu WK, Chen Y, Uras RA, Chang CT (1996) Generalized multiple scale reproducing kernel particle methods. Comput Methods Appl Mech Eng 139(1–4):91–157. doi:10.1016/S0045-7825(96)01081-X
Liu J, Yang T, Liu Y, Zhang H, Wang K, Liu M, Chen G, Xiao X (2012) Krüppel-like factor 4 inhibits the expression of interleukin-1 beta in lipopolysaccharide-induced RAW264.7 macrophages. FEBS Lett 586(6):834–840. doi:10.1016/j.febslet.2012.02.003
Lu D, Kassab GS (2011) Role of shear stress and stretch in vascular mechanobiology. J R Soc Interface 8(July):1379–1385. doi:10.1098/rsif.2011.0177
Malek AM, Alper SL, Izumo S (1999) Hemodynamic shear stress and its role in atherosclerosis. JAMA 282(21):2035–2042. doi:10.1001/jama.282.21.2035
Maurovich-Horvat P, Ferencik M, Voros S, Merkely B, Hoffmann U (2014) Comprehensive plaque assessment by coronary CT angiography. Nat Rev Cardiol 11(7):390–402
Moore JE, Maier SE, Ku DN, Boesiger P (1994) Hemodynamics in the abdominal aorta: a comparison of in vitro and in vivo measurements. J Appl Physiol 76(4):1520–1527
Nazareth J (2009) Conjugate gradient method. WIREs Comp Stats 1(3):348–353. doi:10.1002/wics.013
Paige CC, Saunders MA (1975) Solution of sparse indefinite systems of linear equations. SIAM J Numer Anal 12:617–629. doi:10.1137/0712047
Pereira G (2013) An iterative incompressible immersed boundary method applied to biofluid-structure interaction problems. Biophys J 104(2):508a–509a. doi:10.1016/j.bpj.2012.11.2808
Pereira GC (2016) A multi-layer non-Newtonian model of cardiovascular inflammation. J Biomed Eng Med Devic 1(3):1–6
Pereira G, Krams R, van Wachem B (2014) Modelling the mechanics of the circulation: blood rheology and atherosclerosis. Biophys J 106(2):376a. doi:10.1016/j.bpj.2013.11.2127
Peskin C (1972) Flow patterns around heart valves: a numerical method. J Comput Phys 10(2):252–271. doi:10.1016/0021-9991(72)90065-4
Peskin CS (1977) Numerical analysis of blood flow in the heart. J Comput Phys 25(3):220–252. doi:10.1016/0021-9991(77)90100-0
Provenzano PP, Rutland CJ (2002) A boundary layer model for wall shear stress in arterial stenosis. Biorheology 39(6):743–754
Reese JM, Thompson DS (1998) Shear stress in arterial stenoses: a momentum integral model. J Biomech 31(11):1051–1057. doi:10.1016/S0021-9290(98)00130-4
Rogers SE, Kwak D (1990) Upwind differencing scheme for the time-accurate incompressible Navier–Stokes equations. AIAA J 28(2):253–262. doi:10.2514/3.10382
Roma AM, Peskin CS, Berger MJ (1999) An adaptive version of the immersed boundary method. J Comput Phys 153(2):509–534. doi:10.1006/jcph.1999.6293
Roshko A (1953) On the development of turbulent wakes from vortex streets. National Advisory Committee for Aeronautics, Washington, DC, pp 1–78. doi:10.1016/0045-7930(91)90031-C

Saiki EM, Biringen S (1996) Numerical simulation of a cylinder in uniform flow: application of a virtual boundary method. J Comput Phys 123(123):450–465. doi:10.1006/jcph.1996.0036

Scarborough P, Bhatnagar P, Wickramasinghe K, Smolina K, Mitchell C (2010) Coronary heart disease statistics, 2010 edition. British Heart Foundation, London, p 21. doi:10.1002/hep.24445

Scarborough P, Wickramasinghe K, Bhatnagar P, Rayner M (2011) Trends in coronary heart disease, 1961–2011. British Heart Fou ndation, London

SenBanerjee S, Lin Z, Atkins GB, Greif DM, Rao RM, Kumar A, Feinberg MW, Chen Z, Simon DI, Luscinskas FW, Michel TM, Gimbrone MA, García-Cardeña G, Jain MK (2004) KLF2 is a novel transcriptional regulator of endothelial proinflammatory activation. J Exp Med 199(10):1305–1315. doi:10.1084/jem.20031132

Sherwin SJ, Shah O, Doorly DJ, Peiro J, Papaharilaou Y, Watkins N, Caro CG, Dumoulin CL (2000) The influence of out-of-plane geometry on the flow within a distal end- to-side anastomosis. J Biomech Eng 122(1):86. doi:10.1115/1.429630

Stone PH, Coskun AU, Kinlay S, Clark ME, Sonka M, Wahle A, Ilegbusi OJ, Yeghiazarians Y, Popma JJ, Orav J, Kuntz RE, Feldman CL (2003) Effect of endothelial shear stress on the progression of coronary artery disease, vascular remodeling, and in-stent restenosis in humans: in vivo 6-month follow-up study. Circulation 108(4):438–444. doi:10.1161/01.cir.0000080882.35274.ad

Stone PH, Coskun AU, Kinlay S, Popma JJ, Sonka M, Wahle A, Yeghiazarians Y, Maynard C, Kuntz RE, Feldman CL (2007) Regions of low endothelial shear stress are the sites where coronary plaque progresses and vascular remodelling occurs in humans: an in vivo serial study. Eur Heart J 28(6):705–710. doi:10.1093/eurheartj/ehl575

Taira K, Colonius T (2007) The immersed boundary method: a projection approach. J Comput Phys 225(2):2118–2137. doi:10.1016/j.jcp.2007.03.005

Tedgui A, Mallat Z (2006) Cytokines in atherosclerosis: pathogenic and regulatory pathways. Physiol Rev 86(2):515–581. doi:10.1152/physrev.00024.2005

Thwaites B (1949) Approximate calculation of the laminar boundary layer. Aeronaut Q 1:245–280

Townsend N, Wickramasinghe K, Bhatnagar P, Smolina K, Nichols M, Leal J, Luengo-Fernandez R, Rayner M (2012) Coronary heart disease statistics. British Heart Foundation, London

Tsuji Y (2007) Multi-scale modeling of dense phase gas-particle flow. Chem Eng Sci 62(13):3410–3418. doi:10.1016/j.ces.2006.12.090

Tsuji Y, Tanaka T, Ishida T (1992) Lagrangian numerical-simulation of plug flow of Cohesionless particles in a horizontal pipe. Powder Technol 71(3):239–250. doi:10.1016/0032-5910(92)88030-l

Uhlmann M (2005) An immersed boundary method with direct forcing for the simulation of particulate flows. J Comput Phys 209(2):448–476. doi:10.1016/j.jcp.2005.03.017

Wang X, Liu WK (2004) Extended immersed boundary method using FEM and RKPM. Comput Methods Appl Mech Eng 193(12–14):1305–1321. doi: 10.1016/j.cma.2003.12.024

WHO (2012) World health statistics. WHO, Geneva. doi:10.2307/3348165

WHO (2013) Media centre Cardiovascular diseases (CVDs). Fact sheet No. 317. WHO, Geneva, pp 1–5

WHO (2015) Global status report on noncommunicable diseases, World Health. WHO, Geneva. Isbn: 978-92-4-156422-9

Williamson CHK, Govardhan R (2004) Vortex-induced Vibrations. Annu Rev Fluid Mech 36(1):413–455. doi:10.1146/annurev.fluid.36.050802.122128

World Health Organization (2012) Cardiovascular diseases (CVDs). Fact sheet No. 317, September 2012. WHO, Geneva

World Health Organization (2015) WHO|Global status report on noncommunicable diseases 2014. WHO, Geneva

World Health Organization (2016) WHO|Cardiovascular diseases (CVDs). WHO, Geneva

World Health Organization, WHO Regional Offices for the Western Pacific, World Health Organization (WHO) Global Action Plan (2014) Global status report on noncommunicable diseases 2014. WHO, Geneva, p 176. Isbn: 978-92-4-156485-4

Zastawny M, Mallouppas G, Zhao F, van Wachem B (2012) Derivation of drag and lift force and torque coefficients for non-spherical particles in flows. Int J Multiphase Flow 39:227–239. doi:10.1016/j.ijmultiphaseflow.2011.09.004

Zhao H, Isfahani AHG, Olson LN, Freund JB (2010) A spectral boundary integral method for flowing blood cells. J Comput Phys 229(10):3726–3744. doi:10.1016/j.jcp.2010.01.024

Index

S. Malik (ed.), *Biotechnology and Production of Anti-Cancer Compounds*,
DOI 10.1007/978-3-319-53880-8

MIX
Papier aus verantwortungsvollen Quellen
Paper from responsible sources
FSC® C105338

If you have any concerns about our products,
you can contact us on
ProductSafety@springernature.com

In case Publisher is established outside the EU,
the EU authorized representative is:
Springer Nature Customer Service Center GmbH
Europaplatz 3, 69115 Heidelberg, Germany

Printed by Libri Plureos GmbH
in Hamburg, Germany